Modernization and the Japanese Factory

ROBERT M. MARSH and HIROSHI MANNARI

Modernization and the Japanese Factory

Princeton University Press, Princeton, New Jersey

Published by Princeton University Press
Princeton and London

Library of Congress Cataloging in Publication Data
will be found on the last printed page of this book

Composed and printed in the United States of America
by Princeton University Press, Princeton, New Jersey

ISBN 978-0-691-64427-1

TO
Susan, Eleanor, Christopher, and Diana
AND
Fusako, Noriko, and Takashi

PREFACE

Why are Japanese industrial organizations so successful? Some studies have sought answers to this question in such factors as Japan's national economy, polity, and culture, or its international setting. In recent years, other researchers have focused more on the characteristics of Japanese organizations themselves; we join this group, for this book is most centrally a study in the sociology of complex organizations. Our purpose is to relate three sets of variables—technological modernization, the modernization of social organization, and organizational performance or effectiveness—at the level of the manufacturing firm and its subunits. The main theoretical problem with which we deal is the relative validity of the paternalism–lifetime commitment model of the Japanese factory and the convergence theory of modernization.

In method and exposition, we combine two approaches that are seldom used together. On the one hand there is ethnographic coverage-in-depth that results from some three months' field work in each of the three main firms studied, including quotations from interview protocols and observations of work and non-work settings. On the other hand we also capitalize on the strengths of the method of systematic comparison across firms. A standard questionnaire was completed by employees in each firm (N = 1,695 for the three firms, with a mean response rate of 84 percent). Data from the questionnaire and company records enable us to supplement qualitative description with multivariate statistical analysis. The separate and joint influences of a variety of theoretically-derived independent variables upon given dependent variables are analyzed by means of linear multiple regression analysis. Propositions concerning both intrafirm and interfirm variations are tested with the comparative data from the firms studied.

The reader who takes up a book of this length deserves to know what the fruits of his labor will be. It is our position that the paternalism–lifetime commitment model of Japanese factories exaggerates their uniformity, traditionalism, and "Japaneseness"; that the most successful firms (each of our three firms is among the top five in sales in its industry) have in fact moved in the direction of a more modern organizational structure, and will continue to move in this direction; and that those firms that have not moved in this direction will suffer significant organizational strains. We shall also attempt to show that although a Japanese firm may have a given attribute of the paternalism—lifetime commitment model, the empirical consequences of that attribute predicted by the model are often not, in fact, observed in our data. Finally, our data suggest that the more "Japanese" elements in factory social organization have less impact on the functioning of firms than do the more universal organizational features.

This book has been over six years in the making; it would have taken considerably longer had it not been for the help—intellectual stimulation, introductions, advice, patience, and understanding—of numerous people in the United States and Japan. We are painfully aware that we can acknowledge in particular only a few of the many whose lives crossed ours and that of this book over these years. During the initial planning stage, we benefited greatly from the experience of Professors John C. Pelzel of Harvard University, Richard K. Beardsley of the University of Michigan, Bernard Karsh of the University of Illinois, and Dr. James C. Abegglen of the Boston Consulting Group. Professor Sidney Goldstein, chairman of the Department of Sociology at Brown University at the time, encouraged the project and gave it strong administrative backing.

In Japan, Mr. Yamada Hiroshi, president of the Japan Human Relations Association, was instrumental in paving the way for us to do extended field work in Sanyo Electric Company, and by inviting us to present the earliest results of our research before the members of his Association on several occasions during 1970, provided a setting in which we could receive valuable reactions to our line of thinking. Yamada *san* in the course of time became our mainstay, friend, and patron. A longtime friend, Professor Morioka Kiyomi of Tokyo Kyoiku University, deserves thanks for his advice on several matters.

The study would not have been possible without the financial and other support of the John Simon Guggenheim Memorial Foundation and the Ford Foundation, a sabbatical leave from Brown University for Marsh, and the generosity of our sponsoring institution, Kwansei Gakuin University, Nishinomiya, Japan. Mr. Ishii Tōru, a graduate student in the latter university, performed a valuable role as our principal assistant in field research. In addition, Brown University helped during the analysis phase by providing free computer services and research stipends.

Once the field work began, our indebtedness to the managers, staff, and workers of the Hojō Plant of Sanyo Denki K. K. (Sanyo Electric Co., Ltd.), of the Diesel Engine Division of the Sakurajima Plant of Hitachi Zosen K. K. (Hitachi Shipbuilding and Engineering Co., Ltd., Osaka), and of the Nishinomiya brewing plants of Konishi Shuzō Co. (Konishi Brewing Co., Ltd.) quickly cumulated. These individuals not only admitted us to their midst, but made us feel at home, both at work and in recreational activities. What might have been lonely periods for us on Sundays and evenings were made memorable human occasions by the warmth, humor, and hospitality of these people. In particular we wish to acknowledge our gratitude to Mr. Gotō Seiji, vice president of Sanyo Denki K. K., Mr. Furushita Shozo, division managing director, Hojō Factory, Sanyo Denki K. K., Mr. Sanji Hironobu, former plant manager, and Mr. Nishiyama Takuya, present plant manager of the Sakurajima Plant, Hitachi Zosen K. K., Mr. Matsumoto Teizo, general manager, and Mr. Satō Zenshin, engineering manager, Konishi Shuzō Co., Ltd., Nishinomiya.

Since finishing our field work, we have published several articles and extensively revised the manuscript of this book three times. During the four-year analysis phase, our colleagues at Brown University, Kwansei Gakuin University, and elsewhere have in many ways provided theoretical, empirical, methodological, and bibliographical help for which we are deeply appreciative. If we have not always been able fully to attain their high standards, the book is, we trust, the better for the attempt to meet them. We, of course, must take responsibility for whatever limitations remain. Special thanks are due to Professors Peter B. Evans, Colin Loftin, and Dietrich Rueschemeyer of Brown University, to Professor William Form of the University of Illinois, to Professor Koshiro Kazutoshi of Yokohama National University and the Japan Institute of Labor, to Professor James L. Price of the University of Iowa, and to Professor Azumi Koya of Rutgers University. The members of the Columbia University faculty seminar on modern East Asia: Japan, of which Marsh is a member, offered most searching criticisms of a paper based on this study, and deserve thanks. Finally, our students—both graduate and undergraduate—have given us the benefit of their fresher perspectives on our data and inferences.

We wish to thank the Macmillan Company for permission to quote some materials from James C. Abegglen, *The Japanese Factory: Aspects of its Social Organization*, published by The Free Press, Glencoe, Illinois. Copyright 1958 by the Massachusetts Institute of Technology.

ROBERT M. MARSH and HIROSHI MANNARI

August 30, 1974

CONTENTS

TABLES

FIGURES

Modernization and the Japanese Factory

CHAPTER ONE

Introduction to the Problem

A society may be considered modernized "to the extent that its members use inanimate sources of [energy] and/or use tools to multiply the effects of their efforts" (Levy 1966:11). This definition has at least two virtues. First, it is an objective definition and therefore lends itself to operational measurement: a society's level of modernization is its inanimate energy consumption per capita. Second, it is a deliberately narrow, rather than broadly inclusive, definition. It therefore treats the question of what kinds of polity, family patterns, values, etc. are associated with varying levels of energy consumption per capita as a question for theoretical and empirical investigation, rather than one to be settled by definitional fiat.

In this study we are concerned with the validity of the convergence theory of modernization. This theory holds that as societies become more highly modernized they tend to become more alike in their social and cultural structure. Few would argue that the attainment of high modernization eliminates all differences among societies which stem from their different histories and culture, from whether they are early or late modernizers, and the like. But analysts disagree on the *extent* of similarity that results.

This study attacks the problem of convergence at the level of complex organizations—specifically, industrial manufacturing firms. At this level there is a problem in defining "modernization" in terms of per capita energy consumption that is avoided at the societal level. This is that energy consumption per capita is highly industry-specific. Steel firms consume much more energy than lace textile firms, but are not necessarily more modernized. To handle this problem, we shall use energy consumption per capita as the measure of modernization only for within-industry comparisons of firms; for cross-industry comparisons of firms we shall use the extent of reliance on inanimate, mechanized, and automated processes as the criterion of modernization.

Japan, though still less modernized than the United States and some other Western societies, is narrowing the "modernization gap" between itself and those societies, both at the general societal level and at the level of its larger firms. Moreover, the performance of those Japanese firms (as measured, for example, by their volume of sales) is also tending toward levels as high as those of comparable Western firms. The convergence theory of modernization, as applied to the level of complex organizations, would predict a similar convergence in the social organization of Japanese and Western firms. On social structural and technological grounds, theory would lead us to expect large, successful Japanese firms to be relatively modernized in their social organization, and to be becoming more so. One widely known model of the Japanese firms—the paternalism–lifetime commitment model—denies this, and argues instead for an anti-convergence theory. ". . . [I]ndustrial organization in Japan has followed a different course from that of the United States; yet it has also achieved outstanding results. Indeed, it seems likely that it is as a consequence of having developed a different, Japanese approach to organization that Japan has accomplished the industrial success that it has" (Abegglen 1969:100).

It is our position that the paternalism–lifetime commitment model exaggerates the uniqueness, uniformity, and traditionalism of the social organization of Japanese firms; that the most successful firms have in fact moved in the direction of a more modern organizational structure; and that those firms that have not moved in this direction will exhibit significant organizational strains. We shall also attempt to show that although a Japanese firm may have a given attribute of the paternalism–lifetime commitment model, the empirical consequences of that attribute predicted by the model are often not, in fact, observed in our data.

JAPAN'S RAPID INDUSTRIAL DEVELOPMENT

By almost any measure, Japan's economic and, more specifically, industrial development has been a notable success. Its real annual economic growth rate averaged over 10 percent through the 1960s (Oriental Economist 1969, 1971). Its industrial productivity index number (1960 = 100) was 172 in 1965 and 371 by 1970 (Ōuchi 1971:3). Among the 49 nations for which the United Nations presented data, Japan had the fourth highest rate of increase of industrial production in manufacturing between 1964 and 1969 (United Nations 1971: table 51, pp. 167–77). Even among 10 leading industrial nations, Japan's index number for labor productivity in manufacturing (1963 = 100) in 1968 was second only to that of the United States.

At the level of firms a pattern similar to that at the national aggregate level can be seen. *Fortune* publishes an annual list of the 200 manufacturing and mining firms with the largest sales in the world (excluding the United States). The number of Japanese firms in this group increased from 31 in 1962 to 51 in 1970 (*Fortune* 1963:139–43; 1971:151–54). In 1962, two other countries—Britain and West Germany—had more companies among the top 200 than did Japan, but by 1969 (and again in 1970 and 1971) Japan had outdistanced them and attained the position of the country with the largest number of firms among the top 200 (*Fortune* 1972:153–56). In 1972, *Fortune* extended its list to include the top 300 firms in the world outside the United States. Japan led with 79 firms, followed by Britain with 61 and West Germany with 43 (*Fortune* September 1973:203). The Japanese firms' rate of growth in sales was higher than that of the European firms. When United States firms are included, the number of Japanese firms among the 200 firms with the largest sales in the world increased from eight in 1962 to 14 in 1969. Although 114 United States firms had sales of over $1 billion in 1969, in contrast to 13 Japanese firms, in dynamic terms the striking fact is that the Japanese firms were generally rising in rank among the world's leading 200 firms (mean rise of 11.3 ranks in sales from 1968 to 1969), while the American firms were generally falling (mean fall of 2.3 ranks).

What accounts for Japan's success in industrial development? At the national level, a number of explanations have been advanced, variously emphasizing political, economic, technological, sociocultural, psychological, or other factors (Levy 1953, Lockwood 1954, Bellah 1957, Smith 1958, Nakayama 1964, Allen 1965, Chao 1968, Yoshino 1968, Stone 1969). The Japanese government's Economic White Paper for fiscal 1968 summarized these explanations as follows: "Japanese people's industriousness, their high rate of savings, ample and excellent labor, constant technological innovations, consumption revolution caused by the advent of new consumer durables . . . and the existence of a competitive society based on institutional reforms are the major factors that have supported the nation's rapid postwar economic growth" (*The Mainichi Daily News*, 1969:5). In this book we are concerned with explanations at the level of the industrial firm. As Japan's larger firms begin to overtake their Western counterparts in sales and productivity, it is significant to ask, what are the factors on the level of the firm that explain this success? This takes us at once into what may be called the "Japanese factory controversy."

THE JAPANESE FACTORY CONTROVERSY

The controversy over the Japanese factory concerns whether, given the Japanese sociocultural setting, industrial efficiency is maximized by traditional forms of organization, by some combination of traditional and modern forms, or by the wholesale adoption of modern forms.[1]

[1] Our use of the terms "traditional" and "modern" is analytical rather than historical. Some elements hypothesized as being modern may have existed in the historical past in Japanese organizations or in Japanese society in general; and some elements hypothesized as being traditional may not have existed in Japan's historical past but may, indeed, have been introduced in recent times.

Japanese factory owners, managers, and personnel directors are preoccupied with the rationality and efficiency of their production system. However, they find it difficult to evaluate whether their traditional practices in personnel management are effective in production under changing conditions. Thus, what for sociologists is a central theoretical and research problem is for Japanese industrialists a very practical dilemma.

By the "paternalism–lifetime commitment model" we refer to a variety of similar ideas that have been identified by different terms, such as Whitehall and Takezawa's (1968) "involvement hypothesis," and what Dore (1973) calls being "organization-oriented" rather than "market-oriented." We recognize that these and other Japanese and Western scholars—Matsushima and Nakano (1968), Abegglen (1958), Levine (1958), Matsushima (1962), and Vogel (1963)—do not all regard the thesis as of equal validity.

By the paternalism–lifetime commitment model of the social organization of Japanese firms in this study, then, we refer to the following elements. The relationship between the firm and the employee is one of lifetime commitment (*shūshin koyō*): the employee enters a firm after completing school and remains in that same firm until retirement. Lifetime commitment is supported by Japanese beliefs in "the firm as one family" (*kigyō-ikka*), and "familistic management" (*keiei kazoku-shugi*) (Hazama 1960:137). The basic link between the employee and the firm is more a matter of loyalty and reciprocal obligation than a rational economic calculus (Abegglen 1958:17). There is a seniority system (*nenkō joretsu-seido*) according to which base pay, allowances, and promotions depend more on age and length of service in the firm than on job or performance. Inefficient employees are retained, and recruitment screening pays little attention to skills directly related to work performance.

The model posits further that the strength of lifetime commitment derives not only from the present social structure of Japan, but also its traditional, feudal past. "At repeated points in the study of the factory, parallels to an essentially feudal system of organization may be seen—not, to be sure, a replication of the feudal loyalties, commitments, rewards, and methods of leadership but a rephrasing of them in the setting of modern industry" (Abegglen 1958:131).

It would be one thing if this paternalistic, particularistic, diffuse, and less rationalized form of social organization were found mainly in small-scale, technologically less advanced firms. But the paternalism–lifetime commitment thesis is that these social organizational features are more common in Japan's largest (over 1,000 employees) and technologically most advanced firms. It is this that gives the thesis much of its anti-convergence flavor.

The controversy over the extent of lifetime commitment and other aspects of traditional social organization in Japanese industry has gone on for over a decade now. The paternalism–lifetime commitment model has been challenged by several Japanese and Western social scientists (Tominaga 1961, 1962; Taira 1962, 1970; Odaka 1963; Noda 1963; Tsuda 1965; Shirai 1967; Sumiya 1967; Cole 1971a; Okochi 1972; and Okochi, Karsh, and Levine 1973). These critics argue that lifetime commitment applies to only a small minority of wage earners in manufacturing. They argue further that the situation the model describes as traditional in Japan was in fact a postwar aberration in the capital and labor markets and that during the 1960s Japanese industry began to return to more rational modes. But the model has had more influence than its critics have on social scientists who are not Japan specialists[2], and Abegglen has held to his original position in more recent writings, claiming that the alleged changes in factory social organization are more apparent than real (Abegglen 1969). Moreover, Dore's entering the lists (Dore 1973) will give the model a new lease on life. Dore's version of the model depicts the typical Japanese factory as organization–oriented, that is, employees have a lifetime commitment to one firm, there is a seniority-plus-merit wage system, an individual has a career *within* the firm, is trained in and by the firm, the source of his welfare is the firm, not the government, and a high degree of consciousness of the firm as such is nurtured.

As we subject the paternalism–lifetime commitment thesis to close theoretical and empirical scrutiny in this book, we shall refer most often to Abegglen's version. This is not because we regard it as the best statement; Dore's is superior, but was published too late to be our central referent. It is not our intent to single out Abegglen for *ad hominem* attack, but rather to provide the reader with a specific source for the propositions we shall test. In other words, what is important is the paternalism–lifetime commitment model, and what can be learned from it and from the Japanese experience, not the particular assertions of any of the proponents of the model.

It should be clear that the controversy over the model decisively joins a number of theoretical issues: To what extent can there be a convergence of technology without a convergence of social structure in organizations? What type of organizational structure maximizes performance and efficiency? To what extent can the behavior and attitudes of members of organizations be explained on the basis of the culture

[2] March (1965:xi) has shown that Abegglen's *The Japanese Factory* is among the most frequently cited books in the field of the sociology of complex organizations, where it is usually used as evidence of how formal organizational structure varies from society to society as a result of cultural and historical factors. It is also often cited by those who attack convergence theories of modernization, i.e., by those who argue that highly modernized societies do *not* become more alike in social and cultural structure.

and history of their society, to what extent on the basis of technological and structural characteristics of the organizations themselves?

With these questions in mind, a number of systematic empirical studies in Japanese factories have been undertaken in recent years, by Okochi et al. (1959), Whitehill and Takezawa (1968), Cole (1971a), Dore (1973), and Okochi, Karsh, and Levine (1973); the present book is an addition to this growing literature.

These questions involve the relationships between three sets of variables—social organization, technology, and performance—at the level of complex organizations. We next present a conceptual scheme that defines these variables.

CONCEPTUAL SCHEME

A complex organization is a structure for coordinating the work of many persons that is deliberately established to accomplish certain ends. Firms, hospitals, schools, churches, and armies are complex organizations; families, friendship cliques, and neighborhoods are not. For any of these complex organizations, we may analytically distinguish such components as their environment, technology, social structure, or culture. The core concept of this study is that of the social organization (social structure) of the firm and its subunits. At the most general level, this social organization can be viewed according to its degree of bureaucratization.

In Weber's famous statement, bureaucracy is one type of administration, namely, the modern-rational, as opposed to the traditional and charismatic (Weber 1946). An organization is bureaucratic to the extent that its personnel operate on the basis of norms and values of rationality, functional specificity, universalism, and achievement; authority is monocratic, but functionally specific, not diffuse and paternalistic; personnel are selected on the basis of skills related to specific tasks; promotions, pay, and the like are based on workers' performance, rather than on age or seniority alone. Remembering that these elements are variables, not dichotomous attributes, let us consider them as concepts that disaggregate the master concept of the social organization of a firm.

1. *Formal structure.* Formal structure is defined as "the stable and explicit pattern of prescribed relationships in the organization" (Woodward 1965:10). More specifically, the formal structure refers to the number of hierarchical levels in the organization, the ratio of managerial and staff personnel to all personnel, and the span of control of managers on different levels (Blau and Schoenherr 1971:15).

2. *Systematic procedures.* The tasks performed by personnel follow "general rules, which are more or less stable, more or less exhaustive, and which can be learned" (Weber 1946:198). More recent studies have made a sharper distinction between "rules that carefully codify behavior and sanctions for violation, thus becoming substitutes for close supervision" and "procedures that give guidelines, the standards that are necessary when something occurs frequently" (Azumi and Hage 1972:223–24).

3. *Division of labor.* Formal organizations are by definition designed to produce output relevant to the attainment of a specialized goal. To do this requires that the activities of the members of the organization be coordinated in a division of labor. The job of each member can be categorized according to its complexity—the degree of skill and formal education it requires—and its scope—the extent to which it involves narrow, fractionated, highly repetitive tasks rather than a range of highly diversified tasks.

4. *Status structure.* In a complex organization, the division of labor has "vertical" as well as "horizontal" components. Status structure refers to dimensions of positions that are hierarchically ranked in the organization, including in this study job status, formal rank, education, seniority, and status of the section. In its individual referent, status structure refers to an employee's organizational status, in terms of his job status, rank, and other variables.

5. *Rewards.* In economic or utilitarian organizations, the main incentives or rewards for participation (work) are (a) economic—in the form of monetary remuneration—and (b) influence and prestige—in the form of promotion in the job and rank hierarchy. (Job satisfaction and other elements to be noted below may also be rewards, but we shall consider them separately.) These rewards are distributed differentially. Obviously, not all personnel have an equal say in defining the criteria of rewards. Generally, it is specific personnel at the top of the organization who have the authority to define these criteria and to decide the relative contribution (and therefore rewards) of various other employees.

Pay and promotion are the main objective components of an organization's reward system, but there are also subjective components with which we shall be concerned. In regard to pay, we shall analyze the type of wage system employees prefer, e.g., seniority- or job classification-based wage systems. We shall also try to uncover employees' values concerning the factors they think *should* determine pay. With regard to promotion, we shall introduce two concepts. First, employees' perceived promotion chances. Second, for employees who see their promotion chances as poor, the concept of the legitimacy of the promotion system is important. The more employees think the reasons for their poor promotion chances lie in their own limitations, and therefore think they do not deserve promotion, the more legitimacy the promotion system has. Conversely, the more the organization is seen as to blame for employees' poor promotion chances, the less legitimacy the system has.

Up to this point we have disaggregated the master concept, social organization, or social structure of the firm, in terms of variables central to the theory of bureaucracy. These variables refer mainly to the formal structure and the processes of social differentiation—the division of labor, status, authority, and rewards—in organizations. In contrast, the following variables refer more to the informal structure and the internal social integration of organizations.

6. *Social integration.* Processes of internal integration have the function of coordinating and unifying the differentiated parts of an organization. Specifically, social integration will here refer to the ways in which, and the degree to which, the employee is informally socially integrated into the firm. Because, according to the paternalism-lifetime commitment model, this is the most distinctive aspect of the Japanese factory, it is here that we shall focus much of our attention. The concept of social integration can be disaggregated into seven subconcepts which have special reference to Japanese firms, though they are in principle applicable to any organization in any society.

a) *Employee cohesiveness.* Employee cohesiveness is defined as the extent and intensity of social bonds (friendship, teamwork, etc.) among the employees of an organization. An organization with high cohesiveness is one in which employees' workmates are also their friends, in which employees interact in leisure contexts outside of work as well as at work.

b) *Company paternalism.* Company paternalism may be defined as present when "the managerial element assumes responsibilities for workers over and beyond the basic contractual provisions for wages and routine working conditions. The responsibilities include . . . the practice of carrying workers on the payroll in periods of business decline; housing; religious facilities; and many others" (Bennett 1968:475). In our usage, paternalism refers on the one hand to the extent of benefits the company makes available to its employees, especially in the form of company housing, and on the other hand to the extent to which employees prefer their company and its management to have diffuse and particularistic, rather than functionally specific, relationships with them.

Although we think that social structural factors generally have more influence on organizations than cultural factors, we are aware of the importance of cultural variations in the meaning of "paternalism," and of how this could affect our structural analysis. The Japanese term for paternalism is "*onjō shugi*," literally, "warm heartedness," a term which has very positive connotations. (The more Western, negative connotation of paternalism as "meddling, gratuitous interference in another's life" is expressed in the Japanese term *kanshō seiji*.)

c) *Participation in company recreational activities*. Organizations attempt to integrate their members not only by coordinating their work activities and distributing rewards, but by encouraging them to participate in organization-sponsored recreational activities, such as clubs and sports. The more successful the company is in this area, the more of the total life space of its personnel is encompassed by their membership in the organization; potential competing organizational loyalties can thus be minimized.

d) *Company involvement and identification*. The paternalism–lifetime commitment model posits a relationship of close functional interdependence between company paternalism and employees' involvement in the company. Participation in company recreational activities is only one aspect of employees' involvement in the company; they may also identify with the company's more instrumental goals. We refer to the latter as company involvement and identification. The model we are testing sees the employee's involvement in and identification with the company's performance goals as his way of reciprocating the paternalistic benefits he receives from the company. In its extreme form, the employee's involvement and identification would be such that he would not differentiate his personal identity and problems from those of his firm. The extent to which this is true, and its conditions and consequences, are what we shall study.

e) *Conflict*. Conflict is defined as "a struggle over values and claims to scarce status and resources in which the aims of the opponents are to neutralize, injure, or eliminate the rivals" (Coser 1956:8). We distinguish between dissatisfaction and social conflict. The former is subjective and refers to attitudes of employees; the latter is objective: as in the above definition, conflict refers to overt struggle, which may be verbal or physical or both. Although dissatisfaction is a necessary condition for conflict, there can be dissatisfaction without conflict.

We shall analyze conflict in Japanese firms by drawing on both conflict and consensus models of organization. The conflict model derives from the Marxist tradition. In the present context, the model holds that for any given industrial firm, strata can be distinguished on the basis of the ownership or non-ownership of the means of production, or in a more recent formulation, on the basis of authority differences in an imperatively coordinated organization (Dahrendorf 1959). These strata have opposing interests, which result in social conflict, unless the "real" interests are hidden by the operation of socialization techniques, the mass media, and the like. Japanese scholars' work on Japanese industry is often based on the conflict model.

The consensus model stresses more the identity of interests and common ideology among managers and workers. The human relations tradition developed specific managerial techniques for maximizing this cooperative system and identity of interests.

f) *Interfirm mobility*. According to prevailing views of Japanese organizations, individuals who change firms are less integrated into their present firm and have less lifetime commitment than those who always work for the same firm. In this view, firms can be compared with regard to the integration of their employees by comparing the proportion of employees who have always worked for their present firm with the proportion who have worked for one, two, or more other firms. This is the role behavior side of the concept of lifetime commitment.

g) *Lifetime commitment norms and values*. The core of the concept of lifetime commitment (or "permanent employment") is not simply that Japanese exhibit the role behavior of staying in one firm; it is that they stay because they hold certain beliefs and have internalized certain norms and values. Lifetime commitment beliefs, norms, and values profess that employees *should* stay in one firm and express moral disapproval toward those who voluntarily change firms.

A crucial difference between Abegglen's analysis and our own is that he assumes that the behavioral and the norm-value aspects of lifetime commitment vary together in such a way that those who never change firms are more wedded to lifetime commitment norms and values than those who do change firms. We, on the other hand, treat this as an open empirical question.

The reader should note that underlying all these dimensions of social integration is one fundamental variable: the extent to which members of an organization differentiate their own personality, identity, social relationships, and activities from those of the organization.

This completes the discussion of the social organization of the firm. The next two concepts are job satisfaction, which is more social psychological than social structural, and value orientations, a concept that refers analytically to the cultural system rather than the social system.

7. *Job satisfaction*. Individuals experience different degrees of gratification or deprivation from the specific jobs they perform in an

organization; the attitudes they express in this area are conceptualized as their job satisfaction. The functional role an employee performs in the organization's division of labor and his status and authority interact with his subjective needs and expectations to produce a given degree of job satisfaction.

8. *Work values.* Apart from their attitudes toward their specific work roles, members of an organization have more general work values—conceptions of what is or is not desirable, bases on which they choose among alternative courses of action (Kluckhohn and Strodtbeck 1961). We shall analyze the extent to which employees choose work values in preference to family or pleasure values. Work values have been variously referred to in the literature as the Protestant ethic (Weber 1950, Bellah 1957, Eisenstadt 1968), labor commitment (Moore and Feldman 1960), and the work ethic. Analytically, value orientations are part of the cultural system rather than the social system. Although, like job satisfaction, values are analytically distinct from the social structure of a firm, they interact with it and thus should be taken into account. We conceive of work values as being in part orientations individuals develop from their socialization in the wider society and bring with them when they enter a firm, but we also view the priority an individual gives to work values relative to other values as interacting with his specific experiences throughout the period of his membership in the firm. Specifically, we view these values as being influenced by one's position in the division of labor, one's organizational status, and so on, as well as influencing the social organization and performance of the firm.

The two remaining concepts of this study—technology and performance—are also analytically distinct from the social organization of a firm.

9. *Technology.* This is defined as the techniques and materials an organization uses in its workflow activities, in order to create its output. In a manufacturing setting, technology refers to those sets of person-machine activities that together produce some commodity (Azumi and Hage 1972:138). Types of technology and other conceptual distinctions will be introduced in Chapters 4 and 5.

10. *Performance.* This is defined as the degree to which an organization achieves its goals (Price 1968). An organizational goal is defined, following Etzioni (1964:6), as "a desired state of affairs which the organization attempts to realize." Organizations with high performance are effective (they realize their goals) and efficient (they realize their goals with minimum cost). Further conceptual distinctions concerning performance will be introduced in Chapter 11.

PLAN OF THIS BOOK

Chapter 2 introduces the three Japanese firms we studied. Each of the next nine chapters (3–11) deals with one analytical aspect of social organization, technology, or performance. In each chapter there are parallel analyses of each of the three firms. Although we incorporate the results of the multivariate statistical analysis in the text, the technical details appear in the appendices. There is some explicit comparison of firms in Chapters 2 through 11, but this becomes the main focus only in Chapter 12, when we consider systematically the uniformity and variation in the social organization of the three firms. In the last chapter we also put our findings in a larger context by relating them to data on other Japanese firms and Japanese society in general, and draw general substantive and theoretical conclusions.

CONCLUSION

The aim of this study is to relate three sets of variables—technological modernization, the modernization of social organization, and organizational performance or effectiveness—at the level of the industrial manufacturing firm and its subunits (factories, departments, sections). The core design consists of a comparative study of each of these three sets of variables, principally within Japan. The main theoretical problem involves the relative validity of the paternalism–lifetime commitment model and the convergence theory of modernization. Although this issue ideally calls for a study of comparable data on Japanese and Western firms, we concentrate here on variations among firms within Japan.

CHAPTER TWO

Three Japanese Firms in Their Industry Settings

SELECTION OF FIRMS TO BE STUDIED

We take a middle ground between the older method of the case study of a single organization and the newer strategy of collecting aggregate or structural data from one or a few informants in each of a large number of organizations. Our research questions can only be answered on the basis of a comparative analysis of firms and factories. At the same time, the paternalism model of the Japanese firm which we are testing concerns the properties of individuals as well as those of organizations. To test this model we need data on employees' behavior, attitudes, and values. Our analysis will combine ethnographic coverage-in-depth, based on three months' field observation in each factory, and multivariate statistical analysis. The emphasis on field work limited the number of firms we could compare.

The firms we studied were selected on the basis of the following criteria. First, they should vary in industry and type of technology in order that we might discover the extent to which the paternalism–lifetime commitment type of organization is constant, or varies, as these conditions vary. Testing relationships in a variety of industries and technologies within Japan should also lend somewhat more generality to our findings. Second, the proponents of the lifetime commitment model say it is most applicable to larger firms—those with at least 1,000 employees—and we accordingly sought to select only firms of this size.[1] Third, to be selected a firm had to be among the top five in performance (as measured by sales) in its industry. This criterion was imposed because we wanted to test the assertion that the paternalism–lifetime commitment pattern is characteristic of even the most successful large Japanese firms.

Some of the firms that met these three criteria refused permission or put severe restrictions on the duration of field work or type of data that could be collected. In general Japanese firms are very open to brief

[1] We decided to study Sake Company even though it employed less than 1,000 people. This gave us the opportunity to compare a new, relatively automated plant and three traditional brewing factories, thereby broadening our range of variation in technology.

inspection tours, but are more reluctant to permit research of the depth we requested, involving three months of field work, access to detailed company personnel and productivity records, close observation on the plant floor, and intrusion into work time for interviews. Thus, of the 13 firms we approached, only four fully accepted our conditions; time permitted a full-scale study of three of these.

Therefore, we make no claim that these three firms were randomly selected. The fact that they are leading firms in their respective industries in itself reduces their representativeness. Nor do we assert that they are representative of large, leading firms in general, since we have studied only a small range of industries. The extent to which these firms are representative is not our concern, for our intent is not to generalize from the firms studied to some larger universe of Japanese firms. Rather, it is to describe and explain patterns in the firms studied, as these patterns relate to theoretical issues.

Table 2.1 summarizes some main characteristics of the three firms, and of the branch plant(s) studied in each firm. In each of these plants data were collected from four sources, using parallel procedures and research instruments: (1) company records on personnel and production; (2) observation of employees both in their work and non-work settings; (3) semi-structured interviews with individual managerial, staff, and production personnel lasting on the average one and a half hours; and (4) a self-administered questionnaire distributed to all managerial, staff, clerical, and production personnel. Further details on data collection and field work are given in Appendix A. The questionnaire, which is discussed more fully in Appendix A, was completed by a total of 1,695 employees in the three firms, making the average response rate 84 percent (see Table 2.2). Comparison of the status and job characteristics of respondents with those of all employees in each factory indicates no major differences (see Appendix A).

The remainder of this chapter is devoted to describing the three firms in their respective industry settings.

THE JAPANESE ALCOHOLIC BEVERAGE INDUSTRY

Japan's indigenous alcoholic beverages were all made from fermented rice, and the principal one was rice wine, or *sake*.[2] Sake brewing was one

[2] The material dealing with Sake Company is based on two major sets of data. The first is the field work conducted by three Japanese sociologists in the winter of 1966–67, and published in Mannari Hiroshi, Maki Masahide, and Seino Masayoshi, "Sake tsukuri no rōdō no soshiki: sangyō shakaigakuteki kenyū."(The Organization of Labor in Sake Manufacturing: An Industrial Sociological Study), *Kwansei Gakuin Daigaku Shakaigakubu Kiyō* 15 (December 1967):1–32. Section 1 of this paper was by Maki, section 2 by Mannari, and section 3 by Seino. The first section dealt with the "Roles of Master and Worker in a Traditional Sake Plant," the second with "Innovation and Labor Organization in a New Sake Plant," and the

of Japan's largest industries prior to the industrial revolution in the 1890s. An 1874 source, *Fu-ken bussan hyō* (List of Prefectural Products), tells that the food industry, including alcoholic beverages, manufactured 42 percent of all industrial products in Japan. Breweries alone produced 16 percent of all manufactured products (Sakaguchi 1964:ii). While the

TABLE 2.1
Characteristics of Firms and Plants Studied, 1969–70

Characteristic of Firm	*Electric Company*	*Shipbuilding Company*	*Sake Company*
Industry	Electric machinery, appliances	Shipbuilding	Beverages
Capitalization[a]	¥ 24,192	¥ 18,960	¥ 100
Annual Sales[a]	¥ 215,205	¥ 139,512	¥ 10,727[b]
Rank in sales among manufacturing and mining firms	20th	39th	—
Number of employees	14,838	18,500	525
Sales per employee[a]	¥ 14.5	¥ 7.5	¥ 20.4[b]
Characteristic of Branch Plant Studied	*Electric Factory*	*Shipbuilding Factory*	*Sake Factories*
Main products	Electric motors; electric fans; vacuum cleaners	Diesel and turbine engines; industrial heavy machinery	Rice wine (sake)
Number of employees	1,212	756	68
Percent male employees	50	96	100
Average age of employees	22	34	49
Average length of service in company	4 years	11 years	7 years
Quit rate during previous years	7%	5%	33%

Source: *President* 1970, Nihon Keizai Shimbunsha, *Kaisha sōkan* 1970, and *Kaisha nenkan* 1970.

[a] In millions of yen. In 1969, the official exchange rate was U.S. $1.00 = ¥ 360.

[b] The published figure for Sake Company's sales was ¥ 20,239 million, and for its sales per employee, ¥ 38.5 million, but the sake industry is a government monopoly, and the sales figure in 1969 includes a 47 percent tax. After this is deducted, the sales were ¥ 10,727 million and the sales per employee ¥ 20.4 million.

third with "The Rural Social Situation of Workers in Sake Plants." Because Mannari also collaborated with Marsh in restudying some of the same sake factories during the fall and winter of 1969–70, we have, with the permission of Maki and Seino, translated the 1967 article into English and taken the liberty of reorganizing it along lines that fit our own conceptual and analytical scheme better. The second major set of data is from Mannari and Marsh's field work in Sake Factory in 1969–70. In the following pages it should be understood that we are heavily indebted to the original published paper, though we shall not cite it every time we draw material from it.

TABLE 2.2
Questionnaire Response Rates

	Number of Employees[a]	*Number of Usable Questionnaires*	*Response Rate (Percentage of Usable Questionnaires)*
Electric Factory	1,201	1,033	86.0
Shipbuilding Factory	756	596	78.8
Sake Factory	68	66	97.1
Total	2,025	1,695	83.7

[a] At the time the questionnaire was distributed.

textile and steel industries were still in a handicraft stage in Japan, sake brewing had already attained the level of large-scale, albeit traditional, manufacture.

Centuries ago, the sake guilds were among the strongest such organizations in Japan. "As early as during the *Kamakura* period (1192–1333), the sake brewers formed one of the most powerful *za* or guilds, enjoying practically monopolistic rights in each city, the members further increasing their wealth by lending money at exorbitant rates. Sake and oil were two of the most important commodities consumed in the country" (Casal 1940:72). During the eighteenth century, the powerful sake guilds "seceded from the Yedo-Osaka Shipping Federation to form their own organization, which within a few years boasted of 106 vessels and annually transported up to 260,000 casks between the two cities" (ibid.:42).

The Japanese sake industry was created and has grown as one means of commercialization of the rice crop. Based on the manual labor of peasant owners, it has absorbed surplus labor in slack seasons of the agricultural year. The distinctiveness of the sake industry lies in its dependence upon surplus labor and surplus rice production. In the area of Japan in which our field work was carried out—Osaka, Kobe, and environs—the combination of rice agriculture in the Tamba–Tajima area and the sake breweries in the Nada area is an example of regional functional interdependence.

Sake consumption is increasing in Japan faster than the rate of population increase, partly because women are drinking it more. However, beer is more popular than sake, and whisky, though still much less popular than sake, is increasing the fastest in production (see Table 2.3). The sake industry has long been influenced by government economic policies. In earlier centuries the government rationed the amount of rice that could be used for sake production, because rice was a generalized medium of exchange, and the rice surplus varied from year to year (only recently has the surplus become stable). It was expected that by 1974 all government restrictions on sake production would be removed. But since the number of sake firms is expected to decline (due to rationalization and increasing scale of the firm), many present firms will not survive to enjoy the era of no restrictions.

TABLE 2.3
Japanese Production of Alcoholic Beverages in 1967

Product	*Kiloliters Produced*	*Percentage of Market*	*Percent Increase in Production since 1962*
Beer	2,411,000	56.3	62
Sake	1,402,000	32.7	43
Shōchū[a]	233,000	5.4	—
Whiskey	107,000	2.5	113
Other	133,000	3.1	—
Total	4,286,000	100.0	

Source: Shiba and Nozue 1969:6.
[a] Shōchū is distilled sake with a high alcoholic content.

About 40–50 percent of the retail price of sake reflects the government tax. If there were no tax a regular-grade bottle of sake could sell for about ¥ 450, instead of ¥ 880. Because the tax on alcohol is a source of government revenue, the government in turn assists companies to modernize by taxing that part of a company's profit used to modernize plant and machinery at a low rate.

In 1962 there were 3,913 sake firms in Japan, in 1970 about 3,500. Most sake firms have been and continue to be small. In earlier centuries every village or town in Japan made its own sake. "That *sake* brewing was a home-industry during feudal times, seems obvious from the number of 'brewers' engaged, given as some 27,200 about A.D. 1700 in the *Kwantō* alone, with the *Kwansai* an even bigger producer" (ibid.:42).[3] Sake production is measured in terms of *koku*, and one koku equals 0.18 kiloliters. The smallest-scale firms (150–200 kiloliters, or 825–1,100 koku of sake production per year) are each typically one-family enterprises. In somewhat larger firms (700–1,000 kiloliters, or 3,850–5,000 koku), the owner and his family still form the core of the work force (Mannari, Maki, and Seino 1967:22). In 1969 the company we studied produced 80,700 koku of sake in its eight plants. Although it is still a family-owned firm, with over 500 employees it is obviously no longer a family-staffed firm; not even the core of its employees can be said to be from one family.

Japan's sake industry has a much lower concentration ratio than the other two industries we studied. Let us define this ratio as the share of

[3] Kwantō and Kwansai are the old terms for northeast and southwest Japan, respectively.

total market sales in a given industry that go to the five firms with the largest sales. In 1970, the concentration ratio was 19 percent in the sake industry (Tōyō Keizai, *Tōkei geppō* 1972:8), in contrast to 70 percent in the shipbuilding industry (Kobayashi 1973:204) and 71 percent in the household electrical appliances (*kateidenki*) industry (Komiya 1973:17–18).[4]

Basic comparative data on the five largest sake firms appear in Table 2.4. As of 1969, the sake company we studied (Sake Company B in the table) ranked second in sales, third in sales per employee, second in number of employees, first in age, and fifth in capitalization. (These ranks refer only to position among the five largest firms.)

TABLE 2.4
Selected Comparative Data on Japan's Five Largest Sake Companies

	Sake Co. A	*Sake Co. B*[b]	*Sake Co. C*	*Sake Co. D*	*Sake Co. E*
Sales[a] (millions of yen)					
1965		9,313		8,811	
1966		10,304	12,519	8,208	5,845
1967		14,406	14,566	11,244	8,787
1968		15,720	14,042	10,453	8,827
1969	27,500	20,239	17,943	13,058	11,333
Increase: 1966–69		96.4%	43.3%	59.1%	93.9%
Number of employees in 1969					
Male	420	313	310	248	231
Female	150	212	120	100	85
Total	570	525	430	348	316
Sales/employee in 1969[a] (millions of yen)	48.2	38.5	41.7	37.5	35.9
Capitalization in 1969 (millions of yen)	280	100	350	210	126
Date established	1637	1550	1743	1711	1660
Date incorporated	1927	1933	1897	1935	1908

Source: Nihon Keizai Shimbunsha, *Kaisha sōkan* 1970.
[a] Sales include a government tax on sake.
[b] The firm we studied.

[4] Sales are given in kiloliters rather than in yen in the sake industry. In the shipbuilding industry, if we define the concentration ratio in terms of tonnage of new ships produced, rather than total sales—which includes repairing ships—the ratio is even higher: 76 percent for the leading five firms in 1970.

Each sake firm produces sake in its own plant(s). But in recent years larger firms have been obtaining an increasing proportion of their sake from their subsidiary firm(s) and from a large number of independent plants located throughout western Japan. This sake is blended and then sold under the company's own label.

DEKASEGI: THE INTERDEPENDENCE OF THE SAKE INDUSTRY AND RURAL STRUCTURES

The most outstanding respect in which the sake industry has been and remains traditional is the *dekasegi*[5] pattern of seasonal labor supply. In the 1880 to 1912 period dekasegi workers had their roots in and continuing attachments to their farms, but participated in wage labor during winter months.

At that period, the dekasegi supplied the bulk of the male labor force in the factories (Yoshino 1968:70; Hazama 1964). But the dekasegi pattern had prevailed in the sake industry for centuries prior to that time, and still did at the time of our field work. The work group in a sake plant consists of a master and workers, all of whom are engaged in farming from about April to November, and who come together to make sake in the other months. This calls for a complex set of institutional structures which we shall describe and analyze.

A feeling of trust by workers and their home villagers toward the master (who recruits them annually) still exists (Mannari, Maki, and Seino 1967). There are many proverbs such as "boys will become mature through working in a sake plant away from home," and "when an old man retires, his sons will go to work in a sake plant."[6] The social structure of the sake plant and that of the villages from which workers are recruited are united into one large community. The specific relationship between master and workers is based on the fact that they share "one community" or a "common life" (*kyōdō seikatsu*). The same attitudes can be found among them, because they come from the same villages. The shared feelings of villagers and master, their trust for him, and the master's diffuse authority over the workers are all factors making for the smooth operation of the system.

Thus, workers in a traditional sake plant are not a true proletariat, but rather farmers who work for wages on a seasonal basis. They are still classified as farmers in Japanese society, and it is important to understand not only their role in the factory but their social situation in their home village or town.

[5] *Dekasegi* means migrating away from home in order to work. Okochi (1972:155–62) discusses "dekasegi gata rōdō" as a "feudal irrational type of labor force," comparing three patterns at different stages of Japan's industrialization.

[6] Usually a father and son worked in different sake plants.

THE DEVELOPMENT AND STRUCTURE OF SAKE COMPANY

The history of the company we studied—which we have called "Sake Company"—is inextricably intertwined with that of the development of the Japanese sake industry. Sake first began to be sold outside a local market in the Ikeda and Itami area. Sake Company started manufacturing sake in Itami during the reign of Emperor Gonara, of the Temmon era (1532–1554), Ashikaga (Muromachi) Shogunate. The company shipped sake to Edo (Tokyo) from Itami by horseback, later by small boats, using the Itami–Amagasaki rivers and channels. During the later part of the Tokugawa period, the center of sake manufacturing in Japan moved to Nada go-go—the five villages of Imazu, Nishinomiya, Uozaki, Mikage, and Nishi, in the Nada area.

The *daimyō*, or feudal lord, of the immediate region was a kinsman of the landlord of the area, who founded Sake Company. The daimyo gave Sake Company the protection of the imperial court, which enabled it to enjoy stability over a long period. In good times two or three brothers of the family might divide the company into (1) an original or head or main house (*honke*) sake plant and (2) branch (*bunke*) sake plant(s). Then in subsequent generations, when times were worse, the bunke and honke would merge again into one company. Ever since the sixteenth century there has been the continuity of at least one "Sake Company."

In 1840 *miyamizu*, a special quality water, was discovered in the area and thenceforth used as the best raw material for sake. Sake Company started manufacturing sake in the Nada area (around 1850) by renting three *kura* (plants) in Nishinomiya that another brewing company had operated. Later the other company became a subsidiary firm of Sake Company. Full ownership of these kura—the main ones we studied—came only after World War II, when Sake Company rebuilt them.

Sake Company became known as "Sake Company K. K. (Ltd.)" in 1933. Its stock consists of two million shares at ¥ 50 per share and is not sold on the stock market. The shares are owned by a total of 134 shareholders. The family and an institution representing it control some 79 percent of the stock; former executives in the company, now retired but not members of the family, own three percent; and the remaining 18 percent is owned by about 129 individual shareholders, mostly long-term employees of the company. One becomes eligible to buy company stock after ten years' employment in the firm.

Family ownership is not unique to Sake Company. In the five largest sake firms in 1969, among top management officials listed in Nihon Keizai Shimbunsha, ed., *Kaisha sōkan: mijojo kaisha han 1970*, two of the firms had four individuals with the same surname, two (of which Sake Company was one) had two individuals with the same surname, and the fifth had no individuals with the same surname.

Table 2.4 shows that between 1965 and 1969 Sake Company's sales per year increased sharply. On other economic performance indicators, such as rank in sales among Japanese mining and manufacturing firms and gross capital, it also showed a pattern of success.

Sake company's products

Sake Company has a more specialized line of products than do either Electric or Shipbuilding Company. Its only activity is the brewing of alcoholic liquors (*shurui no jōzō*), including refined sake (*seishu*), synthetic sake (*gōsei seishu*), and *shōchū* a distilled sake with high (45 percent or more) alcoholic content. The company produces one-third of its sake in its own plants, of which about one-half is produced in the traditional plants and the other half in the new automated plant; the other two-thirds of its sake is bought from other local breweries and blended with its own. The company is most famous for its Special Brand sake, known by connoisseurs throughout Japan. Of this Special Brand, a company advertisement says: "Special Brand sake was first brewed in the year 1550 and in the 400 years since, brilliant achievements in the art of brewing and distilling . . . have been made. As a result, Special Brand's present yearly sales are about 11,000,000 gal. and it is ranked among the top class of Japanese breweries. Special Brand sake . . . has an alcoholic content of 16 percent–18 percent. . . . Lovers of Special Brand sake throughout Japan have the following saying, 'As Mt. Fuji is the highest among mountains, Special Brand is foremost among all sakes.'"

In its advertisements, Sake Company gives evidence of at least partial Westernization in that it features mixing its sake in foreign combinations to produce sake martinis, sake on the rocks, sake Manhattan cocktails, sake collins, sake old fashioneds, sake screw drivers, and so on, in addition to sake "Japanese style": "Warm the sake gently, and then sip from tiny, heated cups. This is best enjoyed when served in a Japanese sake set of delicate porcelain."

Market trends, consumer taste, and changing technology are now operating to change the pattern of sake production. More and more consumers want sake with artificially added alcohol (synthetic sake). Only the company's Special Brand uses alcohol processed under the master's supervision; other brands are the result of blending with ingredients bought from other breweries. The new automated plant cannot make the Special Brand, highest grade sake; this can be done only in the traditional plants. As the master of the Number 2 Factory told us: "A year-round operating plant can mass produce sake, but only of one standard quality, not with various tastes, which are still in demand. The best sake can only be made in the winter, in a 'seasonal plant.' But a larger proportion of sake will probably be made in year-round plants in the future."

THE JAPANESE ELECTRICAL MACHINERY INDUSTRY

The growth of Japan's electrical machinery industry since World War II has been spectacular, even compared to the growth of Japanese industry in general.[7] The industry employs about one million workers, ten percent of the labor force in manufacturing. At the beginning of World War II, as a result of mergers a few large firms (Toshiba, Hitachi, Mitsubishi Electric Machine, and Fuji Electric Machine) produced all the diversified products of the industry. Much of the pre-World War II production was of heavy electrical machinery (e.g., generating and transmission equipment), because the government's rush to industrialize made it emphasize capital goods more than consumer goods. However, "following the start of mass production to [*sic*] radio receivers in the 1920's, Matsushita Electric Industries (founded in 1918) climbed to a position ranking next to Toshiba as a maker of vacuum tubes and radio receivers and laid the foundations for becoming a forerunner in the postwar household electrification boom" (Koshiro 1969:5).

The company we studied—which we have called "Electric Company"—owes its success at least partly to this boom. It now produces virtually all the major household appliances that have been a major part of the consumer revolution in postwar Japan: washing machines, television sets, electric fans, etc. The size of the market for these products in Japan alone—not to mention exports to foreign markets—is suggested by data from the Economic Planning Agency. Between 1957 and 1968, the percentage of blue and white collar employees' households with television sets increased from 2.7 to 96.7; households with electric washing machines increased from 8.4 percent to 85.0 percent, and households with electric fans increased from 13.9 percent to 80.1 percent. Electric vacuum cleaners, which were found in 7.7 percent of all employees' households in 1960, were found in 59.8 percent of these households by 1968 (ibid.: Table 1). Although the market for black and white TV was almost saturated by about 1965, a new boom, which started in 1966, was stimulated by sales of color television. By 1969 about one-fourth of all Japanese households owned a color television set.

Japan's electrical machinery industry is one of the industries now characterized by fierce (oligopolistic) competition. Despite the occupation policy, which banned monopolies, by 1955 "the powerful firms had almost completely organized under their control the promising small and medium businesses by means of financial and technical assistance" (ibid.:5).

[7] This discussion borrows heavily from Koshiro 1969.

The growth of Electric Company[8]

Electric Company was founded after World War II. In its first year, seven employees began the production of bicycle dynamo lighting sets. When it became incorporated its initial capital was $55,000. By 1955 the company had, on the basis of a conveyor belt mass production system, achieved 60 percent of the total share of the competitive bicycle lighting field. Between 1950 and 1968 its capital multiplied over 1,000 times—to almost $60 million; sales reached $478 million in 1968. The company has established a number of joint ventures in other countries, e.g., Hong Kong, Malaysia, Taiwan, Ghana, Singapore, and Spain.

Electric Company's rate of expansion is singular even in the Japanese electrical industry, which, as we have seen, is notable for its rapid expansion. The company's success is outstanding relative to other firms, measured by a number of indicators (see Table 2.5). Measured by sales it is one of the 200 largest manufacturing and mining firms in the world outside the United States and among the top 25 manufacturing and mining companies in Japan. As early as 1962, the company had captured nine percent of the total Japanese market in household electric appliances.[9]

Electric company's goals and products

The various products of the electrical machinery industry can be classified as follows (Takeuchi 1966:237, cited in Koshiro 1969:6): (1) electron components, (2) electron tubes and semiconductors, (3) electron

TABLE 2.5
Electric Company's Rank Among 54 Electrical Machinery Companies in Japan, 1964 and 1968

Performance Characteristic	*Electric Company's Rank*	
	1964	*1968*
Annual sales	5	5
Number of employees	8	8
Value added	7	7
Value added per employee	8	11
Sales per employee	4	3
Equipment costs for labor (per employee)	18	11

Source: Tōyō Keizai, *Chingin soran* 1970:156–66.

[8] For a structural history of an American electrical appliance firm, see Downing 1967 on General Electric.

[9] Koshiro 1969:6. Only three firms were ahead of it in this field by 1970: Matsushita, with 28 percent of the market in household electrical appliances, Hitachi (17 percent), and Toshiba (11 percent). Between 1962 and 1970, Electric Company moved from fifth to fourth place in sales in this industry (Komiya 1973:18).

equipment for industry, (4) electric machines for industry, and (5) electric appliances for households. Electric Company has specialized mainly though not entirely in household appliances. Its earliest products were bicycle dynamo lighting sets. It put out a mass-produced radio with a plastic cabinet (rather than the older wooden cabinets), which sold for less than any previous radio. It was the first firm to achieve large-scale production of washing machines. Thus, the company's first image was as a producer of bicycle accessories, its second as a synonym for washing machines. Although it has remained Asia's largest producer of washing machines, its image now is more differentiated. It is no longer limited to consumer electrical appliances, since it also puts out "cold chain" frozen food storage equipment and automatic laundry machinery.

Electric Company's highly diversified products are manufactured in 10 regional plants or affiliates. The main products of Plant Number 1 are compressors and their applications (air conditioners and freezers) and semiconductors (transistors, diodes, and integrated circuits) and products in which they are used (stereos, TV sets); Plant Number 2 produces color and black and white TV sets, radios, and tape recorders; and Plant Number 3 produces washing machines, dry cleaning equipment, and well pumps. Plant Number 4 produces refrigerators and cookers; Plant Number 5 (the one we studied) electric fans, vacuum cleaners, food blenders, exhaust fans, juicers, circulators, and bicycle accessories; Plant Number 6 solid state radios, phonographs, and transceivers; Plant Number 7 gas ranges, heaters, and dry cell batteries; Plant Number 8 color TV sets; Plant Number 9 rechargeable batteries; and Plant Number 10 radios, phonographs, electronic calculators, kerosene stoves, and gas ranges. Other products of Electric Company include: cassette recorders, video tape recorders, microwave ovens, water heaters, percolators, toasters, irons, car cleaners, electric shavers, hair dryers, battery powered children's cars, and electronic components. Since these products appear in numerous models—there are 20 models of washing machines—the company's overall product list runs into the hundreds if not thousands.

Electric factory

The factory we studied is the original and foundation plant of Electric Company. Since 1962 the company has had a division system, in which the factory has become one plant under the Small Appliance Division. Its major goals are to design products, make motors, and assemble parts (many of which are made by subsidiary firms) into finished products. It specializes in the manufacture of small motors and products driven by such motors: electric fans, vacuum cleaners, etc.

The factory had over 1,200 employees in 1969, nine percent of all employees in the company. The value of its products—13,000 million yen in 1967—is about 10 percent of that of the company. Its productivity per employee has been increasing, as Table 2.6 shows.

TABLE 2.6
Productivity of Electric Factory, 1965 to 1969

Year	*Production (millions of yen)*[a]	*Number of Employees*[b]	*Production/ Employee (millions of yen)*
1965	¥ 4,413	958	¥ 4.61
1966	5,576	936	5.96
1967	8,288	1,109	7.53
1968	10,253	1,200	8.54
1969	10,900	1,223	8.91

Source: Business reports for Electric Factory.

[a] Production is defined as total costs and is some 40–50 percent less than sales.

[b] Since the number of employees here is a twelve-month average, while that in Table 3.4 below is for a specific month, the 1969 figures do not agree exactly.

THE JAPANESE SHIPBUILDING AND HEAVY MACHINERY INDUSTRY

The older branches of Japan's engineering industries—shipbuilding, machine tools, textile machinery, and electrical machinery—have expanded rapidly since World War II, but not as rapidly as the newer branches—motor vehicles and household electrical appliance industries.[10] Since 1956, Japan has been the world's largest shipbuilder. By 1969 it had cornered about one-third of the world shipbuilding market, and that year launched nearly half the world's new ships, in contrast to only 11 percent in 1953 (United Nations 1971, Table 132, p. 319). Japan's total tonnage of new ships increased from 1.7 million to 9.3 million gross tons between 1959 and 1969. In 1969 Japan was far ahead of the second ranked nation, West Germany, with 1.6 million gross tons.

Japan has been a pioneer in building large ore carriers and giant tankers; e.g., it built the first supertanker (131,000 tons) in 1962. It has also done a large business in repairing and remodeling ships for foreign and Japanese owners. Because of the heavy capital investment and scale requirements, the industry is highly concentrated. Thus, nine-tenths of all postwar tonnage, including all the large ships, was built in 24 shipyards owned by 19 companies.

Japan lost many of its ships in World War II, and in the immediate postwar years Greek, Norwegian, and other tanker owners were economically stronger than the Japanese. Japanese government loans were

[10] This discussion of the shipbuilding industry draws heavily from Allen 1965.

available to shipbuilding firms after 1945, but with strict regulations attached; as a result, profit margins were small and at times in particular firms even negative. Only since 1967–68 have profits begun to climb somewhat, partly as a result of a deliberate policy to accept only orders with given profit margins (Nippon Kōgyō Ginkō 1972:22).

Another problem facing the shipbuilding industry is its tendency to produce an excess capacity. In 1970 tankers were being built that can carry up to 367,000 deadweight tons. Whereas the ton-miles of oil cargo carried grew 10.5 percent a year for the period from 1960 to 1969, the tonnage available for carrying oil will expand 12.5 percent a year in the period 1970–1973 (United Nations 1970). Among the developments likely to reduce the rate of expansion of demand for tanker space during the 1970s are the greater reliance on new pipelines for overland shipment, restrictions on the sale of oil by Middle Eastern producing nations, and the shortening of shipping routes, by the use of the Northwest Passage from Alaska to the eastern coasts of the United States and Europe.

The growth of Shipbuilding Company

Founded in the early years of the Meiji period, almost a century ago, by a foreign entrepreneur, Shipbuilding Company (the name we shall use for the shipbuilding firm we studied) is now ranked as one of the world's largest. Its position among the world's major diesel engine shipbuilders (in terms of horsepower) rose from eleventh in 1959 to fifth in 1968. Table 2.7 shows the company's rank among Japan's seven leading shipbuilding firms on several economic performance variables. From 1964 to 1970 Shipbuilding Company averaged between third and fourth rank on these eight economic performance measures, with an overall rank of 3.5 on all measures.

Shipbuilding Company's goals and products

The two main products of Shipbuilding Company have always been ships and heavy land machinery for industrial uses. Its shipyards and other plants and offices are now found throughout Japan, with the following specialized lines: Plant Number 1 builds and repairs ships; Plant Number 2 (the one we studied) produces diesel engines, industrial machinery, plants, and steel structures; Plant Number 3 repairs ships and manufactures castings and forgings; Plant Number 4 builds and repairs ships and produces diesel engines, industrial machinery, plants, castings, and forgings; and Plants Number 5 and 6 build and repair ships and produce industrial machinery, plants, and steel structures. Each shipyard specializes in building ships of certain categories, such as mammoth tankers,

TABLE 2.7
Shipbuilding Company's Rank Among Seven Leading Japanese Shipbuilding Firms on Economic Performance Variables, 1964 to 1970[a]

	1964	*1965*	*1966*	*1967*	*1968*	*1969*	*1970*	$\bar{X}$
Sales	3	3	3	3	3	4	4	3.3
Sales per employee	3	5	5	4	3	4	3	3.9
Value added	3	3	3	3	3	4	4	3.3
Value added per employee	3	2	3	3	3	4	3	3.0
Value of equipment investment per employee	7	5	2	2	3	3	4	3.7
Efficiency of equipment investment[b]	1	3	6	4	5	4	4	3.9
Number of employees	3	3	3	3	3	3	3	3.0
Value added as percent of sales	3	1	4	4	6	2	4	3.4
Horsepower of diesel engines produced	3	4	4	4	5			4.0
			1964–1970		$\bar{X}$ =			3.5[c]

Source: Tōyō Keizai, *Chingin soran* 1971: p. 156; 1973: p. 168.

[a] The fiscal year for most of these seven firms ends in March. Since most of the fiscal year falls in the previous calendar year, we use the latter as the year designation. Thus, 1964 in this and other such tables referring to Shipbuilding Company means "the fiscal year ending March 1965"

[b] Computed by dividing the value added per employee by the value of equipment investment per employee.

[c] The overall mean of means, 3.5, includes the mean for horsepower, even though that is based on only the years 1964–1968.

cargo liners, fishing ships, naval craft, and hydrofoil ships. In addition to these six plants, the company also has a technical research laboratory.

Like other shipbuilding companies, Shipbuilding Company has followed a policy of product diversification, in order to maintain an advantageous economic position. It now builds cement mills, chemical recovery plants, sugar refineries, pulp and paper plants, and fertilizer plants and manufactures cooking equipment, water gates, underwater observation towers, and bottle washing machines. The company refers to itself as "one of Japan's most versatile heavy industrial companies." It not only designs and builds such plants, structures, and machines but can send its representatives anywhere in the world to supervise their installation and operation.

In 1969, 60 percent of Shipbuilding Company's products were ships, the other 40 percent industrial land machinery. The latter products have represented a larger portion of the total in recent years, and are expected to reach 50 percent by 1975.[11] In 1947 the biggest diesel engines built by the company had 2,000 horsepower, and the biggest ships for which the company built engines were whaling ships. A few years later, however, the company concluded a sublicense agreement with a European diesel engine designer, and started to manufacture engines on a much larger scale: by 1969, 36,000 horsepower diesel engines for tankers up to 200,000 tons. Rather than investing in research and development on methods of building diesel engines for ships larger than 200,000 tons, the company is returning to the development of an earlier line: turbine engines. Turbines can be built on the basis of existing know-how for ships larger than 200,000 tons. In general, management policy is to maximize efficiency *within* the existing allocation of resources, rather than to emphasize basic research or breakthroughs into new fields in shipbuilding. We shall return to this point below, in the discussion of information and staff and line relationships.

SHIPBUILDING COMPANY'S ENGINE DIVISION

Shipbuilding Company's X Plant has 2,300 employees, of whom 780 are in the Engine Division (*Gendōki Bu*).[12] Our field work was carried out from December 1969 through February 1970 in the Engine Division. The character of X Plant changed during the period 1964 to 1969. In 1964 and 1965 it employed between 3,600 and 4,000 people, and the division that was then called the Shipbuilding Division built both ship hulls and their engines, as well as repairing ships. In 1966, hull and general ship construction and repairing functions were transferred to other plants, along with over 1,000 employees. The division was renamed the Diesel Engine Division (*Nainenki Bu*), and along with the Machining Division (*Kikō Bu*) specialized in the construction of marine engines and industrial machinery. In 1969, the Diesel Engine and Machining divisions of X Plant were consolidated into one division, called the Engine Division (Gendōki Bu). As in the case of Electric Factory, X Plant was the first plant established by the company of which it is a part; it has been in existence almost a century.

The Engine Division is responsible for a number of firsts over the years, such as the production of the world's first large diesel-propelled tanker, and the development of a high output engine which raised the mean effective pressure by a sizeable percentage.

[11] This estimate of 1970 was probably revised by 1973, in view of developments in the international economy. One report noted sluggishness in the machinery division, relative to the shipbuilding division of the industry, and pessimism concerning the demand for industrial machinery (Nippon Kōgyō Ginkō 1972:19).

[12] Strictly speaking, two of the company's shipyards produce diesel engines. We shall use "Engine Division" to refer only to the one studied.

The expanding output and productivity of the Engine Division's two main product lines from 1960 to 1969 are shown in Table 2.8.

TABLE 2.8
Production Record of Diesel Engines and Industrial Machinery and Plants, Shipbuilding Company's Engine Division

		Diesel Engines			*Industrial Machinery and Plants*	
Year	*Number of Employees*[a]	*Number of Units*	*Horsepower (HP)*	*HP/ Employee*	*Output (tons)*	*Output/ Employee*
1960		12	74,000		11,000	
1961		21	143,000		9,000	
1962		16	119,000		7,000	
1963	790	10	94,000	119	10,000	12.7
1964	834	19	211,000	253	17,000	20.4
1965	867	23	303,000	349	11,000	12.7
1966	776	24	374,000	482	13,000	16.8
1967	789	29	397,000	503	22,000	27.9
1968	769	19	230,000	299	31,000	40.1
1969	753	26	305,000	405	35,000	46.5

Source: Shipbuilding Company records.
[a] Includes white collar (managerial, staff, and clerical) and production personnel.

CONCLUSION

The aim of this chapter has been to place each of our three main firms in the context of its industry and market settings. Each firm is among the leading five firms in sales in its respective industry. Both shipbuilding and electrical appliances are more concentrated industries than the sake industry. Electric Company and Shipbuilding Company have more product diversification than Sake Company. Sake Company is more traditional than Electric Company and Shipbuilding Company in that it is still family owned, does not sell stock on the stock market, and is not unionized.

The inclusion of Sake Company has advantages and disadvantages for our research design. The disadvantages are that it is smaller than the size class of greatest interest in the paternalism–lifetime commitment model, and its traditional brewery workers are industrial workers only half of each year. We believe these factors are offset by certain advantages. Some proponents of the paternalism model claim the model existed historically in Japan; if so, we might expect to observe its operation in the centuries-old organization of sake brewing. Moreover, studying Sake Company adds plants using both traditional craft technology and automation, thus increasing our range of variation in technology. This enables us to analyze the relationship between the modernization of technology and the modernization of social organization within a single firm.

The stage is now set for a detailed examination of the several aspects of social organization, technology, and performance.

CHAPTER THREE

Formal Structure

As stated in Chapter 1, as an analytical property of a complex organization, formal structure refers to the number of employees and the scale of operations, the division of labor (role differentiation), the degree of differentiation of productive units, such as plants, the hierarchy of authority, and the span of control.

SAKE COMPANY

Traditionally, a sake factory produced on the average 1,000 koku (180 kiloliters) a year, so the scale of operations and number of employees were small. A plant that produced 1,000 koku of sake a year typically had a master (*tōji*), a head (*kashira*), a subhead in charge of malt (*emon* or *daishi*), a fermentation chief (*sake-no-moto mawashi*), an instrument caretaker (*dōgu mawashi*), a steamer (*kama-ya*), and about 10 upper and middle workers (*jō-bito* and *chū-bito* respectively), totalling around 16 employees. For example, a sake plant in Uozaki village in 1886 had 16 workers; in the Edo era (1600–1868) one plant producing 1,000 koku of sake had only 10 workers (Mannari, Maki, and Seino 1967:10)[1]. The average number of employees in a plant probably remained between 10 and 16 until quite recently, when the modernization of technology and the increasing scale of sake production began to change the situation.

Compared with this traditional base line, two recent changes in formal structure can be identified: the shift from the single-plant sake firm to the multi-plant firm, and the differentiation of roles within individual plants.

In 1969, Sake Company had the following plants and other units: seven brewing plants (kura) with personnel in the traditional hierarchy of positions, three of which we studied intensively in 1969–70; one new, automated brewing factory; one general affairs unit (business unit); one refined sake manufacturing plant; one alcohol factory; three storehouses; one blending (mixing) factory; one bottling factory; one white rice factory; and one products warehouse. The total area of these plants and units was 33,000 square meters in 1969. All Sake Company's brewing plants

[1] Mannari, Maki, and Seino found their figures for the earlier period in *Dōmō sake tsukuri ki* (A Children's History of Sake Making), written in the Jōkyō period (1684–87).

and factories are in two cities near Osaka, although it also has a blending and binning factory near Tokyo. Table 3.1 gives more information on the traditional brewing plants and the automated factory.

TABLE 3.1
Structure of Sake Company's Brewery Plants,
October 1969–March 1970 Season

	Date Established	*Number of Employees*	*Days Worked*	*Koku of Sake Produced*
Traditional Plants in N-City[a]				
Number 1 Kura	c. 1850	38	150	9,000
Number 2 Kura	c. 1850	15	150	5,500
Number 3 Kura	c. 1850	15	150	5,500
Traditional Plants in I-City[a]				
Number 4 Kura	c. 1750	35	150	4,200
Number 5 Kura	c. 1570	15	150	5,500
Number 6 Kura	1940	15	150	5,500
Number 7 Kura	c. 1670	15	150	5,500
Automated Factories in I-City				
Number 2 Factory	1963	28[b]	300	23,000[c]
Number 3 Factory	1969	36	280	40,000

[a] "Traditional" plant means here those plants whose personnel occupy the positions of master, head, blenders, etc., as defined in the long-standing hierarchy of sake brewery work roles and statuses; plants open only about half the year; and plants that are not fully mechanized. The three plants in N-City are the ones we studied most intensively in 1969–70. They had burned down in 1945; Plants Number 2 and 3 were rebuilt in 1947, Plant Number 1 in 1953.

[b] The Number 2 Automated Factory employed 28 workers for 300 days in 1966, when Mannari, Maki, and Seino studied it. In 1969 it stopped operations, and its 28 employees were transferred to Number 3 Automated Factory.

[c] Sake produced in 1966.

With the exception of Plants Number 1 and 4, the number of employees in each position in the traditional plants (kura) had not increased markedly over the traditional base line. Table 3.2 shows that from 1684 to 1687, in 1886, and in 1969 a given kura always had only one or two workers in the position of steamer; blenders had increased from one to about four per kura, and so on. The preservation of essentially similar positional distributions and scale of personnel at the individual plant level is itself a major illustration of the traditionalism of the sake industry and of Sake Company; we shall see later why there has been this kind of stability. As sake firms have expanded, they have done so partly by segmentation: as plants are added, the company continues to replicate in each new plant the same traditional positional hierarchy of master, subhead, etc.; a large company, such as Sake Company today, consists of several kura, each of which performs the same set of functions (steaming, blending, etc.), as shown in Table 3.2. Unlike the other two firms, Sake Company did not give us a table of organization; its closest approximation is shown in Table 3.2.

Company growth has brought inter- and intraplant differentiation, as well as segmentation. Some plants now specialize in brewing, others in bottling, still others in white rice processing. There is technological differentiation of automated brewing plants from semi-traditional, manual brewing plants. A third form of differentiation is at the work role (*yakumei*) level. The eight role or positional levels in Table 3.2 had become differentiated into the 20 indicated in Table 3.3 for Sake Company in 1969. For example, the traditional role of malt head had become differentiated into the two roles of associate master and malt head or malt master; that of fermentation chief had become differentiated into the two roles of upper and lower fermentation controllers.

TABLE 3.2
Number of Employees in Traditional Hierarchic Positions of Sake Plants at Three Points in Time

			Sake Company, 1969				
Position	*1684–1687*	*Uozaki Village Sake Plant 1886*	*No. 1 Plant*	*No. 2 Plant*	*No. 3 Plant*	*Other Plants*	*Sake Co. Total*[a]
Master (tōji)	1	1	1	1	1	4	7
Head (kashira)	1	1	0	0	0	0	0
Malt head (emon, daishi)	1	1	3	1	1	6	11
Fermentation chief (sake-no-moto mawashi)	1	1	8	4	4	25	41
Instrument caretaker (dōgu-mawashi)	1	1	8	3	3	16	30
Steamer (kama-ya)	1	1	1	1	1	7	10
Upper worker (jo-bito)	4	10	8	2	3	10	23
Middle worker (chū-bito)	0		9	3	2	14	28
Total	10	16	38	15	15	82	150

Source: Adapted from Mannari, Maki, and Seino 1967:10.

[a]Sum of columns 3–6. Since Sake Company in 1969 had employees in positions other than these traditional ones and plants other than breweries (e.g., a bottling plant), the company total here is less than the total number of Sake Company employees, which is 656.

TABLE 3.3
Role Differentiation in Sake Company[a]

Role, Position, or Section	*Number of Employees*	*Percentage of Total*
Master (*tōji*)	7	2.3
Associate master (*tōji hosa*)	9	3.0
Malt head (*kōji shunin*)	2	0.7
Upper fermentation controller (yeast master) (*shubo shunin*)	2	0.7
Upper yeast section (*shubo kakari*)	12	4.0
Lower fermentation controller (yeast section) (*shubo kakari*)	22	7.3
Rice steaming section (*jō mai kakari*)	7	2.3
Instrument caretaking section (*seibo kakari*)	30	9.9
A-Section workers (*kaka-in* A)	23	7.6
B-Section workers (*kaka-in* B)	27	8.9
Temporary workers (*rinjiko*)	31	10.2
Alcohol (*shusei*)	11	3.6
Inspection (*kensa*)	6	2.0
Cooking (*suiji*)	4	1.3
Refrigeration room (*kōri muro*)	3	1.0
Bottling (*binjō*)	74	24.4
Mixing (*chōgō*)	26	8.6
Research (*kenkyū*)	3	1.0
Boiler (*boira*)	2	0.7
Transport operators (*untenshu*)	2	0.7
Total	303	100.2

[a] This table understates the degree of differentiation, because, although it includes all personnel in the listed positions or sections in all the company's plants in the two cities in the Osaka area, it omits sales, clerical, and office (except for research) personnel.

ELECTRIC FACTORY

As can be seen from Figure 3.1, which presents Electric Factory's table of organization, the factory as a whole has the status of a division (*jigyōbu*) within the company's 10 geographically separated plants. The division director is officially at the top of the factory's management, but he spends considerable time planning and promoting the sales of his division all over Japan. The actual running of the factory is in the hands of the associate division director, who is factory manager (*kōjōchō*). The organizational structure consists of six departments (*bu*), which are divided into sections (*ka*), which are subdivided into three lower levels: *kakari*, *han*, and *kumi*; three sections (Research, Development, and Standardization) are responsible directly to the factory manager.

FIGURE 3.1

Table of Organization of Electric Factory, June 1969

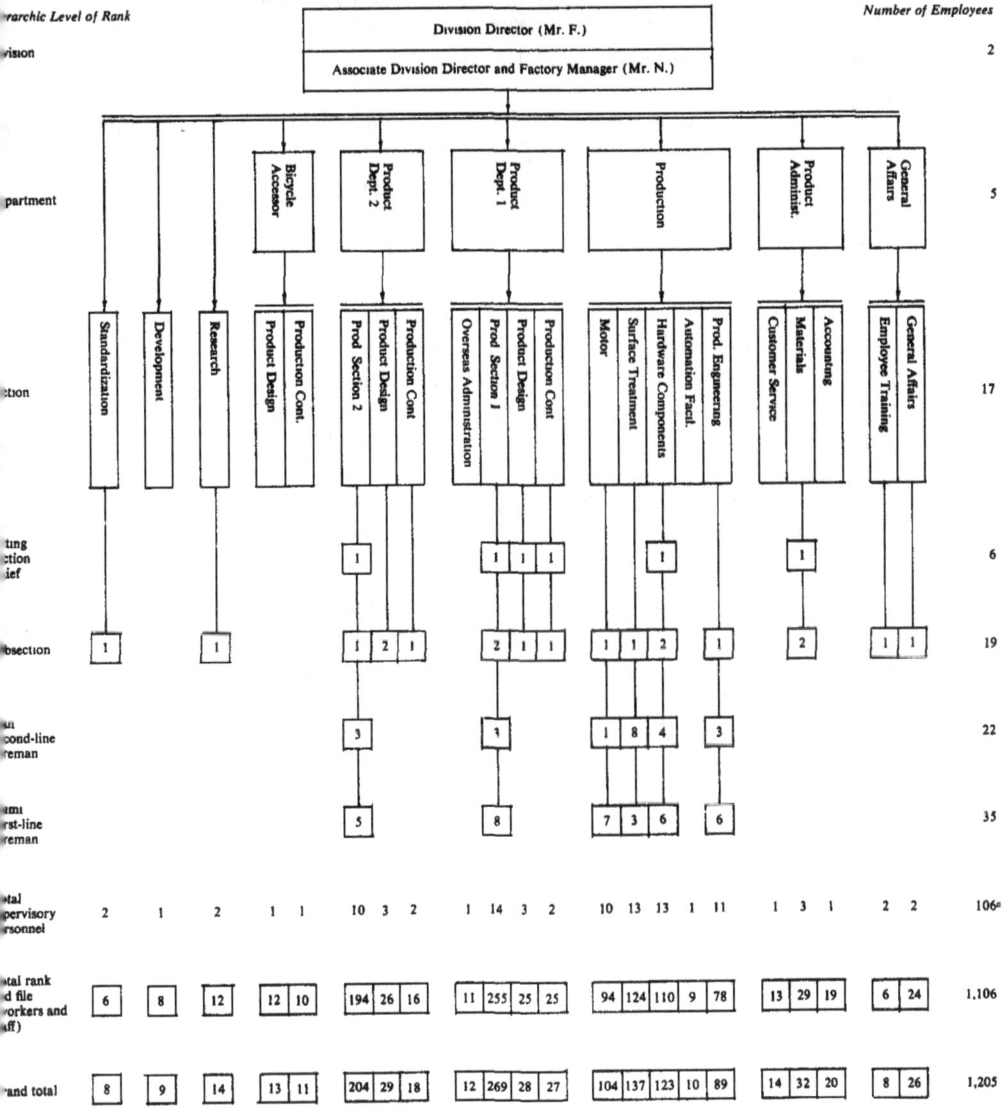

Positions defined in terms of this organizational hierarchy are as follows: division director (*jigyōbuchō*); associate division director and factory manager (*fuku jigyōbuchō*); department head (*buchō*); associate department head (*bu jichō*); section chief (*kachō*); acting section chief (*kachō-dairi*); subsection head (*kakarichō*); second-line foreman (*hanchō*); first-line foreman (*kumichō*); and rank-and-file workers (staff, clerical, and production).

Departments (bu) are generally headed by buchō, but the Department of General Affairs, and Product Department Number 2 were headed by men with the rank of associate department head (bu jichō). Most sections (ka) were headed by kachō, but six are headed by acting section chiefs (kachō-dairi), a position junior to kachō. Although all 22 sections have either a kachō or a kachō-dairi, not all have a fixed number of lower supervisory ranks. Four kakari (subsections) have two kakarichō each, 11 have one kakarichō each; seven kakari have no kakarichō. Only six sections have first- and second-line foremen (kumichō and hanchō). These are the sections engaged in direct production—Hardware, Surface Treatment, Motor, Production Sections Number 1 and 2—and Production Engineering.

The company is now seeking to simplify its organization by reducing the number of hierarchic levels. In 1969 the rank of *shunin*—which had been between kakarichō and hanchō—was abolished, and all shunin became kakarichō. In 1970 it was planned to abolish the rank of acting section chief (kachō-dairi). This will have flattened the organizational hierarchy from 10 to eight levels (including rank-and-file workers) in a two-year period.

Figure 3.1 shows that three departments (Product Departments Number 1 and 2 and Bicycle Accessories) each have a Production Control (*ko-mu*), and a Product Design (*setsukei* or *sekkei*) Section; and Production Departments Number 1 and 2 both have a Production Section (*sagyōka*). These functionally parallel sections are differentiated by product. Thus, Production Department Number 1's Control, Design, and Production sections are concerned with the administration, planning, design, and production of electric fans and air circulators. The parallel three sections in Production Department Number 2 deal with the administration, planning, design, and production of vacuum cleaners, juicers, and blenders. The sections named Control and Design in the Bicycle Accessories Department plan and design bicycle accessories.

Formal functions of major units

Electric Factory specifies in detail in writing the formal functions of each unit and position in its organization. Space does not permit the description of the factory directory of formal job tasks for the 22 sections. Suffice it to say that in this sense Electric Factory is a bureaucratic organization: it stands midway, as it were, between a more traditional, pre-bureaucratic organization with unwritten, diffuse definitions of the functions, units, and roles on the one hand, and a "task force," post-bureaucratic organization, as described in Bennis and Slater (1968), on the other.

The table of organization presented in Figure 3.1 is a first approximation of the formal structure of Electric Factory. A closer approximation is provided by Figure 3.2, which shows the formal work flow among the various sections. Even this, of course, oversimplifies the total formal structure and functions of the units. The exchange of information and the process of decision making, testing, and checking do not proceed only in the one major direction indicated (from top to bottom on the chart). Line sections such as Hardware tell the Design Section the easiest way to make a given part or product; veteran workers in the production sections tell designers that their plans have some deficiencies in safety or efficiency. Thus there is a feedback loop *from* machine workers *to* designers. In general, however, the direction of the work flow is as shown in Figure 3.2. It will be most useful if we postpone further remarks about the functions of specific sections until the next chapter, when we shall deal with technology, the division of labor, and general working conditions in specific sections.

Differentiation of units (sections) over time

Electric Factory was a much more differentiated organization in 1969 than in 1960 (see Figure 3.3). Actually, the differentiation of organizational units took place quite suddenly between July and August 1961, a period that in some ways was a turning point of the organization, and about which we shall have much more to say in other contexts. The number of named units of the factory organization (ka, shitsu, sho, etc.) increased from 15 in July 1961 to 25 in August 1961. Since then, the total number of units has remained between 22 and 28.

Variations in the number of employees

Table 3.4 reveals that the number of employees in Electric Factory declined from 1,053 in July 1964 to 892 at the end of 1965, a drop of 15.3 percent. This was caused by a slump in the electrical machinery industry, which became a recession in 1964–65. The industry had surplus production capacity, output was reduced, and both profits and dividends fell. Since 1966, however, the boom has returned, centered on household

FIGURE 3.2

Work Flow in Electric Factory

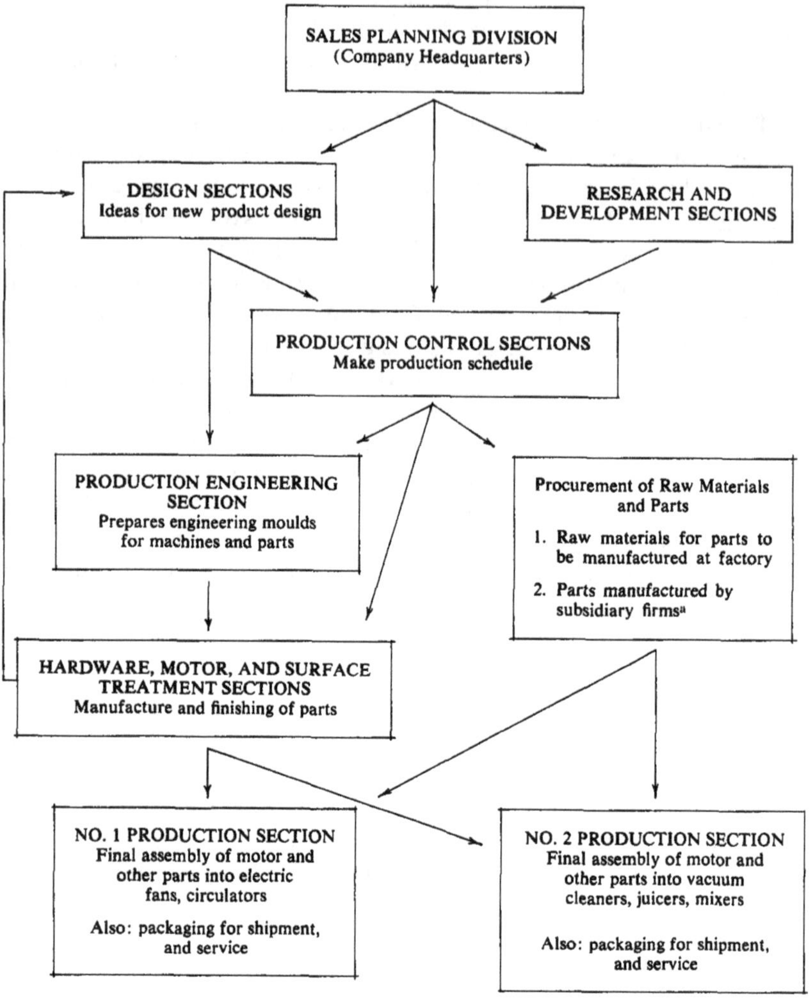

[a] In general, subsidiary firms that produce parts only or mainly for Electric Company are located near the company. Other subsidiary firms which produce parts do work for many other firms as well, and tend to be farther away from Electric Factory.

FIGURE 3.3

Differentiation of Units in Electric Factory

Sections in 1960

1. General Affairs[a]
2. Manufacturing[a]
3. Bicycle Production
4. Electrification Production
5. Technical[a]
6. Engineering (kōsaku)
7. Hardware Components
8. Quality Control
9. Research
10. Service
11. Surface Treatment

Sections in 1969

1. General Affairs
2. Bicycle Production Control
3. Bicycle Product Design
4. Production Engineering
5. Hardware Components
6. Research
7. Service
8. Surface Treatment

New Sections

9. Employee Training
10. Accounting
11. Materials
12. No. 1 Production Control
13. No. 1 Product Design
14. Overseas Administration
15. No. 1 Production
16. No. 2 Production Control
17. No. 2 Product Design
18. No. 2 Production
19. Development
20. Standardization
21. Motor
22. Automation Facilitation

[a] Then a department (bu) rather than a section (ka).

TABLE 3.4
Number and Sex Distribution of Employees in Electric Factory, 1962 to 1969

Date	*Number of Employees*	*Percent Male*
December 21, 1962	1,103	59.2
July 6, 1963	1,094	59.5
December 21, 1963	1,044	61.3
July 10, 1964	1,053	62.2
December 21, 1964	983	64.6
July 5, 1965	943	65.5
December 21, 1965	892	68.0
July 21, 1966	931	64.8
December 21, 1966	952	62.2
July 5, 1967	1,091	56.8
December 21, 1967	1,087	55.7
July 12, 1968	1,195	52.6
December 21, 1968	1,157	46.1
July 14, 1969	1,228	51.3

Source: Electric Factory records.

electric appliances. The company's employees again rose in number, and its other economic indicators turned upward.

SHIPBUILDING FACTORY

As already noted, X Plant is one of Shipbuilding Company's six shipyards and plants scattered throughout Japan. X Plant is headed by a plant director, under whom is an associate plant director. The plant has five major divisions (bu), each headed by a buchō: General Affairs, Management, Engine (the division studied), Machine, and Steel Structure. Within each division, Shipbuilding Company has the same number of formal hierarchic levels as does Electric Factory:

Electric Factory		*Shipbuilding Factory*	
Level	*Rank*	*Level*	*Rank*
Bu	Buchō	Bu	Buchō
Ka	Kachō	Ka	Kachō
Kakari	Kakarichō	Kakari	Kakarichō
Han	Hanchō	Shop	Sagyōchō
Kumi	Kumichō	Han	Hanchō
	Workers		*Boshin* (informal rank) Workers

Historically, Shipbuilding Company had close links with the Imperial Japanese Navy. The term boshin is a Japanese rendering of bos'n, i.e., boatswain, a petty officer.

All personnel in Shipbuilding Company are divided into two groups: B Group, which contains managerial, engineering, and clerical workers and A Group, which contains line production workers and their hanchō or first-line foremen. For convenience we shall refer to these two status groups as "white collar" (or "staff") and "workers," respectively. These terms are not precisely accurate, since the worker (A) group contains 41 hanchō, who are above rank-and-file workers, and the white collar (B) group contains 12 sagyōchō, or second-line foremen, who combine desk work with work on the production shop floor.

Among line personnel, the sagyōchō is the shop leader (*shokuchō*). Each sagyōchō has three or four hanchō under him, and each hanchō supervises between 10 and 20 workers. Below the hanchō is an informal rank, boshin, and this person is the chief assistant to the hanchō. Sometimes, first- and second-grade boshin are distinguished.

The kakarichō is considered to be the lowest management position. Thus management positions include buchō, kachō, and kakarichō; supervisory positions include sagyōchō, hanchō, and boshin (informal).

Figure 3.4 presents the table of organization of the Engine Division. The scale (number of employees) in the organizational units of X Plant that were included in the Engine Division in 1969 has been highly stable in recent years (Table 3.5). In 1970 a fourth section, the Turbine Section, was created, changing the table of organization in the following way from that shown in Figure 3.4:

I. Design Section
II. Engine Assembly Section:
    Service Kakari
    Diesel Staff Kakari
    Production Control Kakari
    Assembly Kakari
    Piping Kakari
III. Machining Section:
    Tool Engineering Kakari
    Number 1 Machine Kakari
    Number 2 Machine Kakari
IV. Turbine Section

This elevation of the turbine unit from the status of subsection (kakari) in 1969 to that of a section (ka) in 1970, reflects the company's adaptations to the exigencies of the market, discussed above. The company's policy is

FIGURE 3.4

Table of Organization of Shipbuilding Company's Engine Division, 1969

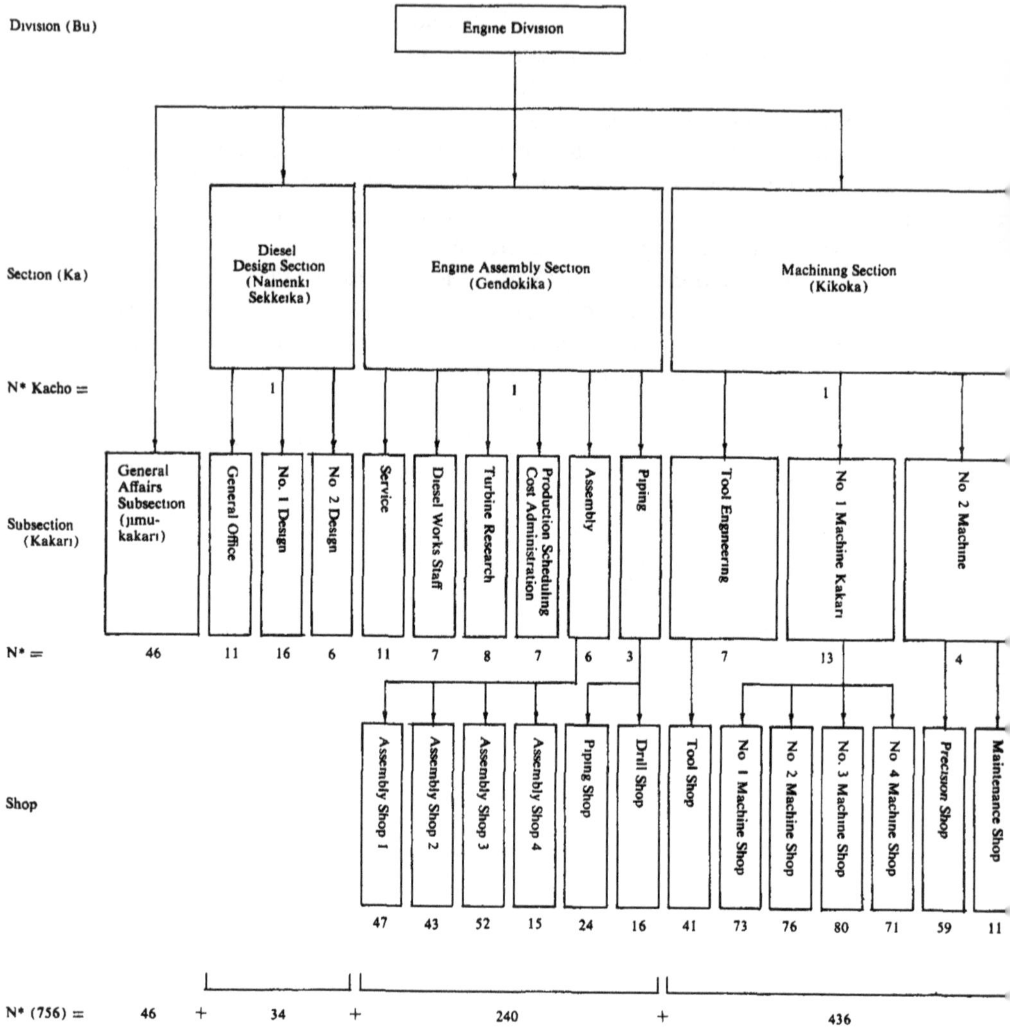

N* includes all regular line, clerical, staff and managerial personnel.

Basic Terminology

Assembly Subsection	*kumi tate kakari*	Machine Subsection	*kiko kakari*
Assembly Shop	kumi tate 1 shop, 2 shop, etc.	Machine Shop	kiko 1 shop, 2 shop, et
Piping Subsection	haikan kakari	Precision Shop	seimitsu shop
Tool Engineering Subsection	kōgu gijutsu kakari	Maintenance Shop	hosen shop

TABLE 3.5
Number of Employees in Engine Division, 1964 to 1969

	B Group (Supervisory and White Collar)						*A Group (Workers)*			
Year	*Kachō and Higher*	*Kakarichō*	*Sagyōchō*	*Ordinary Male*	*Ordinary Female*	*Subtotal*	*Hanchō*	*Ordinary Workers*	*Subtotal*	*Total*[a]
1964	5	15	9	73	10	112	41	561	602	714
1965	6	15	9	80	15	125	45	622	667	792
1966	5	16	10	74	11	116	39	626	665	781
1967	4	15	9	71	15	114	40	631	671	785
1968	4	14	10	65	16	109	45	620	665	774
1969	4	14	12	64	15	109	44	603	647	756

[a] This total may differ slightly from the total in other tables since it is based on different months of the year.

to develop the turbine engine, and produce more of these for industrial machinery, relative to the production of diesel engines for tankers.

STATUS IN THE ORGANIZATION

One of our major independent variables is an employee's status in his company. This refers to the employee's rank or position with regard to various rewards the company allocates to its personnel, including economic and prestige rewards and authority. The specific variables that are conceptualized as components of status in Electric and Shipbuilding factories[2] are: (1) rank, (2) monthly pay, (3) job classification, (4) seniority, (5) education, and (6) informational level of the section in which one works. Each of these variables is trichotomized as shown in Table B.2.

The sixth component of organizational status—the informational level of the section—is itself based on an index, and calls for some discussion. It combines six properties of the section as a unit: (1) mean education of employees in the section; (2) proportion of employees in the section who are college graduates; (3) proportion of employees with a Creative job classification; (4) number of rank-and-file employees per supervisory and managerial employees; (5) proportion of employees in the section who have a Routine job classification; and (6) proportion of employees who are middle school graduates. The first three properties are positive measures of informational level, the latter three are negative measures. The resulting informational index score is obtained by (1) ranking each of the 22 sections on each of the six measures, (2) summing each section's ranks

[2] Because of the much smaller number of respondents in Sake Company we did not attempt as detailed a statistical analysis as in the other two firms. Therefore, no index of organizational status (or any other index), was constructed for Sake personnel.

on the first three measures, (3) summing each section's ranks on the last three measures, and (4) subtracting (3) from (2).[3]

Parsons (1966) develops a cybernetic theory of social systems in which there is a negative relationship between the informational and the energic components of these systems. Using data on electricity consumption (in kilowatt hours per capita) in each of the 22 sections in Electric Factory, for the period January through July 1969, as a measure of energy, we were able to confirm this relationship: the Spearman rank order correlation between informational and energic level of sections is −.610.

Shipbuilding Factory's sections were also ordered in terms of an informational-energic continuum (see variable 88 in Table B. 2).

Table 3.6 presents the matrix of relationships between all pairs of status variables, for each factory. All relationships are significant at the .01 level and either strong or moderate. With the exception of the relationship between seniority and education, all relationships are positive. This one negative relationship reflects the secular trend of rising educational levels in Japan; thus, employees with more seniority represent a cohort that entered the company after completing less education than more recent recruits to the company. Despite this exception, the meaning of high status in each company is clear: the higher an employee's education, job classification, section informational level, seniority, pay, and rank, the higher his overall status in the firm. It is also clear that in general employees in both factories empirically tend to cluster at one of two poles: a high or a low status on these six dimensions.

This being so, we subjected the six status variables to an item-analysis computer program (ITEMA) and constructed an index of organizational status which contains these six variables. By the formula for estimating index reliability from the items' intercorrelations (Nunnally 1967:193),

[3] For example, the index score for the Development Section, −53.5, is obtained as follows:

$$\begin{array}{lcl} \text{Positive measures: 3 ranks} & - & \text{Negative measures: 3 ranks} \\ (1 + 1 + 1 = 3) & - & (16.5 + 20.0 + 20.0 = 56.5) \end{array}$$
$$3.0 - 56.5 = -53.5$$

It will be seen that the larger a section's negative score, the higher its informational level; the larger its positive score, the lower its informational level. Section index scores range from −53.5 through 0 to +47.0. The 22 sections were trichotomized into three groups: high, medium, and low informational level. Sections that are high in informational level are Development (−53.5), Standardization (−51.5), Research (−47.0), Number 1 Product Design (−33.0), Automation Facilitation (−31.0), Number 2 Product Design (−30.0), Overseas Administration (−23.0), Bicycle Product Design (−17.5), Employee Training (−15.5), Bicycle Production Control (−10.0), Number 2 Production Control (−3.5), and Accounting (−3.0). Sections medium in informational level include: Production Engineering (12.5), Number 1 Production Control (16.5), General Affairs (22.0), Materials (25.5), and Customer Service (35.0). Sections low in informational level are: Motor (36.0), Hardware Components (41.5), Surface Treatment (44.5), Number 1 Production (46.0), and Number 2 Production (47.0).

TABLE 3.6
Relationships between Organizational Status Variables[a]

	Variable	84/94	78/91	9/89	13/95	24/88
8/110.[b]	Pay	.86**	.71**	.78**	.31**	.38**
		.76**	.59**	.45**	.38**	.19**
84/94.	Seniority		.57**	.77**	−.42**	.37**
			.47**	.51**	−.41**	.32**
78/91.	Rank			.85**	.44**	.31**
				.79**	.60**	.63**
9/89.	Job classification				.69**	.68**
					.59**	.85**
13/95.	Education					.64**
						.35**
24/88.	Informational level of section					

** Significant at the .01 level.

[a] In each cell, the upper number is the degree of association (C/C_{max}) in Electric Factory, the lower number the degree of association in Shipbuilding Factory. C/C_{max}, our basic measure of strength of relationship between variables, is a correction for C, the coefficient of contingency. $C = \sqrt{\frac{X^2}{X^2 + N}}$. C equals 0 when the variables are independent, but its upper limit depends on the number of rows and columns and is always less than 1.00. C/C_{max} corrects for this by dividing the value of C by the maximum value of C, C_{max}. $C_{max} = \sqrt{\frac{k-1}{k}}$, where k is the number of rows or columns, whichever is smaller (Blalock 1972: 297–98).

[b] The first variable number refers to Electric Factory, the second to Shipbuilding. For a complete list of variables by number, see Appendix C.

coefficient alpha is .78 for this index in Electric Factory (variable 138) and .47 in Shipbuilding Factory (variable 123).[4] This index sums each employee's score for the six status items; the higher one's education, pay, rank, etc., the higher one's overall status score.

What do these data tell us about the sensitivity of specific status items as predictors of one's overall organizational status score? Some proponents of the lifetime commitment model suggest that seniority or education are the most important elements of status in Japanese firms, while some critics of the model argue that pay is most important. If we use the item-total correlation as our criterion of importance, the evidence in Table 3.7 supports the critics more than the proponents of the model. In other words, in both factories, the most sensitive predictors of a person's overall

[4] The measurement and coding of the six status variables are most similar for Electric and Shipbuilding factories with regard to pay, seniority, education, and rank. Jobs cannot be as precisely ranked in Shipbuilding as in Electric, and the ranking of sections is less clear with regard to cybernetic-informational level in Shipbuilding than in Electric.

TABLE 3.7
Status Indicators in Relation to Status Index Score in Electric and Shipbuilding Factories

Electric Factory	*Item-total* r^a	*Shipbuilding Factory*	*Item-total* r^a
Pay	.786	Pay	.456
Job classification	.716	Rank	.361
Seniority	.607	Job classification	.283
Rank	.474	Seniority	.202
Informational level of section	.407	Informational level of section	.173
Education	.206	Education	−.006

[a] The item-total r is the correlation between the item and the total score when that item is not contributing to the total.

organizational status are pay and rank or job classification, not seniority or education. This should not be interpreted as a statement concerning which status variables *cause* which other status variables. Clearly, for example, one's pay does not cause one's education, since education precedes pay. These causal questions will be considered in later chapters.

BUREAUCRATIZATION IN ELECTRIC AND SHIPBUILDING FACTORIES

Our first impression was that Shipbuilding Company had a classical bureaucratic form of organization. Company manuals indicated that Shipbuilding Company was more formalized and bureaucratized than Electric Company. At the level of codified rules and procedures in the company manuals, the contrast is as shown in Table 3.8. With regard to (3), for example, the Shipbuilding Company manual states that sagyōchō and hanchō have the authority to assign a worker a given job, whereas a kakarichō does not have this authority.

Another difference between the two companies' manuals is that Shipbuilding Company defines authority relationships on a division-wide basis: all managers or supervisors at a given level have the same authority, regardless of the section in which they work. Electric Factory, on the other hand, defines the responsibilities of given managerial ranks section by section. The manuals thus give the impression that organizational structure and authority are less highly generalized in Electric Company, i.e., more unit-specific than in Shipbuilding Company.

However, when we posed this contrast to a personnel manager in Shipbuilding Company, he gave a different view of the significance of the 1967 organization manual: "Between 1950 and 1963, we made revisions in the rules of job and authority definition seventeen times. Then, after 1963, we gave up the attempt to rigidly specify definitions of job and authority. One reason for this was that our organization expanded, and new jobs were always emerging."

TABLE 3.8
Comparison of Shipbuilding and Electric Company Manuals

Shipbuilding Company Manual, Engine Division	*Electric Company Manual, Electric Factory*
1. Defines what the company means by such terms as "authority" (*kengen*), "orders" (*meirei*), "instructions" (*shiji*), etc.	1. Does not define such terms.
2. Defines explicitly the responsibilities and functions of each job and rank.	2. Defines explicitly the responsibilities and functions of each job and rank.
3. Defines explicitly the authority relations among levels in the organization. For each level—kakarichō, sagyōchō, hanchō—specifies to whom one reports, whether one makes given kinds of decisions, draws up plans, etc.	3. Does not define explicitly the authority relations among levels. Leaves implicit exactly what the boundaries of authority between levels are.

Shipbuilding Company is in fact less bureaucratized than its 1967 manual would indicate. The personnel manager said: "If we followed the rules in the manual in our procedures, we'd have no time left for work. In practice, we threw out rigid authority over jobs." Although the authority and duties of kakarichō and lower supervisors are still specified, the higher levels of kacho and buchō are no longer given rigid job and authority definitions:

> It's top management of the plant whose authority is more difficult to describe nowadays. . . . The company is moving to eliminate the title of kachō, and substituting the term "team leader" of continuous casting, etc. Thus, when there is an order for us to build a sugar plant, we call on a team leader for this job. People would not then have fixed titles like hanchō, or fixed sections. On the other hand, the production of turbine engines is more constant, so we will keep a hanchō there. In company headquarters, we have abolished the title of kakarichō. We're also setting up the titles "specialist buchō," "specialist kachō" and "specialist kakarichō" [*senmon buchō, senmon kachō, senmon kakarichō*]. Each of these positions would have no subordinates. They are positions especially for university graduates who don't have leadership ability, but have to be given some recognition for their specialist knowledge. These positions are created for the man.

Thus, Shipbuilding is introducing some elements of a task force type of organization, with team leaders and workers brought together on a temporary basis to solve a particular problem or accomplish a particular job, after which they are disbanded. (On task force theory, see Burns and Stalker 1961, Bennis and Slater 1968, and Perrow 1970.) To quote the personnel manager again: "We have a divisional organization, but our organization is more flexible than, for example, electrical appliance firms, because within a division [bu] we have many products, and many are custom made [bespoke]. Each product is made somewhat differently, and our organization therefore has to be more flexible. In contrast, an electrical company can produce thousands of standard television sets without any changes."

The task force pattern is already in operation in parts of the Engine Division. In one han there are 12 workers on paper, but two or more of them may spend much of their work time away from their han. One worker is assigned to the tool shop, one to part-time work in the staff production planning kakari; a third may be abroad for weeks or months, helping to set up a new sugar refinery; and so forth. Also, from day to day, small groups of workers are assigned on an ad hoc basis to work in various locations. In an Assembly han, seven workers worked on a Pattern 84 diesel engine, while three others were detached from the han to work in the pit, building an engine for a sugar refinery.

Another modification in strict bureaucratic organization that one commonly encounters in Japanese firms was verbalized by a Shipbuilding Company worker who was about to retire: "This company has strict rules, but it is still more 'human' [*ninjō*][5] and paternalistic than some of our competitor firms.

In the classical Weberian theory of bureaucracy, the tasks performed by personnel follow systematic procedures; these are widely known and followed by *all* personnel. In Electric Factory, a hanchō of the Hardware Section said: "All work performance here is done by rules." Table 3.9 compares Electric and Shipbuilding personnel on the variable, knowledge of procedures.[6] By this measure Shipbuilding employees are significantly, but weakly, more likely than their counterparts in Electric Factory to say they "always know the measures to take to deal with the trouble" (between-factory $C/C_{max} = .14$). Both factories are partially bureaucra-

[5] *Ninjō* denotes humaneness, sympathy, kindness, and is contrasted with being cold-hearted or heartless.

[6] Note that we are measuring how much knowledge of procedures an employee *claims* to have, not how much knowledge we observed or otherwise independently measured him as having. We cannot say how well the former measures the latter, which would be a more direct measure of the Weberian concept. Knowledge of procedures is variable number 39 in Electric Factory and variable number 18 in Shipbuilding.

TABLE 3.9
Responses of Electric and Shipbuilding Factory Personnel to the Question on Knowledge of Procedures, "In your shop, when there is an accident such as work trouble, or a fire, do you know how to deal with it?"(%)

	Electric Factory	*Shipbuilding Factory*
I always know the measures to take to deal with the trouble	30.8	35.9
I sometimes know, but sometimes do not know	55.9	56.7
I hardly know how to handle it	13.3	7.4
Total	100.0	100.0
N	(1,015)	(584)
$C/C_{max} = .14^{**}$		

** Significant at the .01 level.

tized, in the sense that the majority of personnel know the procedures sometimes, but not always.

What explains the differential knowledge of procedures? We hypothesized that knowledge of procedures is positively correlated with (1) organizational status, (2) age, (3) cohesiveness with fellow employees,[7] (4) job satisfaction,[8] and (5) work values. The theoretical rationale for these hypotheses is as follows. (1) Knowledge of procedures should increase with status in the company because higher status employees have greater seniority, which means they have had more time to learn the procedures; their higher rank and job classification mean greater responsibility and higher expectations that they should know the procedures; their higher pay is more of an incentive for knowing the procedures, as one element in overall performance. (2) Knowledge of procedures should increase with age for somewhat the same reasons. Older employees in general have more seniority, and thus more time to learn the procedures; they are more likely to be expected to know the procedures. (3) Cohesiveness with other employees provides informal system inputs into one's knowledge of procedures, which can supplement formal system sources of knowledge of procedures. That is, one can learn the procedures from fellow employees, in addition to, or instead of, learning them directly from company manuals or supervisors. The weakness of this hypothesis is that it assumes the

[7] The index of employee cohesiveness, described in Chapter 8, combines six measures of social solidarity among employees.

[8] The index of job satisfaction, described in Chapter 5, combines six questions on attitudes toward one's job.

informal system is as oriented to knowledge of formal procedures as the formal system, which may or may not be true. (4) The more satisfied one is with his or her job, the more one will endeavor to know the procedures that are core elements of the job. (5) Employees with "work is my whole life" values are more likely than those who give primacy to either "happy family" or "work is only a means to pleasure" values to have the work commitment that would motivate them to learn procedures.

In Electric Factory a sixth independent variable, sex, is introduced, as will be done in relation to other dependent variables later in this study. Since 96 percent of the Shipbuilding Factory respondents are male, in contrast to only 49 percent of Electric Factory employees, we shall not normally use sex as an independent variable in Shipbuilding Factory analyses.

The results of our correlation and multiple regression analysis are as follows.[9] In Electric Factory we find that knowledge of procedures can be predicted virtually as well on the basis of three of the independent variables—status, sex, and cohesiveness—as on the basis of all six originally hypothesized independent variables. The extent to which one claims to know the procedures increases with status in the company and with cohesiveness with fellow employees; men are more likely than women to say they know the procedures. When one knows these three things about an employee, knowing in addition the employee's age, job satisfaction, and values adds virtually nothing to explaining knowledge of procedures. Moreover, of the three most important variables, we find that knowledge of procedures is somewhat more a result of (high) status in the company than of being male or having high cohesiveness with fellow employees.

In Shipbuilding Factory the results are similar. Insofar as employees profess to know the procedures, it is because they have somewhat higher status in the organization, are somewhat older,[10] and have slightly higher cohesiveness with fellow employees. When these variables are taken into account, the remaining two variables—job satisfaction and values—add virtually nothing to the explanation of the variation in knowledge of procedures.

Electric and Shipbuilding factories are only partially bureaucratized organizations. This is due in part to the fact that their personnel include lower status workers. Such workers lack the prerequisites for detailed and

[9] The more technically inclined reader will want to know the details of the statistical analysis by which these conclusions were reached. These appear in Appendix D, Tables D.1 and D.2, p. 366 and p. 368.

[10] Had there been more female personnel in Shipbuilding, and had we used sex as a variable, it might have been among the most important predictors of knowledge of procedures, as in Electric Factory. This apart, age is less strongly positively related to status in Shipbuilding, and thus operates together with status as an independent cause of knowledge of procedures.

comprehensive knowledge of procedures. Because they have lower status, they are not as likely to be expected to know the procedures. Employees who have low cohesiveness with their fellow employees are less likely to know the procedures, which may be due to the fact that they lack even the informal system's communication network as a means of learning procedures.

CONCLUSION

The recent history of all three firms has been marked by increasing internal structural differentiation: the emergence within each company of more specialized plants, divisions, sections, and roles. A comparable index of organizational status has been developed, in which each employee's status on six hierarchic dimensions within the company is summed. The degree of bureaucratization in each firm was assessed on the basis of its emphasis on written procedures and the knowledge employees professed to have of procedures. Although the available evidence indicates that Shipbuilding Company is somewhat more bureaucratized than Electric Company, basically the same factors determine knowledge of procedures in both factories—status and cohesiveness, and age or sex.

The Weberian theory of bureaucracy assumes that rank-and-file manual or clerical personnel in a bureaucratic organization know the systematic procedures on the basis of which the organization functions. Our findings show that this is less true of lower status than of higher status personnel, thereby suggesting that the assumption needs to be modified. A second qualification is that one source of knowledge of procedures is cohesiveness, part of the informal structure of an organization that Weber de-emphasized.

CHAPTER FOUR

Technology and the Division of Labor

This chapter will describe the production process in each factory in terms of the technology, the work flow, and the division of labor, both manual and non-manual. This discussion is presented in some detail because there are few reports in Western languages, based on direct observation, on what specific Japanese factories are actually like. This chapter will set the stage for the next, in which we shall relate attitudes and values concerning work to the technological and social structure of the work situation.

SAKE COMPANY

We begin with a description of technology and work flow in the traditional plants of Sake Company. After the preliminary work of rice washing, rice steaming, and cooling, there are three important stages: (1) making malt (kōji), (2) fermentation (sake-no-moto or shubo), and (3) making fresh sake (*ro* or *nigorizake*). In the scientific language now used in a Sake Company brochure the sake brewing process is described as follows:

> The raw materials of brewing are rice and water. The first step is the preparation of steamed rice, the second [is] the preparation of Koji (culture of *Aspergillus oryzae* on steamed rice) as saccharifying agent, and then the preparation of Moto, the yeast inoculum.
>
> Sake is brewed in an open fermentation system and all processes in sake brewing are carried out in open tanks at low temperature (app. 10°C) . . . .
>
> Moto is prepared by Koji, steamed rice and water. The first step is spontaneous nitrat-nitrite reduction and lactic fermentation which prevent the contamination by bacteria and wild yeasts. Only then starts the propagation of sake yeast. Such classical Moto is called Ki-Moto or Yamahai-Moto. In the modern method Sokujo Moto, commerical lactic acid is added to an extent of 0.5% of the mash in place of the lactic acid fermentation.
>
> The main fermentation (Moromi) is carried out by adding steamed rice, Koji and water to Moto and these three consecutive steps thereby

increase the volume each time. The fermentation takes about 20 days. Mash is filtered to remove the cake and sake is produced.

It is characteristic for sake fermentation that the saccharification of rice and its fermentation by yeast proceed simultaneously.

Contrary to the case of wort, sake mash is dense and mashy because it contains much solids such as Koji and steamed rice and therefore yeasts are always suspended in mash by these solids during fermentation. These conditions are considered as the main reason for the high alcohol content of sake mash which will become as high as 20–22%.

Work roles

Let us examine the role content of the traditional positions in the hierarchy still found in Number 1, 2 and 3 Plants. In general the master supervises the entire process of sake brewing. The head (kashira) or associate master (tōji hosa) is the master's lieutenant; his main job is planning all the work processes in the plant, and he sometimes represents the master. The subhead (daishi) is the malt master (kōji shujin), i.e., chief of the malt making section. The sake-no-moto mawashi or sake-no-moto ya is chief of the fermentation section (sake-no-moto). The instrument caretaker (dōgu-mawashi) is responsible for maintenance of the instruments of production. The steamer (kama-ya) is in charge of the steaming of rice; usually the chief of the upper workers (jō-bito) takes the title and role of steamer. Upper and middle workers carry out a great variety of tasks under the master's direction, as delegated to the steamer and others of the above chiefs. Heads, subheads, fermentation chiefs, and steamers can be called "independent workers in each division of labor"; upper and middle workers "supplementary laborers" working under the direction of the "independent workers" (Sakaguchi 1964:139). In practice, there is role substitutability, with a given worker being assigned to a job in another division. Independent workers sometimes do the jobs of the supplementary workers, and vice versa. Specialization is not a systematically followed principle within the plant. As we describe these work roles, we shall also show how they have changed in the new, automated plant.

Role of the Master. The master is traditionally referred to as the person who gives the spirit to sake, meaning that it is his skill that determines the quality of the sake. One of the masters in Sake Company echoed this view in 1966: "To make good sake has been the master's duty since a long time ago. If the quality of [our brand of] sake is not good, all responsibility should be shared by the masters working at Sake Company's plants, and they will be accused by the owner of the company, by the workers, and also by their fellow villagers. Even if a master can boast of long experience, he must brush up on his skills like a beginner." Masters still reflect the traditional pattern of recruitment and socialization into work roles by way of a long apprenticeship. The master just quoted is the son of a sake master. He started work in a sake plant in 1923, served a traditional apprenticeship and learned about brewing, and finally mastered his own technique of making sake. He became a master in 1950.

The master's day begins with his examination of the condition of the raw sake and of the chemical analysis he had asked a company scientist to make in the laboratory on the previous day. In the past, the master performed this task without the benefit of modern chemical analysis: he used his intuition, based on his long experience. (Studies in Japan indicate a high correspondence between judgments made by masters on traditional grounds, and those made by modern chemical analysis.) Today, judgment about the condition of the raw sake is a combination of the scientist's and the master's decisions. On the basis of these facts about the state of the sake, the master draws up the day's work plans and informs his heads.

Role of the Head. The head first allocates the day's work to his subordinates according to the master's plan. He then begins one of his own jobs: making fresh sake. In Sake Company increased production has led to a differentiation between upper head (*uwa-gashira*) and lower head (*shita-kashira*). Both upper and lower heads assist the master, but the lower head is a lieutenant of the upper head. The upper head assists in the overall supervision of the work process, but is mainly in charge of the fermentation of the main mash (moromi), a process that takes three weeks. The upper head acts as the head (kashira) in the Number 1 Plant, second only to the master. The lower head is in charge of the fermentation of moto or yeast in seed mash. This process requires two weeks and is done in the Number 1 Plant for the Number 2 and 3 Plants. Upper and middle workers (jō-bito and chū-bito) also work in the main mash section, under the direction of the uwa-gashira.

Role of the Subhead in Charge of Malt. Good malt makes good sake, and the subhead's main responsibility is to make good malt.[1] The introduction of machinery into malt making has eliminated the need for some of the workers who used to assist the malt head in other work, but the malt head's task remains the specialized one of making malt—he does not work in any other stage of the brewing process. In 1970 the malt room had five men who worked there all the time (during the winter); unlike workers in other sections they did not rotate to other jobs.

[1] Malt and moisture operations are now centralized in Number 1 Plant. Prior to 1964 these operations were carried out in all three plants. All other operations are still carried out in a parallel or segmented way in the three plants.

The yeast process formerly took about 44 hours, depending on the speed of the development of moisture. It is important to control the temperature in the malt room, and the installation of a machine that controls temperature now enables the yeast to be developed in 36 hours.

The chief of malt making in Sake Company's automated plant had 17 years' experience in traditional plants. Although the automatic malt-making machine has made this work much easier, some new sources of work dissatisfaction have emerged. A worker in this room said in 1967: "My job is only to control the temperature. It isn't necessary to have long experience and lots of skill. The chief has taught me the technique of controlling the machine. Responsibility for the quality of the malt isn't mine, but the chief's. Another worker examines the quality of the malt. I am only a caretaker of the machine."

Role of the Fermentation Chief. The yeast master is in charge of fermentation. There is an upper and a lower fermentation chief (uwa- and shita-moto mawashi, respectively), both under the lower head (shita-kashira). Both these individuals work as yeast developers; the upper fermentation chief is more experienced and gets higher wages than the lower fermentation chief.

In the new, automated plant, the basic process of fermenting has not changed. But new technology and mass production have been introduced. Refrigeration units have been installed in the buildings and in the fermentation tanks. The scientist controls the temperature of each tank. In the new plant, then, the fermentation room is operated on the basis of scientific methods.

Role of the Steamer. The quality of sake depends on the quality of malt, and it is the steamer's (kama-ya) skills that decide the quality of the malt. The processes applied to rice as a raw material are rice washing, rice steaming, and cooling. In the traditional plants, these operations are performed daily. Steamers start work at 3:00 a.m. Their tools are wooden tanks and wooden boxes. All these early morning steaming operations are manual. The steamer examines the softness of the steamed rice. Long experience and intuitive judgment are necessary in steaming rice with the traditional tools. All the elements that influence steaming—the quality of the rice, water temperature, and other climatic factors—must be considered. The steamer has close contact with the master, who inspects the rice he has steamed as part of a ceremonial procedure.

By contrast, in the new plant, an automatic rice washer washes the rice, and automatic rice steamers and coolers do their respective jobs. A conveyor belt then carries the rice to the malt-making room. All these processes can be controlled by six or seven valves on the machine. There is little need of manual labor. Most of the steamer's traditional jobs and skills have become obsolete. Work on the steaming machine does not need to begin until 8:00 a.m. and ends at noon, with the afternoon devoted to taking care of the rice-washing machine.

The Role of the Instrument Caretaker. The roles of instrument caretaker (*dōgu-ya* or dōgu-mawashi) are differentiated into those of inner- (*uchi-*) and outer- (*soto-*) instrument caretaker. The latter are charged with keeping every part of the factory clean. The former take care of the various instruments used in the plant. Cleaning and dusting the machines and the factory must be done before each phase of brewing begins. The outer-instrument caretaker must prepare and arrange the various instruments for their operation, under the directions of the head.

The Roles of Upper and Middle Workers. Whereas most of the roles discussed so far are relatively specialized, that of upper worker (jō-bito) is not. His main job is tank work (*oke-kake*) and straining lees (*kasu-hagashi*). However, since these tasks are performed in the afternoon, upper workers spend their morning moving about the entire plant, assisting others in miscellaneous work. The boundaries of their role are thus not clear, though the core of it is tank work.

Middle workers (chū-bito) also engage in miscellaneous tasks around the plant (Oshikawa et al. 1962:382–383), assisting other workers in every phase of brewing. Some middle workers are obliged to cook for their fellow workers as part of their job.

The Apprenticeship System. Underlying all these roles is the traditional pattern of socialization through the master-apprentice system. (For a recent historical survey of the role of apprenticeship in modern Japanese firms see Okamoto 1972.) Sake making was a handicraft industry in which skills acquired through long apprenticeship were handed down from generation to generation. Following the custom of his village, a boy 15 or 16 years old would go away to work in a sake plant. At first he would be a servant, arranging meals for the other workers and doing miscellaneous jobs. He would be successively promoted to lower worker, middle worker, and upper worker. After 20 or 30 years of further experience as a steamer, captain, instrument caretaker, blender or subhead, a few would rise to the exalted position of head or even master, the apex of the hierarchy. The case of the master of Plant Number 2 illustrates this: "I was born in 1925 in the Sasayama area, in the same town as the master of a sake factory. I started brewing work at age 16. With the introduction by a person from my village I started to work in a Kobe brewing firm. I worked there three winters, one winter as a cook—the lowest job—and two winters doing kōji [malt] work."

This apprenticeship system was in a state of decline by the late 1960s. Only 42 percent of the workers studied in 1966–67 had begun work in a sake plant before they were 19 years old and could be said to have had a traditional apprenticeship.

The Daily Work Schedule. A combination of scientific advances in biochemistry and technological innovations has made for a substantial shortening of the work day in recent years. The work day in the traditional plants is now from about five in the morning till four in the afternoon; at night there are shifts. Although these hours are still long compared with other Japanese industries, the total actual working hours are down to about eight hours a day. In the new plant—and in automated plants in other sake firms—the work day is 8:00 a.m. to 5:00 p.m.—similar to other industries. Thus even a traditional industry does not escape the general trends of industrial modernization.

Technological modernization and social change

Sake Company's Number 1–3 Plants are only partially mechanized, whereas its new plant is highly mechanized and automated. Major changes in the more traditional plants, introduced in the early 1960s, include the following: enameled steel tanks for storing sake (instead of wooden barrels); automatic facilities for washing, steaming, and cooling rice; a hydraulic presser to squeeze steamed rice; and an automatic malt-making machine with an air conditioner. The enameled steel tank is not only useful in making raw sake, it also preserves the freshness of the taste of sake, and it is easier to wash (Oshikawa et al. 1962:456).

In other respects the traditional plants are still non-modernized in their technology and work patterns. No fundamental changes have taken place in the preliminary stages of brewing: manual operations are still the rule. In two of the traditional plants the raw materials (rice and water) are delivered by hand, and although the rice is squeezed by a machine, the workers have to throw the raw sake into tanks and do the blending with their hands. The straining of the lees after brewing is also done manually. The state of the raw sake depends on the temperature, which is still determined by the outdoor temperature and other climatic conditions. The master must therefore still attend closely at regular intervals to the condition of fermentation and of the raw sake. The steamer makes the rice cakes which the master examines for quality.

Thus, partial mechanization and manual labor coexist in the traditional plants. But it is not a stable, long-term coexistence. Major steps toward mechanization of operations only began in the early 1960s; the process is likely to continue until the plants are highly modernized technologically. As the master of Plant Number 2 put it: "With mechanization, sake manufacturing changes every year." By modernizing some of their transportation facilities the three traditional plants have increased their productivity, but their productivity per worker is still below that found in more modern sake plants because mechanization has been only partially realized in the brewing process itself.

Even this partial technological modernization has brought in its wake a number of social changes. Some jobs—particularly at the lower end of the traditional hierarchy—have been eliminated altogether. For example, the installation of mechanized malt-making equipment with air conditioning has eliminated the job of the head (kashira) and his assistants who used to make malt with their own hands. Furthermore, there is less dependence upon human experience and intuition.

The basic processes of brewing in the new automated plant are not so different from those in the traditional plants. The major changes are rather that transportation facilities have been modernized and biochemical theory has been applied to all stages of the brewing process. The new plant is one of the most highly mechanized in the entire sake industry. The white rice is transported to the third floor of the building through a pipe, and from there the material goes through the processes of rice washing, rice steaming, rice cooling, malt making, and squeezing by means of a conveyor belt and other machines. At the end of the process the sake is stored in tanks on the first floor of the building.

The visitor used to the sights of a traditional sake plant is struck by the sharp contrasts presented by Sake Company's new automated plant. The plant floor is dominated by machines. There are large banks of computers in the control room.[2] Although 28 people are employed during the November to March period, only seven workers are needed to operate the factory between April and October. Our visit to the new plant in 1969 was in the summer, and although the plant was running, few workers were in sight.

In the West the emergence of refrigeration and air conditioning facilities had profound effects on the brewing industry. Japan first attempted to introduce air conditioning into sake plants around 1925, but it is only since 1962 that it has been in general use. This one technological innovation made possible year-round sake brewing, after centuries of winter-only operations. Now the problem is not how to control the natural conditions, but how the workers are to adjust in the summer months to the contrast between the cold plant and the hot, humid outdoors!

[2] A small Shinto shrine was located high on the wall of the computer control room, superficially at least providing another instance of the coexistence of traditional and highly modern elements. We did not investigate the role of religion in Sake Company, so we do not know what functions and meanings are attached to this shrine.

Figure 4.1 shows the table of organization in the automated factory. When it is compared with the traditional craft hierarchy of a sake plant (Table 3.2), dramatic changes are evident. The number of levels in the organization has been reduced to three in the automated factory. Table 4.1 compares specific work roles in the traditional plants with their counterparts in the automated plant. This table shows that the status of master has disappeared in the new plant; it is now differentiated into the two statuses of scientist and foreman. Responsibility for brewing and authority in personnel administration have passed to the scientist. The foreman has become a first secretary to the scientist, as well as the chief of the fermenting process.

In 1967 all but four of the employees in the new plant had previously worked in the traditional plants, which means they were faced with the problem of relearning roles, or role transition. Those in the position of foreman and chief in the new plant had been masters or heads in the traditional plants. They had to learn new techniques of mass production, to supplement those of their traditional skills that had not become obsolete. In the new plant, the scientist role is the essential role, and traditional apprenticeship skills are no longer essential.

The supplementary workers (in raw materials, malt making, and other processes noted in Table 4.1) have no training in modern technology.

FIGURE 4.1

Organizational Hierarchy in Sake Company's New Factory

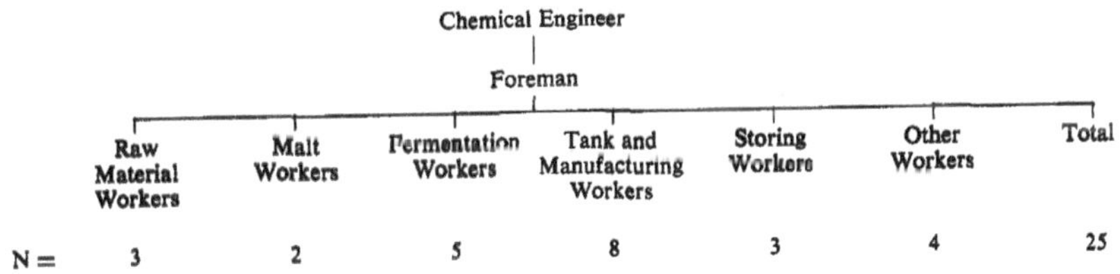

Source: adapted from Mannari, Maki, and Seino 1967:14

TABLE 4.1
Comparison of Work Roles in Old and New Plants of Sake Company

Traditional Plants	*New, Automated Factory*
Master (Tōji)	Chemical engineer
Staff:	
Head (Kashira)	Foreman
Subhead in malt (Daishi)	Chief of each process
Steamer (Kama-ya)	Workers in raw materials
Subhead in malt (Daishi)	
Malt worker (Muro-ko)	Workers in malt making
Fermentation chief (Sake-no-moto-mawashi)	Workers in fermentation Workers in storing
Captain (Sendō)	Workers in squeezing

Source: Mannari, Maki, and Seino 1967: 15–16.

Their function is only to operate the machines under the supervision of their chief. They have little area for initiative, and as a result are said to lack the positive attitude toward their work that could be found in sake workers in traditional settings.

CYBERNETICS AND INFORMATION: THE ROLE OF SCIENCE

Historians of the hoary art of Japanese sake making have discovered that remarkable progress in the understanding of the science of microorganisms had been realized prior to the introduction of Western science into Japan. It is now believed that sake brewing was traditionally one of the most important areas of application of indigenous scientific developments. The knowledge applied was purely empirical, yet, when explained in terms of modern biochemical theory and the analysis of the functions of malt bacteria, yeast, and fermentation, the traditional lore makes considerable sense.

Since the Meiji period, thermometers and hydrometers have been used in sake making. Modern biochemical knowledge has eliminated the putrefaction of sake. These and other scientific applications have had more far-reaching consequences in Sake Company's new plant than in its older ones. But even in the older plants, the master has had to become more technically competent by adding biochemical knowledge to his experience and intuition. He has had to relinquish at least some of his authority to the university educated scientist. In the traditional plants, the master and workers still take the initiative in brewing work: human skills and craft knowledge operate side by side with modern biochemical knowledge. The master and the scientist keep in close touch. The scientist reports the results of his analysis of the condition of the sake to the master. Decisions are made on the basis of the discussions between the master and the scientist: there is an interplay between the latter's biochemical analysis and the former's experience and intuitive judgment. To make good sake it is believed that there must be a close relationship between these two men. The master may consult with the scientist even about many matters outside the scientific field. The master of Plant Number 2 described his relationship with the scientist as "intimate, with good communication."

Data from our interviews illustrate this sharing of knowledge and authority by the master and the scientist in the traditional plants. The scientist, a kachō of the Technical Section in the company, said: "The staff scientist [*sensei*] examines the sake, but is not supposed to give orders; only the master can do that." The master of Plant Number 2 declared: "The job of master isn't much changed, except that after twenty days I used to use my eyes and taste to decide if a bottle of sake was good. Now we take a sample and do analysis *as well as* still using our eyes and taste." The master of Plant Number 1 still had some reservations about the practical efficacy of scientific analysis: "In the summer I attend a seminar to learn the new methods. But I hesitate to use them in practical situations. It's hard to judge whether the new way is better; there's no decisive way to produce good sake. Rice and all other conditions influence making good sake."

In the new, automated plant, the scientist is much more powerful. There is no master (tōji), and the scientist has overall supervision of the plant. He makes the annual plan for the plant's operation. He makes every important decision. Although the new plant has adopted the modern system of staff and line, in reality the staff—the scientist—has the commanding authority. The foreman represents more the methods of the traditional plants, and feels there is a gap between the new technological-scientific system and his own ideas. And since most of the workers have worked previously in the traditional plants, some traditional customs have inevitably been imported into the new plant. These old customs seem to be obstacles to the smooth operation of the modernized factory.

The social composition of the plants also reflects this gap. Workers are more educated in the new plant: 60 percent are high school graduates, in contrast to only 10 percent in the three traditional kura. The heads of each work group in the new factory were recruited from the traditional plants, and are older, as well as less educated. The younger workers studied in polytechnical agricultural high schools. Fermentation workers studied agricultural chemistry in high school; others were recruited to the new factory because of their training as electrical technicians. Although most of these younger recruits had had some years of prior experience in traditional breweries in other companies, what distinguished them from the older men transferred from within Sake Company was that they knew more about making sake with experimental instruments. These younger high school graduates, recruited on the basis of competitive examinations rather than apprenticeship, resented the older sake workers who were above them. This situation generated more conflict in 1966 than in 1970, because the chiefs of each process in the new factory are gradually being appointed from the younger group.

ELECTRIC FACTORY

We shall first describe the production sections; among these we shall concentrate on Hardware Components and Number 1 and 2 Production Sections; then we shall describe more briefly the non-production (staff) sections.

Hardware components section

Physical appearance. One's first impressions when entering the large rooms that house this section's 121 employees are the relatively high noise level from the large machines,[3] the complexity of the machines, and the preponderance of male workers (many of whom have a specific machine assigned to them, and have their name on it). The section has two large rooms, one for pressing and welding, the other for the more complex operations of cutting (lathe work). The machine jobs are organized in terms of four han, two of which do press work and two of which do mainly lathe work.

The section contains 166 machines. A conveyor belt transports parts from worker to worker, and carts transfer raw materials and finished parts. More modern transfer machines have not yet been introduced to this section. The Motor Section is more modernized in this respect. Recently, more lathe than press machines have been introduced into the Hardware Section.

With only 121 employees to handle the 166 machines in the Hardware Section, workers are trained to handle more than one kind of press and lathe machine. A worker is assigned to different machines in the course of time. Although in general lathe work is more highly skilled and complex

[3] Working conditions were worse from 1958 to 1961 than at the time of our field work: more dangerous, dirtier, colder, and noisier. The present Hardware Section chief has improved physical conditions and reduced worker grievances by more skilled handling of human relations. The installation of air conditioning throughout the factory was scheduled for completion in 1971. In the hot, humid summer months there was one electric fan for every two workers in most sections; in Hardware there was one fan per worker. Despite the greater number of fans, it is said that when temporary workers (*rinjiko*) leave Hardware to work in an air conditioned shop, they never want to go back to Hardware.

than press work, and although there are other variations in skill within Hardware (and Motor) Sections, these sections are more diversified in their work—both in an overall sense, and for given workers—than are Number 1 and 2 Production sections. Hardware does cutting, pressing, and welding for parts of fans for Number 1 Production Section, as well as for vacuum cleaners, and mixers for Number 2 Production Section, and also works on bicycle accessories for the Bicycle Production Department. When new products are being considered, it is Hardware that tests them. This enables Hardware workers to be in the vanguard in knowing about new products.

We spent numerous hours observing the spatial layout and the organization of work into assembly lines according to product and han and kumi. Figure 4.2 shows these aspects for the press room, as of July 31–August 1, 1969. The work of lines one through five as of August 1 was as follows: production line 1 was working on cases for window vent circulator fans; production line 2 on the body for vacuum cleaners; production line 3 was doing press and cutting work for cases for fan motors; and production line 5 was doing press and cutting work for bicycle headlight cases.

Description of Division of Labor in Line 1. On July 31, 1969, one of the researchers observed the actual work flow and division of tasks in line 1 (first han, first kumi) of the press subsection. The identification number and sex of the workers in that line are shown in Figure 4.2. On that day the line was making circulators (*junkansen*) for air conditioners and heating equipment. The circulator moves air around so that all the hot air will not stay at the top of a room and all the cold air at the bottom. Here is what the 19 workers present on line 1 that day were actually doing:

Worker 1: takes flat metal piece, inserts it in a machine, which bends two edges.

Worker 2: inserts same piece in second half of same machine to bend its third edge. Workers 1 and 2 must insert the piece into the machine at exactly the same time. The bending is done when one of the two workers steps on a floor bar which lowers the cutting edge to do the bending.

Workers 3 and 4: also work at two halves of another machine, taking the metal piece from worker 2 and, by the same foot pressing, coordinated operation, make the piece into a three-sided tray shape.

Workers 5 and 6: their chairs and space along the line are empty; they are either absent or loaned to another assembly line.

Worker 7: attaches a strip, by soldering, to the outside of the bottom of the circulator.

Worker 8: does same job of soldering a strip, but in this case, the strip is attached to the inside of the bottom of the circulator.

FIGURE 4.2

Hardware Components Section, Room 1, Press Subsection, as Observed on July 31, 1969

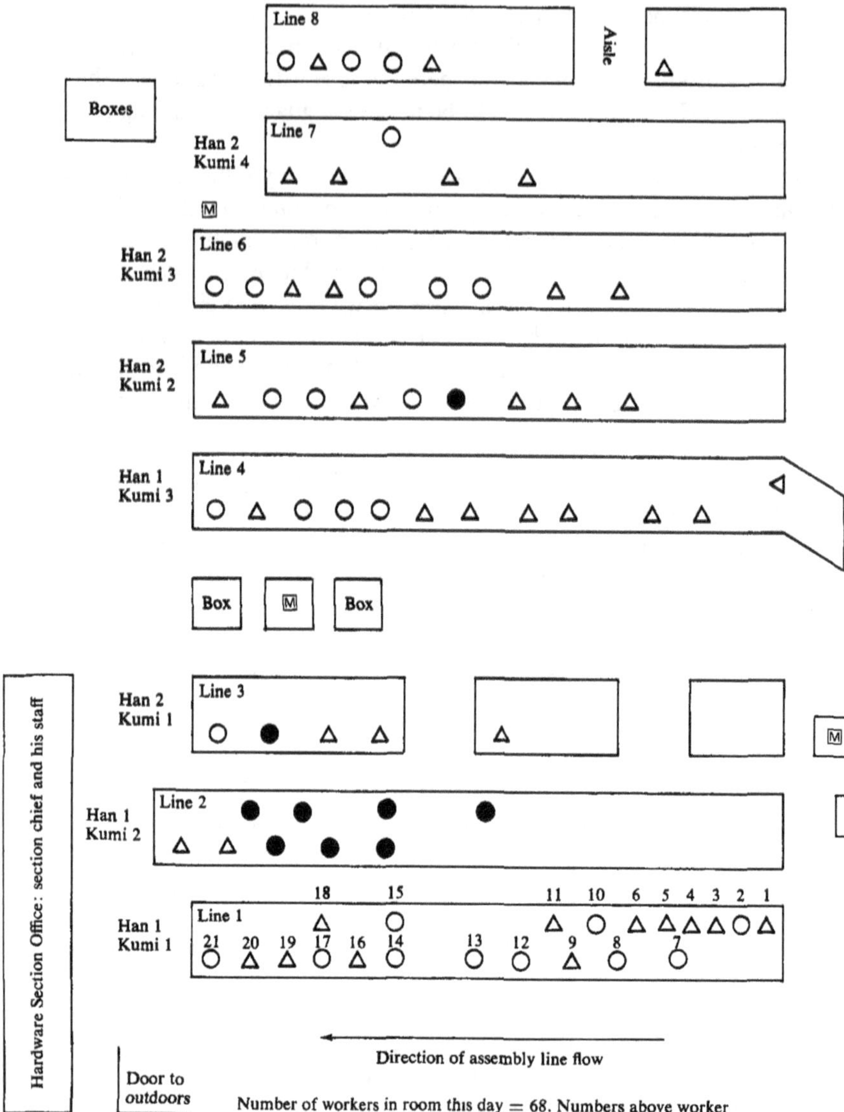

Number of workers in room this day = 68. Numbers above worker identify worker whose specific job is described in the text

Legend: ○ = woman △ = man [M] = machine ● ▲ = absent

Worker 9: using a model bar, attaches two more strips to the inside, to connect the three sides of the circulator.

Worker 10: inserts a piece to the left inside part of the circulator.

Worker 11: takes a metal piece from a box on the floor, solders it to the back of the circulator.

Worker 12: uses a machine to connect the bottom and the back sections of the circulator.

Worker 13: attaches a piece (similar to that done by worker 10) to the right inside part of the circulator.

Worker 14: attaches a third piece to the middle of the bottom piece of the circulator.

Worker 15: attaches a fourth piece to the far left side of the circulator.

Workers 16 and 17: insert a long strip piece into the circulator.

Worker 18: takes piece attached by worker 11 and uses machine to drill one hole. Then, instead of returning the piece to the assembly line (as other workers, above, do), places it in box on the floor.

Worker 19: uses machine to hammer long strip piece firmly onto the circulator.

Worker 20: drills hole and pries up two clips on either end of the circulator.

Worker 21: (a middle aged woman, whereas all other workers are either men or young women) manually bends the bottom of the circulator and then places it in box on the floor.

A transportation worker then hauls away the boxes of finished circulator parts.

This was the situation on July 31. The next day line 1 produced an entirely different part: a case for a window vent circulator fan. However, each worker was doing basically the same kind of job as on the previous day, at the same place along the line.

Other Aspects of Work in Hardware. When a Hardware worker rotates among machines he may at the same time move from his own han to aid another han in the section. Each worker writes a daily work report, which lists the pattern and machine he used, the items being produced, the number of pieces completed, hours and minutes worked, and number of mistakes. The first-line foreman (kumichō) tabulates these reports from all his workers by 10:00 a.m. the following day, and gives it to the kakarichō.

Hardware begins work on a given part usually two weeks ahead of the assembly sections (Number 1 and 2 Production sections). Work scheduling is highly differentiated. For example, the fourth assembly line of the press subsection produces 30 different parts, some of which require only one press machine for two or three hours a month, while others require several machines to be in operation almost every day. The goal is to have 85 percent of the man-hours worked devoted to direct production, and only 15 percent to preparation (changing moulding for pattern, etc.) and miscellaneous. Table 4.2 shows the distribution of work time in Hardware at two points in time, December 1968 and June 1969. The overall man-minutes worked in a month are divided into (1) direct production work, which increased from 79.3 percent to 81.1 percent between December and June; (2) indirect production work, e.g., supervision and switching jobs or patterns, which decreased slightly, from 14.9 percent to 14.3 percent of work time; and (3) lost time, resulting from having to redo work, which decreased from 5.8 percent to 4.6 percent of work time. When the production goal cannot be met by the deadline, workers do overtime work.

The Hardware Section, then, is under continual pressure to reduce lost time and time spent in indirect production. Another problem grows out of the nature of the manual tasks, which are repeated every few seconds, by the press and welding workers who make up the majority of personnel in Hardware. A foreman told us how he handles this exigency: "If the worker continues to operate the same machine for two hours he becomes tired, accident-prone, and he makes defects. So I sometimes let the worker stop the conveyor and shift to other work, such as keeping boxes of parts

TABLE 4.2
Distribution of Work Time in Hardware Section (%)

	December 1968		*June 1969*	
Direct Production		79.3		81.1
Indirect Production				
Supervision	7.6		7.0	
Switching jobs or patterns	4.6		4.7	
Testing patterns	0.3		0.5	
Daily report tabulation	0.8		0.5	
Inventory summary	0.4		0.7	
Conferences	0.7		0.3	
Other	0.5		0.6	
Subtotal		14.9		14.3
Lost work time				
Redoing work	2.0		1.3	
Transporting parts	1.4		0.7	
Repairing and correcting machine patterns	1.5		1.3	
Getting boxes	—		0.4	
Sweeping	0.8		0.6	
Other	0.1		0.3	
Subtotal		5.8		4.6
Total		100		100

Source: Electric Factory Records.

in order. If possible I also change a worker's position along the assembly line so that his operations are more varied. This kind of press work becomes difficult even for workers over 40 years old; for those over 50, they can't do it more than 70 percent of the time." In general, though, workers take very good care of their machines "since they are like their own hands and fingers."

The company's goal of 30 percent increase in labor productivity per year requires either increasing efficiency or reducing the number of workers. The factory introduced an automatic conveying machine for the 200-ton press, following a decision made by the committee on rationalization. This made it possible to eliminate one step in the work process, but not an entire worker.

The stages of socialization into the work role in Hardware are as follows. The new worker is first taught safety in handling the machine and tools. He then assists in boxing the parts and carrying material. Next he begins to do simple press machine work. A kumichō told us he tries to make the worker an independent operator as soon as possible. "I tell him," he said, "the capacity of the machines is such and such, so try to reach that limit. By trial and error, try to come as close as possible to the maximum capacity." After learning how to operate the press machine, they are taught how to remove the moulding for one part and put in a moulding for another part.

The case of Mr. M., an R-3 worker in the fourth assembly line of the Hardware Section, illustrates the process of socialization into the work role. Mr. M. entered the company in 1961, from junior high school. He completed senior high school nights. Now 23 years old, with eight years' seniority, he can handle a 100- and a 200-ton press. Here are excerpts from our interview with him:

> When I entered the company eight years ago, I did odd assisting jobs for one or two years, for example, packing boxes and putting oil on raw materials. Now raw material is sprayed with oil automatically. There were no women in the section eight years ago. Now new workers operate press machines within one year, or within six months, and even girls operate the 30-ton press machine soon after entering the company.
>
> During my first year I wanted to use the machines but wasn't permitted to. In the second year I was given a 30-ton press machine. Soon after I was transferred to a 100-ton press machine. At first, I felt the machine speed was fast; actually the rotation of the big 100-ton machine was slow, but I was scared for one week. . . . The method of changing the metal mould [or pattern] started for me in the third year. Workers had to learn how to use the moulding. Now the change of mouldings is done mostly in overtime [early evening] work. Before, it was done

during normal daytime work hours. I learned by watching senior workers, and they helped me when I couldn't do the work. In the early period I sometimes broke the moulding. Now I know how to use almost all the 15 press machines in my kumi. But if I go to the next han to work its press machines, I'm not familiar with them. I can operate then, but not repair them.

[Q. Does time pass quickly or slowly?]

A. Until noon, quickly; slower in the afternoon. If repair of a moulding is the work, time passes faster. If a moulding isn't straight I report to the kumichō, and he decides whether to stop the machine; I don't decide. I only decide myself on the quality of the parts I manufacture. I want to concentrate on press work, but also learn lathe work so I can become an all-round craftsman.

[Q. Do you like your job?]

A. Press work is a man's work, worth doing. I don't want to change my work, but I want to learn lathe work. One thing I dislike is that now temporary workers (*rinjiko*) operate the press machines, and my work is therefore regarded as simple.

Number 1 production section

The two Production sections (Number 1 and 2) present a sharp contrast to the Hardware Section. Number 1 is housed on two floors of a large, hangar-like building, Number 2 on the second floor of another similarly-shaped building. These two sections, which are the final stage of the production process at the factory, are as much a "woman's world" as Hardware is a "man's world." One is struck in Number 1 by the cleanness, lightness, piped in music, and long lines of uniformed women engaged in the minutely specialized work of assembling electric fans. The fan has gone through Hardware, where press work is done, and Surface Treatment, where the exterior of the fan cover and the guard have been spray painted; only then do its parts enter Number 1 Production, on the continuous overhead conveyor system.

Some 14 makes or styles of fan are assembled. On the first floor the parts—of which there are 170 in major subunits of motor, stand, guard, and fan blades—are placed on trays on a moving belt which carries them upstairs to the second floor where, in three assembly lines, called A, B, and C, final assembly is carried out. Fans for export are made on the first floor, those for the Japanese market on the second floor. Each of the lines (A, B, C) has its own conveyor track for moving parts from the first floor to the second.

Our observation was concentrated on the second floor where there are, depending on the day, four or five long assembly line tables, each consisting of a conveyor belt moving at a fixed speed. Most women are seated along these tables; a few stand. Thus, on August 28, 1969, the three assembly lines' work was as indicated in Table 4.3.

Work in both Number 1 and Number 2 Production sections represents an application of the basic theory of assembly line and time-and-motion studies.[4] A typical worker in the section performs the same motions 3,000 to 4,000 times a day. Consider for example soldering operations. The average fan has about 20 such operations, in contrast to 100 for a television or radio set. Therefore, soldering is done by hand, with each solderer in Number 1 Production Section performing about three operations. In Electric Company's other branch plants, where television sets are produced, the more numerous soldering operations have been automated.

TABLE 4.3
Goals and Output of Assembly Lines in Number 1 Production Section, August 28, 1969

Assembly Line	*Number of Workers*	*Day's Goal*	*Number of Fans Produced at 1:35 p.m.*	*Percentage of Goal Reached at 1:35 p.m.*
A	50	1,750 electric fans model 6EN[a]	1147	65
B	35	2,000 electric fans model 6DA	1297	65
C	30	650 circulators model 8 DR	422	65

[a] Model 6EN was the new model for sale in 1970. Model 6 DA was the 1969 model.

Each fan is tested while running. The time for testing each fan is 5.5 seconds, a pace dictated by the fact that each assembly operation takes on an average 15.6 seconds. Since there are three assembly lines feeding into the same testing line, the latter must move at about three times the pace of the former. This does not mean the entire testing operation is limited to 5.5 seconds per fan. Testing is done at several points along the assembly lines; it is only the fully assembled fan that is tested at the point indicated in the above description.

There are a number of ways to match a worker's job with her interests as far as degree of specialization and repetitiveness are concerned. The hanchō tries to handle complaints about monotonous work by (1) shifting the worker to a different job in the assembly section (job rotation), and (2) giving the worker a combination of two jobs, thus achieving some

[4] Ranked by percent of personnel in Routine (R) and Training (T) jobs combined, among line production sections: Number 1 Production 89.4 percent, Number 2 Production 88.8 percent, Motor 82.7 percent, Surface Treatment 81.8 percent, and Hardware 81.2 percent.

degree of job enlargement. Fans for automobiles are smaller, and the result of letting each girl assemble several parts—job enlargement—was higher productivity and fewer errors. Among the more demanding jobs in the section is testing, and this is another possible assignment. A kumichō told us he tried to overcome the women's dissatisfaction with simple, repetitive tasks by stressing that quality control (QC) is complex and important, and that they should try to find satisfaction in being good in that area, no matter what stage of assembly they work on. As an experiment he assigned a combination of about 10 testing operations to each woman who worked on testing. He found that this job enlargement, or reduction in specialization, led to a decline in efficiency. The same kumichō expressed the belief that "since assembly line work is against human nature, management in this area calls for close leadership." Another kumichō has a different approach to the problem of monotony. "The work is so simple that we place friends side by side along the assembly line. We used to prohibit talking at work, but now allow it to offset the monotony of simple jobs. We put friends and girls of the same age next to each other. . . . I started working here in 1949 assembling bicycle lamps; I have done assembly line work all my life, so I have experienced 'a taste of pain' [*kurushimi*] from repetitive work." The same foreman said that the assembly line workers adjust to their repetitive jobs "by thinking about non-work matters while working: sometimes fantasy, sometimes realistic matters."

Another function performed in Number 1 Production Section is repairing fans for customers. A girl who has already done assembly work may be shifted to repair work for variety, if she has the ability. A kumichō told us how this works: "When a worker repairs an old fan we can evaluate his or her ability to repair. Some repair only the outside of the fan, but I train them to repair both the inside and the outside. Some girls are weak in mechanical skill, so boys have to cover for them. Once a worker can change [replace] the rotor of the fan motor, he is independent as a repair person." When all is said and done, however, the work in this section is simple, highly specialized, repetitive, and monotonous. In addition, as one kumichō told us: "Compared to Hardware, no one in Number 1 Production feels he can acquire any skill that is useful outside" (that is, marketable elsewhere).

As in other sections, there is a strong emphasis on performance evaluation and productivity. Every March workers can be rotated among jobs. Those who have been incompetent are shifted at that time to simpler, lower status jobs. But they are not fired unless their attendance record is very poor. Because the assembly line is so specialized, the efficiency of individual workers doesn't affect total productivity. If there is a slow worker there is usually another, even simpler, task to which she can be shifted. Moreover, delays in the work schedule are not so much the fault of the assembly lines (Number 1 and 2 Production sections) as of delays in receiving parts. When an assembly line has to be stopped because of parts delays, the kumichō uses the time to talk to the workers about "zero defects" (ZD). To stimulate interest in output, one kumichō tells his workers about the statistics on zero defects (i.e., what percentage of fans have been produced free of any defect). He also puts a counter on the women's machines so that they can see how many pieces they have turned out, and praises them accordingly. Sometimes workers get a half hour or more overtime pay to attend a ZD conference.

Excerpts from an interview with Miss N. illustrate some of the conditions of work in Number 1 Production. A high school graduate in a Section where 74 percent of the employees are only middle school graduates, Miss N. is 19 years old, with 1 year and 5 months in Electric Company. She is a group leader in the Zero Defects Movement.

> I have worked in A line. My first job was using the mechanical screw driver. The second and third jobs were other kinds of screwing work. For the last year my job has been to check for defects at the end of the A line (but before testing). It is a difficult job: if I talk to others I may miss defects; I have to stand up, so I get tired. Time passes slowly: faster in the morning, but slowest from 1:00 to 3:00 p.m.
>
> [Q. Can you use your own judgment in your job?]
>
> A. I get a sample of the defects that can occur in a product, and look for these on each fan. But the actual defects often differ from those in the sample, so I have to decide whether to send the fan back; and this is difficult. Sometimes the test line people tell me my judgment was wrong, and sometimes when I think something needs to be repaired, the kumichō comes and decides instead to run it as it is.
>
> There is a schedule to complete. The present goal is to make 1800 model 6 EN electric fans a day. We can make slightly over 1800. There is no machine to count how many. Workers do not seem to mind how many they have to complete. I am always anxious not to make a mistake; I am afraid that workers will come from the testing line.
>
> My present job seems to have no meaning in my life. It is not worthwhile. I wonder whether my job fits me or not. My present job is so simple. I don't know anything good about my job. A bad thing is that it is my job to find out the defects of others.
>
> Four girls entered this company with me from [the local high school]. Two have moved into a clerical job. I used to resent it that I hadn't, but now I forget it.
>
> A good thing about work here is the five-day week. The wages are better than other companies. I can save money for marriage. Life is

stable and the company's welfare is good. My family members think it is good that I work here.

Overtime work varies from month to month. This month we had 10 hours overtime. Defects were found so we unpacked the box and did them over again. We had two hours overtime a day. We are free to refuse overtime. I took one day's absence in the last year and five months. It was a rice planting day. We are free to take time off, but if many workers take time off there would be trouble for production. If we take many absences, it is said that it influenced our bonus.

[Q. Do you make mistakes?]

A. Yes, I made a big mistake. I thought I had checked fans for defects, but I hadn't; all had some defects and we had to do about 100 fans again. I was not scolded, but the kumichō told me to be careful. Usually the hanchō and kumichō don't tell me whether my work is good or bad; I have almost no contact with the kachō and the kakarichō.

There are few changes of products in line A, and we use the same pattern. Sometimes the design changes. B line changes patterns.

There are good workers as well as those who often make mistakes. We help others who are too busy in their work; it's not a shop where people don't help each other.

I don't think while I work: I might make a mistake.

We would be happier here if there were half men and half women employees.

Among the women who actually perform the highly specialized assembly line tasks there is by no means widespread dissatisfaction or alienation. Miss N. is a high school graduate. Until recently, assembly line jobs were done only by middle school graduates. The kumichō in Number 1 Production agreed that middle school and high school graduates "have a different way of thinking." The former are less bothered by assembly work.

One Number 1 Production kumichō said:

My methods of discipline must be specific to each group [high school and middle school graduates]. Both groups start work full of expectation, but after two or three months there are differences. . . . The high school girls see it as more simple and repetitive work, and when their friends move to an office job, they get dissatisfied. The company promises high school graduates that whenever there is an office vacancy they can move, but such vacancies are rare, so many high school graduates must stay on the assembly line. The morale of such girls isn't good. After two years on the job, they forget their original ambition.

Another Number 1 Production kumichō had a similar story: "For high school graduates, after one year's work they begin to quit, because of the monotony of the job—it isn't worthwhile for their life. . . . After three or four years' work here they change their minds because they talk with girls in other companies, and because after all they have to do some kind of work. So after three or four years they accept this work more."

NUMBER 2 PRODUCTION SECTION

The second floor, which houses this section, contains seven long lines of assembly workers. On July 23, 1969 three different items were being assembled by the approximately 200 workers (80–90 percent of whom are women) in the section: motors for vacuum cleaners, cases for vacuum cleaners, and juicer-blenders.

Observation of the work flow and division of labor concentrated on the first two lines, consisting of 24 women, engaged in assembling the motor for vacuum cleaners. Like the others on this floor, each woman had a fan blowing on her for ventilation. Piped in music was difficult to hear because of the moderate noise level. Men present were first- and second-line foremen who circulated in order to replenish the supply of parts some women had near their bench, or to move boxes of parts, and the like. The 24 women whose work will be described were all young, except for two women in their fifties. Three or four wore white socks, the rest hosiery; all wore the company uniform, which they changed into when they arrived at work.

The following description is keyed to Figure 4.3 which identifies each of the 24 women by number and position. Here, then, is what they were actually doing on July 16, 1969, the first day we observed them.

Worker 1 (woman in her fifties): takes wired starter A (*sutata*) out of a box on the floor, pulls out its wires to straighten them, and puts starter on a platter attached to moving conveyor belt. (This is the beginning step of the assembly process for these 24 women. Because the conveyor is continuous and circular, worker 1 has two tasks: to start the assembly in the way just described, and to check the finished motor which has been assembled by the next 23 women.) After checking the finished motor worker 1 puts it on the pulley which carries it to the ceiling conveyor belt.

Worker 2: takes narrower case B starter from box on floor, uncoils some of its wires, heats them over a flame, cools them by immersion in water, then returns the starter to another box.

Worker 3: performs first operation on worker 1's starter A: takes cap from box on floor, fits it onto starter A, presses it down with a machine, then puts new combined motor back on the assembly conveyor belt. Also puts ring and rubber coil on the motor.

Worker 4: inserts a green rotor A into starter, using a couple of screws,

FIGURE 4.3

Assembly Line for Vacuum Cleaner Motors, Number 2 Production Section

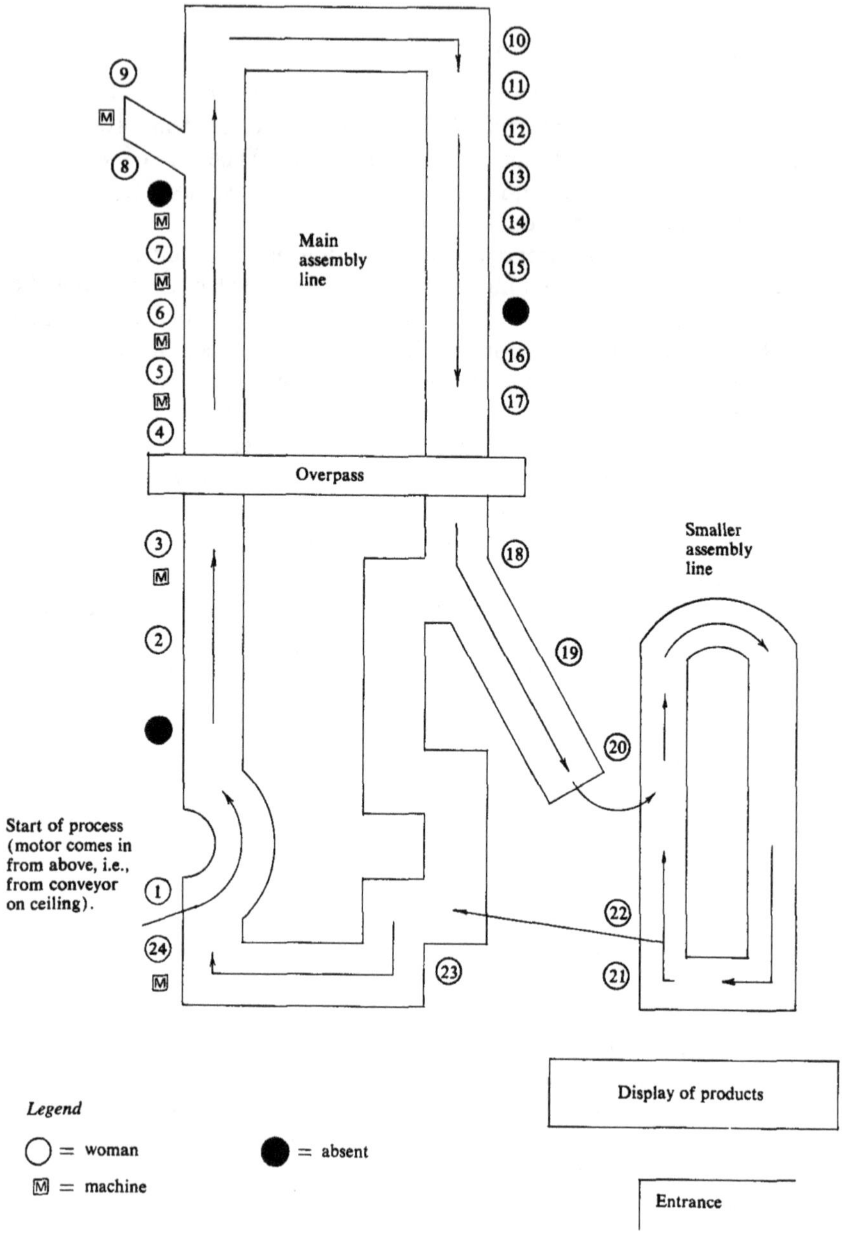

and bangs them together by hand. Motor now has three major parts.

Worker 5: first task is to put a brush holder combination (which she assembles) onto the platter carrying motor along the aseembly line. If she misses some platters the next two workers (6 and 7) do them for her. Second task is to take case A from box on floor and, using machine, insert starter; then place this on same platter with motor leaning against it.

Worker 6: first task is to add brush holder unit to platter (like worker 5), and assemble these while waiting for case A. Second task is to take case A from tray on assembly line, and by means of a machine insert a center piece, place this part on the end of the motor, and return the motor to the assembly line.

Worker 7: first task is to add brush holder, like workers 5 and 6. Second task is to insert screws by machine to fasten together the motor and the case A.

Worker 8: inserts one screw into motor and returns motor to the assembly line. Also removes brush holder from main assembly line and puts it on tray moving toward worker 9 (see Figure 4.3).

Worker 9: inserts brush holder into right side of motor and screws it on by machine. Also inserts another plug (from a box near her seat) onto left side of motor and screws this on by machine. Returns motor to assembly line.

Worker 10: using pliers, winds wire from brush holder around screw on side of motor.

Workers 11 (second woman in her fifties) and 12: insert three wires protruding from motor into screw holes and "paint" these with a brush. Each woman works on every second motor that come along.[5]

Workers 13, 14, and 15: solder a coiled wire hung above the bench onto points on motor where workers 11 and 12 attached wires. Solder five different spots on each motor.

Worker 16: tests current of motor on electric ampere gauge, bends protruding wires, turns motor around to face opposite direction.

Worker 17: inserts washers and fan sleeve and bolts sleeve (by hand) to motor.

Worker 18: takes motor which is on tray separated from the main assembly line, and puts it in a holder. While motor is in holder,

[5] Even while the conveyor is moving, there are times a woman can rest, as when the conveyor belt is empty for a long stretch. There is one five-minute rest period in mid-morning and another from 3:00 p.m. to 3:05 p.m. At that time the women drink milk or cold tea which are prepared for them and dispensed in bottles (milk) or from a machine (tea) near their bench. Immediately after the afternoon break workers sweep their own work bench and floor area.

attaches another fan cover A and puts washer onto it. Holders on this tray pop up every 10 seconds.

Worker 19: inserts one more fan cover washer and screw; watches lights on a gauge.

Worker 20: connects main assembly line with smaller one (see Figure 4.3); takes motor off former, attaches generator (+ and – circuit) plugs to motor.

Worker 21: tests motor on gauge (amps, watts, kilowatts), as each moves along assembly line. Doesn't touch motor unless gauge shows some defect.

Worker 22: same as worker 21, but after testing motor on gauge, returns it to main assembly line.

Worker 23: checks motors given her by worker 22, and after tightening screws returns it to main assembly line.

Worker 24: picks motor off assembly line, puts a cap on it; machines bang cap down. Sprays oil on screw in center and returns motor to assembly line. Also cuts off little pieces of metal before putting cap on.

Worker 1: receives finished motor from worker 24, checks it, and puts it on pulley which carries it to ceiling conveyor.

From this description, it can be seen that the principle of the division of labor is—as in Number 1 Production Section—the classical one of minutely subdivided assembly line tasks. Some variety is introduced by changes in parts to be assembled. On the second day of observation of the same line (July 18, 1969) the women had been moved from assembly of vacuum cleaner motors to assembly of vacuum cleaner cases. This meant they were seated in the adjacent assembly line, and their work places shown in Figure 4.3 were temporarily vacant. This kind of group job rotation occurs when there are seasonal changes in the item being produced, or shifts in particular product orders or deadlines. The kachō of Number 2 Production Section told us his girls like job rotation within kumi, on the same line, and to another assembly task on the same product (the vacuum cleaner). They dislike being moved from the vacuum cleaner to the juicer or mixer assembly.

On a third visit for observation (July 23, 1969) everyone had returned to her job of July 16, as described above, with two exceptions. Worker 1 now had a helper, who took finished motors off the assembly line and stored them in a box on the floor. The helper was worker 2 in Figure 4.3. Consequently the tasks described for worker 2 above were not being performed on July 23.

Kumichō and other supervisory personnel recognize that there are dexterity differences among workers, and at least one kumicho in Number 1 Production felt that "it is hard to place the right person in the right job."

Observation of other sections

Although the time spent observing other sections was limited, the following details may be of interest.

Surface Treatment. This is another direct production section (see Figure 3.2). Because it has machines for heating the surface of parts of the fan and other products, it is very hot. About one-third of the employees are women; 92 percent did not go beyond middle school. Work consists mainly of simple hand assembly operations. As in Number 1 and 2 Production sections, the assembly line here has parts coming down on moving belts from the ceiling. Charts on the wall show the daily output, by number produced, percent change in output, and by the number produced per minute, for each part (e.g., the 30 cm. fan guard A, the 20 cm. guard, etc.). Male workers oversee the automatic dipping of each fan guard into a paint process.

Motor Section. The motor section is housed in an enormous two-story high, hangar-shaped building. It is air conditioned and has large machines, worked mainly by men, one to a machine (with the worker's name on each machine). Piped in music is not clearly audible, because of the relatively high noise level (higher than in Number 2 Production). Tasks here are more highly automated than in Number 1 and 2 Production: smaller machines (run by women) include the automatic slotting machine, the automatic connecting machine, the automatic wedge inserting machine, the automatic winding machine, and the automatic insulation machine.

Production Engineering. This section is housed in two rooms. In the first, a quiet, unairconditioned room, men with blueprints work at desks. The second room is long and somewhat narrow, and contains both desks and tools and machines for constructing the moulds and patterns for the production machines used in other sections. All employees save one are men. Piped music can be heard, though there is a moderate noise level. There are many Zero Defect award placards and pictures of the men of the section hung on the walls.

Number 1 Production Control and Overseas Administration Sections. These are housed in one medium-sized room. The room is replete with reference materials: wall maps of the world and books and other materials on patents, licences of various countries, etc.

Number 1 and 2 Product Design Sections. Designs are drawn for electric fans and circulators in Number 1 and for vacuum cleaners, juicers, mixers, etc. in Number 2 Product Design section. A small number of men work in a small room, sitting at desks, some of which are tilted, drafting-style desks.

Automation Facilitation Section. In this small section eight or nine desk workers work on methods of making machines, tools, and patterns more mechanized and automated. It is housed near Production Engineering, and is a purely staff (not a line) section.

Accounting and Materials Sections. These are also staff sections situated in one medium-sized room, with numerous desks. The only machines are desk calculators or adding machines. Because of this there is less noise than in the production sections, and it is easier for employees to talk, which they do much more here. Accounting has about 30 percent women, Materials about 60 percent. Materials has the task of purchasing parts and raw materials. Adjacent to this room is the computer room, with seven women doing key punching and processing data. These rooms are air conditioned.

Development Section. This section contains a library of technical journals and desks seating eight employees (including one woman). The following tasks are performed: two people work on patents, two on testing, one on development proper, one on electronic data processing, and one (the woman) on general staff work. Next to the library and office is a room equipped with machinery for testing products. When we paid a visit here, the kachō was reading a technical journal, and three of his employees were together at one desk consulting about drawings.

Research Section. This section is housed in a small laboratory which does testing, experimenting, and other kinds of research on the design of the whole range of Electric Factory's products—fans, vacuum cleaners, juicers, mixers—including parts of these products, such as the control circuits of the motor, the main overloads of the motor, and dynamos for bicycles. It also performs these tests on competitors' products.

Standardization Section. This section, which was eliminated the year after our field work, was a small staff section charged with a variety of standard-setting and standard-maintenance operations. It planned standards for production control. drafted methods for measuring the effects of changes in production, gave technical guidance on production management to the company's subsidiary firms, and managed the factory committee on rationalization of production. It also cumulated monthly and annual data on labor productivity. It managed the suggestion system, sought ways to reduce production costs, supervised aspects of quality control, education, and public information and was responsible for overall coordination of Zero Defects activities.

General Affairs Section. This section is responsible for a rather heterogeneous set of administrative tasks, such as safety and hygiene, construction and repair of the physical plant, handling employees' stock subscriptions, purchasing, sales promotion, guarding property and security, and operating factory automobiles.

Employee Training Section. This section handles life insurance and the sickness insurance union, administers company housing and the dining hall, maintains personnel records of all kinds, drafts plans for personnel management policy, and administers hiring, promotions, salaries and bonuses, retirement allowances and pensions, personnel evaluation, and labor union policy.

Conclusion

We have seen that although Electric Factory is moving in the direction of greater mechanization and even automation, at the time of our field work it remained dependent upon classical assembly line mass production technology. To a large extent there is manual control of tools, and the majority of workers (women especially) work at highly specialized, fractionated tasks. Operations are highly repetitive and skill requirements (in terms of formal education or on-the-job training) are low. The system is one of large batch production. At the same time we have seen that there is considerable variation in these aspects of technology from section to section and from worker to worker. With regard to production workers, at the opposite extreme from the short-term female employee on the assembly line (whose work fits the above generalizations) is the male employee of long tenure, running a complex lathe or other production machine, or engineering patterns; his work is much more varied, demanding in skill, and in some ways approximates the craftsman's conditions of work.

As in any factory, work flow and output in Electric Factory are influenced by a number of functional exigencies: calendar deadlines, delays in delivery of parts or raw materials, and the degree of conflict or cooperation between line and staff. One dimension of the relationship between line and staff was made clear to us by a section chief in Design: "The hanchō of the Hardware Section are veterans, rich in work experience. The pattern designers in the Engineering Section are young and less experienced. The judgment of veteran workers and of staff designers is different. There are many cases in which veterans point out some shortcomings concerning safety or efficiency in the work of the designers. Veterans can't do the design work themselves, but they have opinions that are supported by experience. Most of the designers' ideas come from information the machine engineers give them."

Change in technology and work flow is, of course, incessant.[6] The long-range policy of Electric Company is to make parts at Electric Factory but

[6] In recent years Electric Factory and Electric Company have attempted to increase rationalization (*gorika*) by greater mechanization of production and by computerization of

to send them to subsidiary firms for assembly. This is because it is difficult to increase productivity in assembly line work. Even press work will be contracted more from subsidiary firms. Thus, jobs with high labor intensiveness but little value added per worker will be gradually phased out, as a result both of the trend toward greater automation at Electric Factory itself, and of more subcontracting of assembly operations to smaller firms.

SHIPBUILDING FACTORY

PHYSICAL PLANT

Since the typical product manufactured by Shipbuilding Company's X Plant is much larger than that manufactured by Electric Factory, it is not surprising that the area covered by the land and buildings of X Plant is three times larger than that of Electric Factory. The high, long buildings of the Engine Division, in which diesel engine parts are machined and assembled, contain, among their main equipment, machine tools, casting and forging equipment, electric apparatus, and transportation equipment.

WORK FLOW

The typical flow of work can be described in terms of the 1969 table of organization (Figure 3.4). It begins in the Diesel Design Section's General Office Subsection (*sōkatsu gakari*), where negotiations are carried out with customers and with firms that do production subcontracting. The customers may be ship owners, dockyards, industrial firms, or government agencies. Orders for special, custom-built products are taken to specialists in the two Design kakari. In general, the General Office Subsection acts to articulate the work of (1) the Production Scheduling Subsection in the Assembly Section and (2) the two Design subsections. A staff person in the General Office Subsection described production schedules as follows: "I deal with customers who want to order the Pattern 62 diesel engine. Its actual production takes only one month now, but the overall process from writing the customer's order to ordering parts, casting, cutting plates, machining and assembly, takes about eight months."

From the Design subsections the work flow moves through the Machining Section, where the parts are produced, to the Engine Assembly Sections. Here they are assembled into a complete engine (or other machine), tested, inspected for the final approval of the customer's

---

some management and staff functions, as well as by the introduction of more modern personnel practices. Our questionnaire contained an open-ended item, "In this factory there has been a strong effort to advance rationalization. Frankly, what do you think about rationalization?" For an analysis of responses to this question, in relation to technological implications theory, lifetime commitment, and the rate of technical change, see Marsh and Mannari 1973.

engineers, and disassembled for shipment either to a dockyard where the engine will be reassembled in the actual ship that will use it, or to an industrial plant.

Division of labor: job titles

In 1968 Shipbuilding Company introduced the first step in a new job control system (*shokkan seido*), in which 187 job titles in the company as a whole were defined. Of these, 31 job titles exist in the Engine Division. Table 4.4 lists these job titles, the intermediate job classifications under which they fit, and the number of employees in each. The sections and shops in which specific jobs are done are shown in Table 4.5

We shall next describe technology and employees' attitudes and behavior in the work setting, following the work flow sequence from Design to the Machining Section, and then to the Engine Assembly Section.

The diesel design section

The major finding concerning the Engine Division is that most workers' jobs are neither narrowly specialized nor highly repetitive. This fact struck us again and again throughout our interviewing and observation in the division. The Diesel Design Section's college graduate engineers strongly eschew narrow specialization. As one of them put it: "Specialization in the Design kakari is disadvantageous in the long run. Men shouldn't know how to design only one part, such as a piston. If they're sick, the job can't be done. Work here is differentiated both by size of engine—Pattern 82, Pattern 62, etc.—and by parts (turbocharger, cylinder, etc.). But design people are *rotated through these different jobs*. Normally, we spend a year or two learning how to design for one size engine or one type of part, then we change to a different size engine or different parts." It is said to take three years for a college graduate to master the designing of the ten parts of the diesel engine.

At the same time, there are pressures within the Diesel Design sections toward expertise of a kind that requires less frequent job rotation, and greater specialization. The same staff person acknowledged that: "I have been designing turbochargers, remote control systems, and engine exhaust control for the last five years, so I'm expert in these areas."

The company's laboratories do some basic research and gather basic data. There is a laboratory in X Plant, administratively outside of the Engine Division. When there is an accident involving one of Engine Division's parts or products, the Design Subsection asks the laboratory to make tests to help them determine how to handle the problem.

Designers derive satisfaction not just from the diversity of their work, but from another important aspect of their professional role: the company

TABLE 4.4
Job Titles in Engine Division of Shipbuilding Company

Job Title	*Intermediate Job Classification*	*Number of Employees*
Direct Production Jobs		
1. Drilling	Drilling	13
2. Pipe processing	Piping	8
3. Pipe connecting and finishing A		
4. Pipe connecting and finishing B		
5. Pipe connecting and finishing C		
6. Pipe bending		
7. Machine marking	Marking, measuring	23
8. Lathe operator	Machining	288
9. Nakaguri boring machine operator		
10. Horizontal, vertical lathe operator		
11. Milling machine cutter		
12. Drilling machine operator		
13. Cutting, grinding operator		
14. Gear cutting (hagiri) operator		
15. Machine finishing	Finishing	294
16. Tool finishing		
17. Precision finishing		
18. Electric rigging and installation	Electrical	29
19. Frame making	Frame work	1
20. Special indoor sweeping	Sweeping	1
21. Processing by heating and cooling	Heat treatment	4
	Subtotal	661
Indirect, Support Jobs		
22. Packaging for shipment	Transport	134
23. Operating crane		
24. Weighing and measuring	Weights and measures	9
25. Maintaining tools	Tools	34
26. Repairing tools		
27. General inspection	Inspection	4
28. Production process control	Staff jobs	34
29. Procurement of materials		
30. Teaching engineering skills to new employees		
	Subtotal	215
	Total[a]	876

[a] Includes some employees from other divisions of X Plant.

TABLE 4.5
Organizational Units of Engine Division in which Jobs are Performed

Unit	Jobs
Engine Assembly Section	
Staff Kakari	Production process control (28)[a]
	Procurement of materials (29)
	Teaching engineering skills (30)
Assembly Kakari	Machine finishing (15)
	Packaging for shipment (22)
	Operating crane (23)
Piping Kakari	
Pipe Shop	Pipe processing (2)
	Pipe connecting and finishing (3, 4, 5)
	Pipe bending (6)
Drill Shop	Drilling (1)
	Frame making (19)
Diesel Design Section	
Design Kakari	Production process control (28)
	Procurement of materials (29)
	Teaching engineering skills (30)
Machining Section	
Tool Engineering Kakari	Weighing and measuring (24)
	Maintaining tools (25)
	Repairing tools (26)
	Tool finishing (16)
	Cutting, grinding operator (13)
Number 1 Machine Kakari	Machine marking (7)
	Lathe operator (8)
	Nakaguri boring machine operator (9)
	Horizontal, vertical lathe operator (10)
	Milling machine cutter (11)
	Drilling machine operator (12)
	Cutting, grinding operator (13)
	Gear cutter operator (14)
	Packaging for shipment (22)
	Operating crane (23)
Number 2 Machine Kakari	
Precision Shop	Precision finishing (17)
	Heat treatment (21)
Maintenance Shop	Machine repairing (17a)[b]

[a] Number in parentheses refers to number of job title in Table 4.4

[b] The job title "machine repairing" is actually the 31st title, but since we lack data on how many workers do this in the Engine Division, we have omitted it from Table 4.4 and here call it job title 17a, to avoid confusion in the numbering system.

delegates them considerable authority in the engineering area. As the above-quoted staff man told us: "I can design what I want the way I want to."

The machine section

There is a great variety of machines in this section. Machine Shop 2 alone has 30 separate machines for its 77 workers. Each shop and each han tend to specialize by type of machine: for example, Shop 2, han 2, has seven kinds of milling machines (*furaisu*[7]).

Job Diversification. A dominant theme among Machine Section workers was a preference for "multi-skill jobs," (*tanō shokuka*). "Skill" does not mean narrow specialization; quite the opposite. A Machine Shop sagyōchō stated company policy: "Formerly our policy was to attach one man to one machine. [8] Now our unit of thinking is 'finish a certain piece of work,' and if this requires shifting men to other machines, we do it. For example, worker X can do job Y on machine Z the best of all, so when job Y needs to be done, worker X is moved to the machine to do it. At first there was resistance to this change; now it's more accepted." A worker in Shop 2[9] expressed the same policy in these words: "The policy is to train Machine Section workers to work on many machines, to assist the *oyakata* on big machines and learn them, but also to assist workers on smaller machines."

The training period for each worker is defined as a time to familiarize him with a variety of machines. According to a worker in Shop 2: "During my period as an assistant (*sakite*) I had experience with all sizes of boring machines."

Decisions to transfer or rotate workers between shops are made by kakarichō, between han by sagyōchō. The main reasons for rotating and transferring workers were elaborated by one worker as: "To give workers broader skills. Or to fill in needed night shift workers. Also, crane operators and *tamakake* [workers who guide the overhead cranes from the work floor] can be exchanged across shops or han, and tamakake and machine operators can help each other by teamwork." Rotation among machines thus depends both on the policy of the hanchō and on the desire of many Machine Section workers for job diversification. As one Machine

[7] "Fraise" in English: a milling machine, cutter, or cutter for correcting inaccuracies in the teeth of a wheel.

[8] One of the characteristics of a traditional craftsman's role in the past was the "culture of secrecy." The esoteric knowledge of his craft was jealously guarded. A Machine Section worker recalled this period in the history of the company, only to note that the culture of secrecy is now giving way to a culture of open exchange, at least within the organization.

[9] In selecting workers to be interviewed, we decided to sample on a social organizational rather than an individual basis. Hence, we interviewed 10 employees from the same shop (Shop 2), instead of only two or three from each shop.

Shop worker puts it:

> My *rigide* machine is a kind of milling machine. Sometimes I come to work on the night shift and work on the boring machine as a trainee [sakite]. The rigide machine is easier to use than milling machines. I'd like to learn how to use two other kinds of machines in my shop [Shop 2]; and also to learn to use a lathe in Shop 3. I like lathes because they require more thinking than milling machines.
>
> I prefer to know how to use several machines in my han: large machines like the boring machine, and smaller ones like milling machines.
>
> In my eight years here I've learned to use all 15 machines in Shop 2; milling, boring, planing, and other machines. Those people who like to work on one machine all the time can do so; others who want to switch can.

Each of the larger machines has the name of one worker on it. This practice was established about 1966 so that one worker would have ultimate responsibility for each of the large machines. But this name doesn't necessarily correspond to the name of the worker who is using the machine at any given time: this varies from the night to the day shift, and also depends on changing job assignments.

Workers who derive more satisfaction from specializing on one machine may do so: "Most workers know several machines, but I've always worked on boring machines. I'm a boshin of Number 2 Han. It's all right to be this specialized, because the hanchō and others really defer to me on problems related to boring machine work."

Division of Labor by Age. While there is much job diversification and job rotation, the age of the worker sets some limits on the range of jobs to which he may be assigned. A machine worker reported: "My shop has boring machines, milling machines, cutting machines, and marking machines. It has older and newer machines. Older workers can do better jobs on the older machines. Also, the basic principles of boring machines haven't changed over the years, so older boring machine workers can fairly easily learn to use the newer boring machines."

A sagyōchō told us how a certain range of job diversification is reserved especially for older workers: "If older workers get too weak for the strenuous jobs, they can be shifted to drawing work, tool shop work, or materials management work." Older workers are also channeled into frame work, sweeping, heat treatment, drilling, and piping work.

Marking (shindashi) Work. One of the relatively few jobs in the Machine Section that does not involve using a machine is that of marker (*shindashi*).

This job involves reading a blueprint, which tells where an object—e.g., the iron base section of an engine—should be marked for later cutting. Using special measuring devices and an iron pointer, the worker centers the object and then plots points and lines on it. These marks on the object tell the machine cutters where it should be cut. As a shindashi described his job: "The main job of a shindashi is to find the center of the part that is to be cut. Once that is marked, all the other specifications can be found and marked." The work of shindashi is the basic stage of all machining work. As one marker put it: "Marking (shindashi) accuracy determines the accuracy of the next stages in machining." According to another: "Shindashi is more primitive work than machining because the latter continues even when the worker just stands there by his machine. But if the shindashi stops work, work stops completely. So my work [as a marker] is more physically burdensome. Shindashi work more often has to be learned on the job: "More polytechnical high school graduates know how to use lathe or boring machines when they start work than know how to do marking." Marking is so demanding on the eyes that it cannot be done on the night shift; when it is, more mistakes are made. Thus if a marker does night shift work, he works on a boring machine instead of doing his regular daytime job.

Day and Night Shifts. For the day shift, the daily work schedule is: enter the gate before 7:50 a.m., start work at 8:00; lunch from 12:00 to 12:50; work from 12:50 to 3:50; if there is no overtime work, the seven-hour day ends at 3:50 p.m., and the worker leaves the gate at 4:00. This holds for six days of the week. Overtime starts at 3:50 p.m. and continues until 6:50; supper is from 6:50 to 7:50; and from 7:30 to 9:30 some workers stay for more overtime. At 9:30 p.m. all day-shift work stops.

There is a night shift in the Machine Section, which begins at 6:42 p.m. and runs to 11:30, at which time there is a one-hour period for eating and rest; night work and overtime then continue from 12:30 to 6:00 a.m. Of the 647 line production workers (A Group) in the Engine Division in 1969, 347 (mainly in the Assembly Section) worked only on the day shift; the other 300 (that is, those in the Machine Section) alternated between day and night shift work, on a three-week cycle. They were divided into three groups of 100 each and were cycled as follows:

Week	*1*	*2*	*3*	*4*	*etc.*
Group 1	night	day	day	night	
Group 2	day	night	day	day	
Group 3	day	day	night	day	

Research has shown that in the same setting, the work pace and other factors may vary considerably as between day and night shifts (Georgopoulous and Tannenbaum 1957). Interviews with Engine Division night shift workers brought out several differences between the night shift (*yakin*) and the day shift. On the night shift there is only one hanchō for the entire factory, and no higher level managers, so the ratio of supervisors to workers is much lower than on the day shift. At night it is the more automated machines that are run, and workers act more as monitors of the machines. These facts have both functional and dysfunctional consequences for the individual worker and for the organization. A boshin declared: "On the night shift there is only one hanchō . . . so I worry about mistakes that can be irrecoverable. A night shift worker, whose own daytime hanchō isn't on the night shift, may make mistakes. In my han the night shift is *more efficient* than the day shift, because there are fewer distractions: no work conferences that hanchō and sagyōchō have to attend. Also, night shift work is all scheduled, whereas when there is a question about the blueprint on the day shift, they may stop the machines. And the day shift is when time is taken for repairing machines. The physical burden of night shift work is heavier."

Engine assembly section (gendōki-ka)

Parts that are finished in the Machine Section are transported by cranes of various tonnage-lifting capacities, which move along tracks located high above the work floor. Their destination is the Engine Assembly Section, where the final stages of production work in the Engine Division are carried out.

There are several contrasts between the Machine and the Assembly Sections. The Machine Section is higher in a cybernetic informational sense than is the Assembly Section. Machine workers need more skill and receive longer training. Assembly work is more manual, Machine work is relatively more mechanized. Work is more specialized and individualistic in the Machine Section, involving, as we saw, basically a man and his machine. In Assembly, on the other hand, there is more need for teamwork. Also, Assembly Section work is more standardized or routinized. As a Machine worker described this difference: "Particularly the first time we produce a part in the Machine Section we have blueprint problems and other special problems to work out." The contrast is apparently that while those in the Machine Section may have to adjust to new dimensions of cutting or boring of the parts, there is little variation in how the parts are assembled into the finished engine. A piston, for example, may be machined in a quite special way, but the basic assembly of a piston in an engine remains unchanged.

Tight Production Scheduling. Although all phases of production in the Engine Division are scheduled according to a "standard time" system (*yotei jikan seido*), scheduling of the work flow is tightest in Engine Assembly. The standard time production schedule for the assembly of a Pattern 62 Engine in 1968, for example, was 7,912 man-hours. Each step has a scheduled number of hours, and the actual number of man-hours taken is recorded alongside the scheduled number.

More detailed scheduling, based on linear programming, is done at meetings held every Friday. Having estimated the time required to cut and frame the steel plates and do the marking and boring, the delivery date is set. If work is behind schedule, night shift work is done. Despite this weekly adjustment in the programming of the schedule of work, there is a large amount of "emergent," unanticipated work, which may disrupt the schedule.

Block Assembly and Testing. Two stages in the assembly process can be described briefly, block assembly and final testing of the assembled engine. The block is assembled separately from other parts. Then two 100-ton cranes lift these parts and transport them to where they are assembled. Cylinders are also assembled separately, so they can be kept closer to the ground and avoid making workers climb up and down scaffolding that is several stories high. A hoist positions the cylinder block in its place on top of the bed of the engine, and a cylinder cover is placed on top of the cylinder block. The cover controls the explosions of the cylinder. Similarly, the pistons are built in the block, near the floor, before the block is hoisted atop the bed of the engine.

In the assembly of any engine, the climactic moment is when the fully assembled engine—as high as or higher than a three-story house—is actually run and tested. Han 1 of Shop 3 always stays in the location at which this testing takes place, and has the task of watching all the meters and gauges as the motor runs and making final adjustments. Thus, at 10:45 a.m., on December 2, 1969, we observed the testing process. Here are some of our field notes:

> On the control floor—where the master instrument panel is—are eight Assembly workers, and nine representatives from the company that is buying the ship for which this engine is being made (visitors are easily distinguishable from employees because they wear different colored hard hats). Everyone watches as the engine is put through its paces. A big wall chart and papers in everyone's hands tell the sequence of tests through which the engine is going, e.g., "dead slow," "half," "full speed." The sagyōchō gives the order to the hanchō to move the lever from one speed to another. The visiting engineers and representatives of the buyer, are, of course, very alert. Some of the Assembly

workers, on the other hand, are more relaxed. The sagyōchō has the main responsibility during this demonstration: he is at the center of the group of visitors. Some of the latter climb up to the second or third level of the scaffolding around the engine, to observe something more closely.

The demonstration is being conducted in a completely automated way: though there are a couple of Assembly workers on the second or third levels of the scaffolding, everything is controlled from the master instrument panel.

Job Diversification in the Assembly Section. Assembly work is carried out not only in different shops and han, but also in different "factories" (*kōjō*) within the Assembly kakari. ("Factory" refers to a given *line* of workers: one of the large, hangar-shaped buildings houses three such lines—factories 1, 2, and 3—and another building houses factory 4.) Factory 1 does general assembly of Pattern 62 and 84 engines; factory 2 assembles Pattern 74 engines (column, jacket, connection rotor, piston rotor, and cylinder cover); factory 3 assembles Pattern 62 and 74 engines; and factory 4 works on Pattern 84, 62, and 74 cams and does repair work. In addition, certain functions—the operation of equipment for testing, transportation by crane, and ground directions (tamakake)—are performed in all four factories, although the size of the cranes varies from factory to factory.

As in the Design and Machine sections, so also in the Assembly Section there is an emphasis on job diversification, rather than on narrow specialization for each worker. This came out clearly in our interviews with men in the section. As already pointed out in Table 4.5. the job titles in the Assembly kakari are unspecialized: everyone is either a "finisher," a crane operator, or other transport worker. Among finishers, workers distinguish "block assemblers" and "general assemblers," but this exhausts the terms used for job specialties.

Assembly work is all round, not specialized work. One assembly worker described the variety of his jobs: "I've worked with six other workers in setting up two cement mills and one chemical plant. We assemble the parts for these engines. The parts were made by the Machine Section. This is the work I do with N. han. My other work is repairing ship engines. I did repair work on one ship while it was sailing to Hiroshima. It was an 84-Pattern, 12-cylinder, 160,000-ton ship. . . . I'd like to do top-to-bottom jobs of general assembly: the crankshaft on Pattern 84 engines, the columns on another type of engine, the cylinders on a third type. Then over time I'll learn everything about general assembly."

Another worker, aged 19, had entered the company only eight months earlier. His work was already diversified: "My present job is in engine assembly. I work on all stages of assembly. The whole process takes one to two months on one engine, so my work cycle is quite long. I like muscular work best, such as tightening bolts. I'd like to become a master of all engine work. I now work on Pattern 62 engines; Pattern 84 would be almost the same, except that the size of the engine is larger. When younger, I want to learn the engine, and the best way to do this is in all round general assembly work, rather than in block assembly or more specialized stages of work."

Diversification was the explicit reason one worker sought to work in the Assembly Section, when he applied to the company: "I applied to general assembly because that's the way I could learn faster the whole production process here."

There is, of course, some division of labor: "In Pattern 84 work, there is no sharp division of labor, but generally, one group works as a team on the base of the engine, one on the column, one on the piston, etc. Then we sometimes combine." The distinction between block assemblers and general assemblers is related to age and skill. General assemblers are on the average 25 years old and less skilled than the block assemblers, who are on the average 32 years old and have been with the company longer. But here again there are no hard and fast lines. Block work for a given engine peaks one month before general assembly work. When work is in the general assembly stage, several block assemblers move to general assembly. Thus, depending on the work flow, general assembly men may work with block men and vice versa.

Because the work cycle of finishers is a month long or more, there is not the feeling of monotony that narrow assembly line jobs produce in other industries, e.g., the automobile industry. Yet it is no accident that finishing jobs are in general now performed by younger workers. As a hanchō with a long time perspective on changes in the job of finisher explained: "Finishing work in the Assembly Section used to be done by real craftsmen. Now more of this is done in the Machine Section and in subsidiary firms, and we really do more assembly of parts than 'finishing' work. We're still called 'finishers,' but a lot of our finishing functions have been lost." Thus, we must distinguish two types of diversified manufacturing jobs. (1) The traditional craft type, in which jobs of greater and lesser complexity are integrated into an overall product that reflects high individually-skilled workmanship. (2) The newer kind of diversification, as found in Shipbuilding Company's Assembly Section. Here, an individual performs a diversity of jobs, but each job is simpler, and the overall product reflects the integration of a large number of relatively simple operations.

Systematic data from the questionnaire confirm our anecdotal findings from each section of the Engine Division that the majority of workers prefer multi-skilled jobs.

"In every shop in this factory there are multi-skill jobs being carried out. Do you like multi-skill abilities, or do you prefer a specialized job?"	
Prefer multi-skill	54.0%
Pro and con	24.6%
Prefer specialized jobs	21.4%
	100.0%
N =	(594)

The following factors were found to be significantly related to these preferences: (1) *Section* (C/C_{max} = .20). Employees in Assembly and Turbine sections are most likely to prefer multi-skill jobs (69.4 percent), followed by those in Design, Staff, and General Affairs sections (54.6 percent), with those in Machining least likely (48.9 percent). The same percentage (24) are likely to prefer specialized jobs in Design, Staff, and General Affairs as in Machining, in contrast to only 12 percent in Assembly and Turbine sections. Since Assembly workers in fact appear to perform a greater range of jobs than do those in Machining, preferences would seem to fit actual behavior and work demands in these two main sections. (2) *Age* (C/C_{max} = .20). Younger employees prefer multi-skill jobs, whereas older ones are more likely to prefer specialized jobs. (3) *Seniority* (C/C_{max} = .18). The greater the seniority, the greater the expressed preference for specialized rather than multi-skill jobs. (4) *Education* (C/C_{max} = .17). Preference for multi-skill jobs increases with education: 50.4 percent of middle school graduates, 57.5 percent of high school graduates, and 68.4 percent of university graduates prefer multi-skill, rather than specialized jobs. Preference for multi-skill versus specialized jobs varies independently of one's organizational status index score.

The Piping Kakari. The piping shop employs 24 men, about half of whom are from subsidiary firms and work as *shitauke*[10] in Shipbuilding Company. Piping is in a separate building from the Assembly kakari. (Drill workers are officially housed with piping workers, but in fact often go elsewhere to do their drilling.) On the day we observed this shop (December 15, 1969) there were about 20 workers present, engaged in drilling holes in pipes and cutting, bending, and welding pipes. Piping work has the reputation of being the simplest work in the production process, and a boshin in Assembly told us: "Those who started work in this company with me (in 1962) and were assigned to Engine Assembly work or design work got interesting and varied work. Those who got this

[10] *Shitauke* means subsidiary contract, and refers to the subsidiary *firm*, rather than to that firm's workers. For convenience, however, we shall use the term shitauke to refer to both.

assignment have stayed in the company. But those assigned to piping have left the company because the work was too simple."

Social Relations in the Assembly Section. We observed social relationships among workers in one han of Assembly Shop 3, and our notes contain the following description:

> 1:00 p.m.: Five men stand idly near the stove for one half hour. Then their hanchō comes over and takes one man away. One of the remaining men is a tamakake, and has a whistle to attract the attention of the overhead crane operator (13-ton hoist). He whistles for the crane to hoist a gear. After this is done, the men in the group go to work to attach the gear to a huge cam. Two workers from another shop work with them.
>
> 2:10 p.m.: Six men from the han are again idle. They indulge in some horseplay—feigning punching each other in the arm, twisting each other's arms. When the hanchō comes over to talk to them, they form two lines facing him.
>
> It is noticeable in the cylinder assembly shop how much of the time workers are idle. At a given time perhaps only two out of six men are doing any work. The other four are: two hanchō talking together; a signaller to the crane operator, who stands idly between signals, and another worker waiting for a job. In general the very young and the very old workers appear to work the hardest.

It is often asserted, by both Japanese and Western students of Japan, that Japanese are especially diligent workers. Our observations from one factory cannot, of course, conclusively affirm or deny this view. But we do suggest the view needs qualification: the pacing of work and idleness in Japanese work settings, as elsewhere, is situationally variable. At a minimum, it appears that Japanese workers do not go to great lengths to *avoid* periods of idleness on the job. Shipbuilding workers do not try to make work during these lulls.

Another characteristic of social relationships in the Assembly section, as already noted, is teamwork. The work itself provides many occasions for teamwork. For example, it takes three workers to hold a bulkhead section and guide it into position while it is suspended from the overhead crane. Camaraderie can be frequently observed; e.g., when not busy, workers put their arms around each other's shoulders. The stoves located at various points on the work floor are favorite hangouts.

Relations between the Assembly Section and Other Units

The Assembly Section has boundary exchanges with a number of other units, both inside and outside X Plant. It receives such preassembled parts as pistons, cams, and blocks from other han in the plant. But an even larger number of parts—valves, pumps, etc.—come from subsidiary firms.

Indeed, one worker estimated that: "ninety percent of our parts are made outside the company, in subsidiary firms. Therefore our main work is designing [before parts are made] and assembling, [after parts are made]." Because Japanese shipbuilding firms have a surplus of orders, plans are being made to turn over some orders to Singapore dockyards. The specialization will take the form of Singapore's building medium-sized vessels and tankers, while Japan concentrates on the largest ships. Singapore has the advantage that labor costs are only half those of Japan.

But it is not only parts and products that are received by the Assembly Section from these other units; it also receives the services of a special group of workers, the shitauke. About 20 percent of the workers in the Assembly Section come from other sections (e.g., to paint the outside of the engine), or from subsidiary firms. The latter are the shitauke. When the company has a busy season, it uses these shitauke; when work slackens, these men go to another company as shitauke. The subsidiary firm subcontracts to do a specific job, on a piece work basis, for Shipbuilding Company. But the subcontracting firm's shitauke are paid on a regular daily rate, not a piece rate. Although they lack job security and retirement benefits in Shipbuilding Company, these shitauke may often receive a higher daily wage than Shipbuilding Company workers. (Shitauke average about 22 percent overtime work.) One such shitauke performs so well in his Assembly work that Shipbuilding Company would like to hire him on a regular basis, but cannot because his daily wage rate is higher than that of regular employees. (There is also a norm against raiding subsidiary firms.)

The shitauke should not be confused with temporary workers (formerly called rinjikō, now "associate employee" or *jun shokuin*). The temporary employee is working for the company in which he has that status, whereas the shitauke is, as it were, *in* the company but not *of* it. His pay comes from a subsidiary firm in which he is a regular employee. In accounting terms, the subsidiary firm would bill Shipbuilding Company for the services of the shitauke just as it bills the company for the parts it produces for it.

The employment of shitauke has several functions for Shipbuilding Company. Shitauke are willing to do dirty work provided they get higher pay. They perform odd jobs in the construction of engines, which makes it unnecessary for Shipbuilding Company to hire regular workers to do these jobs. For example, the Engine Division has only one regular worker with the job title of sweeper, and he is the supervisor of sweepers, who come as shitauke from a subsidiary firm that specializes in sweeping. This farming out of odd jobs enables Shipbuilding Company to rationalize the jobs of its own regular personnel. The shitauke also form a buffer group who, in slack periods, can be let go without violating the Company's lifetime commitment to its regular personnel. In this sense, the company can provide stability to its own regular employees and at the same time benefit from the inputs of the shitauke. For the shitauke themselves, the pattern has the function of providing stability (in their own firm) and interfirm mobility (in firms to which they are farmed out) *simultaneously*.

Crane and tamakake workers

Both the Machine and the Assembly sections use employees in the role-set (Merton 1968:422–38) crane operator and tamakake, the men who direct the overhead crane operators from the ground. In the Assembly section these roles are the main work roles apart from that of finishing. Extensive use is made of the overhead cranes because most of the objects worked on in the Engine Division are too heavy for men to lift and carry. Cranes vary in the weight they can lift from two to 100 tons. Large tracks run along the side walls, near the ceiling, along which moves the bridge-like span from which hooks, blocks, and hoists are suspended. The raising and lowering of these, and the swift movement of the bridge section, is under the control of the crane operator. He sits in a booth attached to one end of the bridge span, and looks down on the plant floor from his windows. As stated earlier, one building houses three factories, and each of these has two or more overhead cranes.

To the newcomer, such as the researcher, the movements of these crane lifts and of the sometimes colossal objects they are transporting are rather frightening. Since parts and other objects are stacked all over the work floor, the crane has to lift its object and then carry it over the tops of the objects on the floor to its new destination. The speed and skill of the crane operators lead them at times to take what appear to the uninitiated as breathtakingly dangerous chances. Instead of first completing the lifting phase and making sure the object carried is higher than anything on the floor over which it must pass, and only then beginning to move the object over the floor horizontally, the crane operators often combine the vertical and horizontal movements. This means that the object is still gaining altitude as it moves forward or backward above the work floor. Like a plane in extra fast takeoff, the object often clears by what appear to be only inches those things on the floor near its point of origin. Again, the speed with which the cranes move pieces weighing perhaps nearly 100 tons is astonishing.

The roles of crane operators and tamakake have changed over time. Until about 1959 tamakake and crane drivers rotated jobs. Then the cranes got larger, and it was felt that crane driving should be specialized, so rotation ceased. But there were unanticipated dysfunctional consequences, as a tamakake told us: "After trying this [change] we realized that *communication* between crane drivers and tamakake was crucial. No matter how good a driver's individual judgment was, the specialization was preventing the necessary communication between him and the tamakake. So in 1968 we went back to a rotation system, between drivers and tamakake. . . . Now there is rotation of that sort and also rotation among cranes of different sizes."

Crane operators are different in several ways from other workers. They are older, with an average age of over 40. Because crane work involves sitting and is less physically demanding than many other assembly jobs, it tends to be reserved for older workers. Crane recruits are mostly *chūtō saiyōsha*, that is men who have already worked in other companies. It is said that if a man is ambitious, he will prefer machining work and will think crane driving is too simple. Crane operators rotate jobs with tamakake every two weeks in factory 1, but do so less often in factories 2 and 3. When we asked the Assembly kakarichō why this was, he replied: "When we tried to rotate them [in factories 2 and 3] there was strong opposition. We still have some resistance from them. The crane operator has pride. Ten years ago, men were interchangeable between crane driving and tamakake jobs, but there were accidents, so we made them specialize. Now we want to restore the older system of exchange of jobs. From the crane operator's point of view, he thinks the tamakake is working for the assemblers, on the ground, whereas he is more independent of the assemblers. So crane drivers don't want to come down and rotate with the ground, tamakake people." It was also alleged that the social isolation of crane driving influenced the drivers' personalities: since they spend all day alone in their crane driving booth, high above the work floor, they develop "egocentric personalities" and consequently have "bad human relations" with fellow workers when they do come down from their perch.

Because there are varying amounts of work for the different cranes, the operators rotate among them. If a man always worked in one of the harder-worked cranes, he would complain. Another problem for crane operators is that at times they get two requests to lift something from two locations at the same time.

Tamakake work in groups when there is a large object to be raised or lowered. We observed four tamakake, one at each corner of a piece being lowered by a 100-ton crane. Because of the danger of injury and damage, communication is highly patterned. Three of the tamakake signalled the fourth, "It's clear in my location," but only the fourth gave whistle or hand signals to the crane operator overhead. Each hand position, each motion means a specific thing to the crane operator who takes his cues from them. Only certain workers are permitted to attach the wires or cables to the object to be lifted, since there is a special way of doing this to prevent accidents. The tamakake must understand engines well enough to know how to balance them. One tamakake described his job as follows:

> If I master how to wrap wires around the objects to be hoisted, then the rest of the responsibility is the crane operator's. My second job is easy: learning the correct hand signals.

About nine of the 16 men in han 4 of Shop 3 are crane operators, the rest are tamakake. But there aren't nine cranes, so some of them work as tamakake. Pay is similar for crane operators and tamakake. The crane operator has to get a license for both crane driving and tamakake work; the tamakake only has to get a tamakake license. Twelve of the 16 men here have crane licenses. The requirement for a tamakake license is on-the-job training in the factory; for a crane operator license, a government examination.

SUMMARY

Sake brewing is a centuries-old handicraft industry in Japan, and in the traditional kura we studied, hoary manual arts and partial mechanization coexist. But this is not a stable, long-term coexistence. Modern scientific knowledge and technology—biochemistry, hydrometers, and the like—are constantly eroding the older handicraft procedures. The sheer fact that labor productivity increases as a function of modern technology is another impetus to the shift from manual to automated brewing processes. Even in the brewing plants that are only partially mechanized, a number of social organizational changes have taken place; these are seen in a more advanced form in the new, automated plant.

Electric Factory is moving in the direction of greater mechanization and even automation, but at the time of our field work its technological core remained that of classical assembly line, large batch production (Woodward 1965). The majority of its personnel—especially the women—work at fractionated, repetitive tasks with more specialization than is typical in Shipbuilding Factory.

We shall not state larger conclusions from our analysis of technology and the division of labor at this point for at least three reasons. First, the intent of this chapter is not so much to test specific propositions as to describe technology and the division of labor in three Japanese firms. This should provide data for other students of organization, whose perspectives and hypotheses reflect the new wave of interest in the technology of organizations. Second, we have drawn some conclusions within this chapter about the meaning and significance of technology and the division of labor, which the careful reader will have noted. Finally, this chapter and the next are really parts of a larger whole—the relationship between technology, the social organization of work, and job satisfaction and other attitudes and values concerning work. Consequently, some of the "payoff" of the present chapter should become clearer as we discuss job satisfaction and other more subjective aspects of work and responses to the technological situation in Chapter 5.

CHAPTER FIVE

Job Satisfaction and Work Values

Several items in each questionnaire were conceptualized as measures of employees' attitudes toward their own specific job, or of more general values concerning work, as distinct from performance in the job. These attitudes and values are viewed as subjective responses to the more objective conditions of work described in the previous chapter.

ATTITUDES TOWARD THE JOB

The 1969-70 Sake questionnaire contained six job attitude questions, responses to which appear in Table 5.1. These can be summarized by an ordering in terms of the degree of expressed favorableness toward the job. The success of the three traditional brewing plants in structuring their jobs in ways that elicit favorable attitudes toward them is more conspicuous with regard to (1) enabling employees to make use of their experience and ability and (2) workers' perceptions that jobs are interesting rather than boring or monotonous. Success is lower, though still high, in (3) overall job satisfaction, (4) autonomy, i.e., perception that one can use one's own judgment on the job, and (5) feeling that the physical burden of one's job is "just right," neither "too heavy" nor "too light." The company elicits the least favorable responses in the area of job variety: four out of 10 employees think their job does not provide enough variety.

In addition to the sources of job satisfaction noted in the previous chapter, some other sources in Shipbuilding Factory should be discussed before we consider the questionnaire items in Electric and Shipbuilding factories. One source of satisfaction is a sense of achievement. In the words of a Number 2 Machine Shop worker in Shipbuilding Factory: "I've reached the peak of my machine's efficiency and of my manual dexterity on the machine. So my hanchō told me, 'Now use your head in production—by trying to find ways to reduce the seven steps of the rigide machine process to five steps, in producing the turbo-prop fan.' "

One worker expressed his satisfactions in classic "instinct of workmanship" terms: "A real job satisfaction and motivation for us in the Machine Section is when the Assembly Section people say, 'The fit and measurements of the parts you finished are perfect.' This, more than pay, is our motivation to do well."

TABLE 5.1

Job Attitudes (%)

Variable[a]		Electric	Shipbuilding	Sake
25/2/2.	(Interesting job) How do you feel about your present job?			
	1. My job is always interesting	17.5	29.8	55.4
	2. My job is interesting most of the time, but sometimes is boring	51.3	54.0	35.4
	3. My job is sometimes interesting, but mostly is boring, monotonous	26.9	13.7	4.6
	4. My job is completely monotonous and there is nothing interesting about it	4.3	2.5	4.6
		100.0	100.0	100.0
	N	(1,025)	(593)	(65)
	Shipbuilding > Electric: .27**; Sake > Electric: .40**; Sake > Shipbuilding: .25**			
26/3.	(Job and desire) Is your present job identical to your interest and desire?			
	1. Exactly identical	6.0	12.0	
	2. Fairly identical	41.0	54.2	
	3. Not very identical	43.4	28.6	
	4. Not at all identical	9.5	5.2	
		99.9	100.0	
	N	(1,027)	(594)	
	Shipbuilding > Electric: .27**			
27/4/4.	(Variety) Do you think your present job in the factory has enough variety?			
	1. Yes, there is enough variety in the job	42.2	52.3	59.7
	2. No, the job doesn't have enough variety	57.8	47.7	40.3
		100.0	100.0	100.0
	N	(1,025)	(591)	(62)
	Shipbuilding > Electric: .14**; Sake > Electric: .12**; Shipbuilding/Sake: N.S.			
29/6/3.	(Job and ability) Is your present job too high or too simple for your ability and skill?[b]			
	1. Job demands (requires) more ability than I have	21.9	21.4	28.1
	2. Job is just right for my ability	51.7	58.8	65.6
	3. Job requires (demands) less than my ability; it's too simple	26.4	19.8	6.3
		100.0	100.0	100.0
	N	(999)	(592)	(64)
	Shipbuilding > Electric: .11**; Electric/Sake: N.S. Shipbuilding/Sake: N.S.			

TABLE 5.1 (*cont.*)

Variable[a]		Electric	Shipbuilding	Sake
30/7/5.	(Work load) Is the work load (both mental and physical) of your job too heavy, or not?			
	1. Too heavy	30.8	24.4	23.8
	2. Just right	62.5	65.8	69.8
	3. Too light	6.7	9.8	6.3
		100.0	100.0	99.9
	N	(1,013)	(591)	(63)
	Shipbuilding > Electric: .11**; Electric/Sake: N.S.; Shipbuilding/Sake: N.S.			
34/13.	(Move jobs) How often do you think of being moved (want to be moved) to another job (job category) in the factory?			
	1. Often think about it	12.6	13.7	
	2. Sometimes think about it	53.5	47.7	
	3. Rarely think about it	25.7	19.9	
	4. Never think about it	8.2	18.7	
		100.0	100.0	
	N	(1,028)	(593)	
	Electric > Shipbuilding: .23**			
35/14/9.	(Job autonomy) Do you use your own judgment (make your own decisions) in performing your present job?			
	1. Always by myself	16.1	24.7	20.0
	2. Sometimes by myself	59.1	55.9	50.0
	3. Rarely by myself	19.5	15.5	13.3
	4. Never by myself	5.3	3.9	16.7
		100.0	100.0	100.0
	N	(1,024)	(588)	(60)
	Shipbuilding > Electric: .16**; Electric > Sake: .16**; Shipbuilding > Sake: .23**			
36/15/6.	(Job satisfaction) As a whole are you satisfied in your present job, or dissatisfied?			
	1. Very satisfied	4.8	8.6	12.9
	2. Mostly satisfied	47.7	52.0	62.9
	3. Somewhat dissatisfied	39.7	33.0	24.2
	4. Very dissatisfied	7.8	6.4	0.0
		10.00	100.0	100.0
	N	(1,029)	(594)	(62)
	Shipbuilding > Electric: .14**; Sake > Electric: .18**; Shipbuilding/Sake: N.S.			

** Significant at the .01 level; the measure of strength of relationship is C/C_{max}.

[a] When more than one variable number is given, the first refers to Electric Factory, the second to Shipbuilding, and the third to Sake. For a complete list of variables by number, see Appendix C.

[b] In the Sake Company questionnaire the three possible answers were (1) I can show my full experience and ability; (2) I can mostly show it; (3) I can hardly show it at all. Because of this variant wording we do not compute χ^2 or C/C_{max} for Sake versus the other two firms on this item.

Fatigue is obviously an element in how much satisfaction one's work brings: "I like to work on my big machine. It has two workers, so the pace is slower for each of us. Small machines, with only one worker, require intense concentration. I especially like to work with the big boring machine. The work is worthwhile, but physically exhausting; I have to climb up a height of five meters, with rope around my belt, to do work on the columns." Actually, since many of the machines we observed operate automatically, workers are often able to sit down, talk, and relax while their machine cuts or bores automatically. One often sees two workers stationed at a machine, simply watching it operate.[1]

There are, of course, dissatisfactions among Machine Section workers. Some of these were stated by a 55-year-old hanchō who was about to retire and who could view the work of a milling machine cutter (furaisu) in the context of the long period (1935–1969) during which he had done that kind of work: "One change in milling machine work is that senior workers are less powerful now, because the apprenticeship system has declined. Another is that the machines are larger now, yet training is shorter, more superficial. The job has become simpler. Today, management thinks *machines* control both the speed and the precision of production. Workers think that this downgrades their own skill, and therefore don't have as much pride or responsibility in their work as in the past. But workers talk more today about 'workers' rights."

We also encountered this instinct of workmanship among some of the Assembly workers in Shipbuilding: "When I'm just tightening bolts on a pipe, my work is monotonous. But as a whole, my work is worthwhile, because the engine is the heart of the ship, and therefore the most important part. When an engine we assembled passes the test, I'm very satisfied."

We now summarize the responses to eight job attitudes questions asked in both the Electric and the Shipbuilding factory questionnaires (see Table 5.1). In Electric Factory, intrinsic job attitudes, although on balance more favorable than unfavorable, cannot be said to be a major area of strength (relative to other aspects to be dealt with later, such as employee cohesiveness). Particularly important are the relatively large numbers of employees dissatisfied in general with their job (variable 36) and who want to change to another job in the factory (variable 34). Dissatisfaction with the lack of variety in the job (variable 27) and the lack of fit between the job and one's interests and desires (variable 26) is also relatively high. If the production process works smoothly and efficiently, in accordance with management plans, it cannot be said to do so on the basis of overwhelmingly positive worker job attitudes.

[1] Another form of relaxation is the daily exercise period at about 2:30 p.m. This practice is apparently plant-wide, for we observed it both on the production floor and in the offices. Workers array themselves near their machine or desk and a leader puts them through about five minutes' bending and stretching.

The aspects of the job in Shipbuilding Factory that are most favorable from the point of view of employees are the interesting nature of the work (variable 2), the autonomy workers feel they have (variable 14), the fact that the job is seen as a challenge to their abilities (variable 6), that it meets their interests and desires (variable 3) that the mental and physical work load is just about right—neither too heavy nor too light (variable 7)—and that they are in general relatively satisfied with the job (variable 15). Less satisfactory is the relative lack of variety in the job (variable 4); over half the employees sometimes or often want to be moved to a different job in the factory (variable 13). As in Electric Factory, this last aspect is empirically related both to dissatisfaction with one's present job and to the belief that the present job lacks enough variety. This is why we can interpret the relatively large proportion of Shipbuilding employees who want to change jobs as a manifestation of dissatisfaction with their present job. On balance, however, Shipbuilding personnel are more favorable with regard to these aspects of the job than are Electric Factory personnel, a difference we shall examine further in Chapter 12.

THE INDEX OF JOB SATISFACTION

To the extent that the eight job attitude questions asked in Electric and Shipbuilding factories are intercorrelated, data reduction in the form of index construction is possible. Table 5.2 presents the matrix of associations (C/C_{max}) between these job attitude variables for both factories. There are strong or moderate ($\geqslant .30$) positive relationships among the first six job attitude variables in the table. In the language of factor analysis, responses to these six questions all have high loadings on a single factor, which we might label job satisfaction. This factor has a positive and a negative pole of responses; a given individual's responses tend to be either positive or negative to all six questions.

Positive Pole	*Negative Pole*
Job is exactly or fairly identical to my interest and desire	Job is not very or not at all identical to my interest and desire
Very satisfied or satisfied with my job	Somewhat or very dissatisfied with my job.
Job is interesting always or most of the time	Job is dull and uninteresting most of the time or always
Job already has enough variety	Job doesn't have enough variety
Job requires more than my ability or is just right for my ability	Job requires less ability than I have
Rarely or never want to be moved to another job in the factory	Sometimes or always want to be moved to another job

The last two variables in Table 5.2—work load and job autonomy—are at best only weakly associated with the first six. Despite the emphasis on

TABLE 5.2
Relationships among Job Attitudes in Electric Factory and in Shipbuilding Factory

	Variable	*36/15*	*25/2*	*27/4*	*29/6*	*34/13*	*30/7*	*35/14*
26/3.	Job and desire	.74**	.69**	.62**	.55**	.55**	.24**	.24**
		.69**	.66**	.58**	.47**	.57**	.29**	.26**
36/15.	Job satisfaction		.64**	.54**	.52**	.61**	.30**	.20**
			.57**	.52**	.45**	.66**	.28**	.28**
25/2.	Interesting job			.68**	.56**	.44**	.18**	.28**
				.59**	.50**	.46**	.28**	.31**
27/4.	Variety				.58**	.38**	N.S.	.26**
					.47**	.36**	.26**	.18**
29/6.	Job and ability					.37**	.46**	.19**
						.32**	.42**	N.S.
34/13.	Move jobs						.30**	.16*
							.27**	.24**
30/7.	Work load							N.S.
								N.S.
35/14.	Job autonomy							

NOTE: In each cell, the upper number is C/C_{max} for Electric Factory, the lower number, C/C_{max} for Shipbuilding Factory. The first variable number refers to Electric Factory, the second to Shipbuilding. For a complete list of variables by number, see Appendix C.

* Significant at the .05 level.

** Significant at the .01 level.

job autonomy in the literature on the sociology and social psychology of work, in our Japanese data perceived autonomy does not appear to have much to do with other measures of job satisfaction.

All eight job satisfaction variables in Table 5.2 were subjected to an item-analysis program (ITEMA). As expected, the items with the lowest item-total correlation on the first two trials were work load and autonomy; these were eliminated from the index of job satisfaction, which contains the remaining six items. The estimate of index reliability from the items' intercorrelations, coefficient alpha, is .82 for Electric Factory and .76 for Shipbuilding.[2] The higher an individual's total score across these six times, the higher his or her job satisfaction. Henceforth in the analysis we shall use this index (variable 139 in Electric Factory and variable 124 in Shipbuilding) rather than the specific items.

WHAT DETERMINES JOB SATISFACTION?

What accounts for the variation in job satisfaction among employees? On theoretical grounds we hypothesized that job satisfaction is a function

[2] See Appendix B for the distribution of scores and other details on the index.

of the following eight variables. Job satisfaction is positively related to the employee's (1) organizational status, (2) age and (3) sex—men are more satisfied than women. (4) The better one's perceived promotion chances in the company, the higher the satisfaction. (5) The more one's work load is "just right," as opposed to "too heavy" or "too light," the higher the job satisfaction.[3] We also hypothesized that job satisfaction is positively related to the degree to which the employee is socially integrated into the company; thus it should vary positively with (6) employee cohesiveness and (7) frequency of participation in company-sponsored recreational activities, and negatively with (8) previous interfirm mobility (number of previous jobs)—that is, employees who have always worked for Electric Company should have higher job satisfaction than those who previously worked for one or more other firms.

Hypothesis 8 calls for comment. It is not a strong hypothesis, for even within the lifetime commitment model there are some grounds for expecting job satisfaction to vary positively with previous interfirm mobility, or even to vary independently of it. An employee who had moved from a small, less promising firm to a large, stable, promising one, might for that reason have higher job satisfaction than another employee who had always worked for Electric Company. On the other hand, the model implies that lifetime commitment to one firm is so normal and positively value-sanctioned that the mobile person experiences more negative sanctions than the non-mobile person. This would provide a negative contextual influence that might lower one's job satisfaction. Empirically, these counterpressures stemming from previous interfirm mobility may well result in a non-relationship between mobility and job satisfaction.

Our test of these hypotheses takes the form of a multiple regression of job satisfaction (index scores) on the above eight independent variables.[4] In Electric Factory we found that the main determinant of job satisfaction is status; other determinants are cohesiveness, promotion chances, and sex. Job satisfaction is greatest when one has higher status in the company, perceives one's promotion chances as good, is male, and has cohesive relationships with other employees. These four variables together account for 38 percent of the explained variance—$(.62)^2$—in job satisfaction. In Shipbuilding, the relationships were somewhat different. Shipbuilding employees are satisfied with their jobs not so much because of their status in the company as because they are older, enjoy cohesive

[3] From what we have seen in Table 5.2, it is unlikely that this hypothesis will be confirmed. However, we formulated our hypotheses prior to the item-analysis or the correlation and regression analysis, and, as will be our practice throughout this study, we shall not reject them before specific tests are performed.

[4] Details on the regression analysis appear in Appendix D, Tables D.3 and D.4, p. 369–p. 370.

relationships with fellow workers, and see their future promotion chances as good.

Thus, two of the most important causes of satisfaction—cohesiveness and promotion chances—are the same in Electric and Shipbuilding factories, while status and age play somewhat different roles in each factory as influences on satisfaction. In both factories, three of our initial hypotheses were disconfirmed: perceived work load, participation in company recreational activities, and previous interfirm mobility have virtually no effect on job satisfaction, when the other, more powerful independent variables are held constant. Less of the variance in job satisfaction is explained by our independent variables—$(.46)^2$ or 21 percent—in Shipbuilding Factory than in Electric Factory.

TECHNOLOGY AND JOB SATISFACTION

An approach known as "socio-technical theory" or "technological implications theory" has made an important recent breakthrough in organizational analysis. Woodward conducted a comparative study of 100 industrial firms in South Essex, England (Woodward 1965), posing the same question we do: what are the social organizational conditions of organizational performance? She began by testing the classical unity of command theory of organization. But early in the research she made the disconcerting discovery of a "lack of any interrelationship between business success and what is generally regarded as sound organizational structure" (Ibid.:34). Specifically, she found that one could not predict business success on the basis of either the classical theory of unity of command or the human relations theory of "participative management" and workers' morale. Only when she arranged her firms in terms of technological type did she begin to get clear correlations between organizational structure and success.

The basic argument of socio-technical theory is that production technology has gone through a number of historical stages and that contemporary firms can be classified in terms of these stage-types. Herbst (1967) has specified the major variables that underlie this classification:

1. Type of control: manual vs. mechanized
2. Degree of integration of the manufacturing process: low to high
3. Degree of specialization of the worker
4. Degree of knowledge and skill required by the worker.

Herbst proposes an evolutionary classification of levels of production technology:

1. Traditional craftsmen's organization
2. Early mass production

3. Advanced mass production
4. Mass flow production
5. "Short" automated transfer lines
6. "Long" automated transfer lines
7. Automated flow process.

The causal model posited by technological implications theory is shown in Figure 5.1. At present our interest is in how type of technology and organizational imperatives influence job attitudes, specifically job satisfaction. Blauner (1964), Woodward (1965), and others have reduced technologies to three basic stages or types: (1) unit and small batch production, or craft technology; (2) large batch, assembly line, mass production technology; and (3) continuous process, automated technology. Each type of technology is associated with certain organizational imperatives, which, at the level of jobs, make for variations in task specialization, skill levels, autonomy, and the like. Thus, it is asserted (with supporting evidence in Blauner 1964, Ford 1969, Shepard 1970, 1971, and Davis 1971) that assembly line technology, with its narrow task specialization and low skill levels, is associated with lower levels of job satisfaction than are technologies that provide less specialized and more skilled jobs (craft and continuous process technology). The work situations in Blauner's study of printers (craft) and petrochemical workers

FIGURE 5.1

Causal Model of Technological Implications Theory

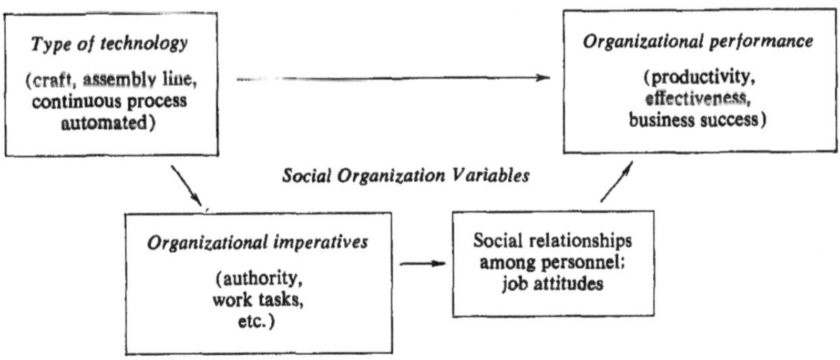

(continuous process) were more conducive to job satisfaction than was that of automobile workers (assembly line).

Electric Factory in general clearly belongs in stage 2, and it is revealing to compare Electric Factory responses with Blauner's American data on three relatively comparable job satisfaction questions (see Table 5.3). By reading across the rows of the table, a curvilinear pattern can be seen for all three questions. The proportion who give negative statements about their job is low in the craft industry, rises in industries with mass production, assembly line technology like that in Electric Factory, and declines again in the most advanced technology, automation.

TABLE 5.3
Job Attitudes, by Type of Technology and Industry, U.S. Factory Workers and Electric Factory in Japan

			Type of Technology and Industry				
			Craft	*Assembly Line Mass Production*			*Automated*
Blauner Appendix Table Number	*Job Attitude*		*U.S. Printing*	*U.S. Textiles*	*U.S. Automobiles*	*Japan Electric Factory*[a]	*U.S. Chemicals*
34.	Percentage of factory workers whose job "leaves you too tired at the end of the day"[b]		12	38	34	31	19
		N	(113)	(387)	(178)	(717)	(77)
42.	Percentage of factory workers whose job is "mostly or always dull"[c]		4	18	34	40	11
		N	(115)	(406)	(174)	(720)	(77)
43.	Percentage of factory workers who feel their jobs are "too simple to bring out their best abilities"[d]		16	23	35	34	21
		N	(118)	(397)	(180)	(700)	(78)

Source: United States data from Blauner 1964, Appendix.

[a] To make Electric Factory more comparable with United States data, which are for factory workers, only personnel in the five production sections are included.

[b] Electric Factory: percentage of production workers who say their work load is "too heavy" (v. 30).

[c] Electric Factory: identical question used in Blauner.

[d] Electric Factory: percentage of production workers who say "my job requires less than my ability" (v. 29).

Does this pattern also hold *within* Electric Factory? Although this factory generally uses assembly line production, some of its workers operate in technological conditions that at least approximate the craft and automated types. These variations can be measured in two ways: job classification and level of mechanization of the section. (1) Workers with an Operative job classification (O) approximate craft skills as operators of complex machines, whereas those in Routine and Training jobs (R, T) approximate the more fractionated, less skilled assembly line jobs; this is especially true of Routine-2 workers, whose jobs are described as requiring simpler methods than Routine-3 workers (for the full job classification hierarchy, see Table 6.3 below). (2) Section variations in level of mechanization per employee measure the extent of manual labor versus inanimate energy used in the prevailing technology. Mechanization is operationally defined here as the amount of electricity consumed per employee in each section (in kilowatt hours, for the period January through July 1969): (1) Number 2 Production, 839 KWH/employee; (2) Number 1 Production, 952; (3) Hardware, 3,329; (4) Motor, 6,581; and (5) Surface Treatment, 12,178.

We hypothesize that among production personnel in Electric Factory's five production section (1) Operatives will have significantly higher job satisfaction index scores than Routine and Training personnel, and (2) the difference in satisfaction will be greater between Operatives and Routine-2 personnel than between Operatives and the more skilled Routine-3 personnel. (3) The level of job satisfaction in these five sections varies positively with the section's level of mechanization. Before testing these hypotheses, we must consider how the sections vary on these two independent variables (see Table 5.4).

Table 5.4 shows that the two main independent variables—job classification and mechanization—are somewhat related. Sections with higher

TABLE 5.4
Job Classification, by Level of Mechanization of Section, in Electric Factory (%)

	Level of Mechanization of Section[a]				
Job Classification	*No. 2 Production (839)*	*No. 1 Production (952)*	*Hardware (3,329)*	*Motor (6,581)*	*Surface Treatment (12,178)*
1. Operatives	7.2	8.0	14.7	12.1	14.6
2a. Routine	48.8	36.1	39.2	39.6	59.2
2b. Trainees	38.6	53.4	43.1	42.9	23.1[b]
2a + 2b. Routine and Trainees	87.4	89.5	82.3	82.5	82.3
3. Routine-2	47.0	31.5	5.9	9.9	23.8

[a] Numbers in parentheses refer to electricity consumption of the section, in KWH/employee.

[b] Column totals for rows 1 + 2a + 2b do not equal 100 percent because some job classifications are omitted.

levels of mechanization tend to have a larger proportion of skilled Operatives and a smaller proportion of relatively unskilled trainees; moreover, of the Routine workers, almost all are less skilled Routine-2 workers in Number 1 and 2 Production sections, whereas the other three more mechanized sections have a larger proportion of their Routine workers in the relatively more skilled Routine-3 job classification. In this sense, the proposition from technological implications theory that level of mechanization-automation is (positively) correlated with skill-complexity level of the work is supported. At the same time, we note that the relationship is less than perfect: the variation in the proportion of Operatives across sections is not large, and the most mechanized section—Surface Treatment—has a relatively large proportion of less skilled, Routine-2 workers.

Hypothesis 1—Operatives have higher job satisfaction than Routine and Training personnel—is confirmed (Table 5.5). Hypothesis 2—the difference in job satisfaction between Operatives and Routine-2 workers is greater than the difference between Operatives and Routine-3 workers—is also confirmed (Table 5.5). Thus, among production workers, job satisfaction is a direct function of job classification. The higher one's job classification, the greater one's satisfaction with the job.

TABLE 5.5
Relationship between Job Classification and Job Satisfaction in Electric Factory's Five Production Sections (%)

Job Satisfaction	*Operatives*	*All Routine and Training Personnel*	*Routine-3*	*Routine-2*
High	20.0	2.4	9.8	1.6
Medium high	62.7	27.2	42.0	24.7
Medium low	17.3	53.1	35.7	57.1
Low	0.0	17.2	12.5	16.5
	100.0	99.9	100.0	99.9
N	(75)	(580)	(112)	(182)

Operatives vs. All routine and training personnel: $C/C_{max} = .52^{**}$
Operatives vs. Routine-3: $C/C_{max} = .46^{**}$
Operatives vs. Routine-2: $C/C_{max} = .68^{**}$

** Significant at the .01 level.

Hypothesis 3 is supported only in a limited way. The level of job satisfaction is lower in the sections with less mechanization, i.e., more dominated by narrow, assembly line technology (Number 1 and 2 Production) than in those with more mechanization, and somewhat more craft or automated technology (Hardware, Motor, and Surface Treatment). But the relationship does not hold in detail:

Mechanization of Section	*Percentage of Employees with Medium High or High Scores on Index of Job Satisfaction*
Number 2 Production (lowest)	31.8
Number 1 Production	25.9
Hardware	59.8
Motor	44.2
Surface (highest)	45.8

This suggests that other factors—job classification, sex, and seniority—are at work. We accordingly regressed job satisfaction on these three variables and section simultaneously, for the 682 workers in the five production sections. To the extent that section mechanization level and job classification are more powerful predictors of job satisfaction, technological implications theory will be supported. Conversely, to the extent that sex and seniority are more powerful, the lifetime commitment model will be supported.

As before, the details of the regression analysis appear in Appendix D, Table D.5, p. 371. Here, we only summarize the results. Among production section workers job satisfaction is more a function of sex and seniority than of section level of mechanization. One key variable in technological implications theory—level of mechanization—thus fails to explain satisfaction. Job satisfaction for these workers is influenced more by seniority and sex than by the mechanization level of the section in which they work. The second variable in technological implications theory—job classification—is so highly correlated with seniority in our data ($r = .84$) that the precise influence of these two variables on job satisfaction cannot be disentangled.

Thus, men with more seniority are more satisfied than women with less seniority, and when these differences have been taken into account, the section in which one works makes little difference. This suggests that higher seniority men would be virtually as satisfied in Number 1 and 2 Production as in Hardware, Motor, or Surface Treatment, and that low seniority women would be as dissatisfied in the latter three sections as in the former two. This is precisely what we find (see Table 5.6).

From this we conclude that while technological theory has some predictive value with regard to job satisfaction in the five production sections, when variables outside that theory—especially seniority and sex—are held constant, most of the variance in satisfaction is attributable to them, not to the section differences in mechanization and job classification. In this specific sense, the lifetime commitment model is supported more than the technological implications theory. Our findings are, however, not definitive disconfirmations of technological implications theory. The

TABLE 5.6
Job Satisfaction of Workers in Electric Factory's Five Production Sections, by Seniority, Sex, and Section Level of Mechanization (%)

Job Satisfaction	*High Seniority Males (12+ years)*		*Low Seniority Females (0–3 years)*	
	More Mechanized Sections[a]	*Less Mechanized Sections*[b]	*More Mechanized Sections*[a]	*Less Mechanized Sections*[b]
High	21.8	13.0	0.0	0.0
Medium high	46.0	61.1	26.9	15.8
Medium low	25.8	18.5	55.8	60.1
Low	6.5	7.4	17.3	24.1
	100.1	100.0	100.0	100.0
N	(124)	(54)	(104)	(253)

[a] Hardware, Motor, and Surface sections
[b] Number 1 and 2 Production sections

variation within Electric Factory in the technological variable so central to that theory is quite possibly insufficient to provide a true test of the theory.

VALUES CONCERNING WORK

Thus far we have been concerned with attitudes toward quite specific aspects of one's job. The sociological literature has also found that people hold more general values toward work which are at least analytically distinguishable from job attitudes. These have gone by a variety of names: the Protestant Ethic, work ethic, labor commitment, and the like. Students of Japanese culture have uncovered a variety of value orientations (Bellah 1957, Nakamura 1964, Dator 1966, Berrien 1966), and Caudill and Scarr (1962), following Kluckhohn and Strodtbeck (1961), have attempted to measure the hierarchy of Japanese value orientations in terms of dominant and variant orientations.

Three of the values that have been identified as prominent in contemporary Japanese culture and society were included in one of our questionnaire items. The results appear in Table 5.7. The differences between Electric and Shipbuilding employees are significant, but weak. In both factories the modal value orientation is a primacy of pleasure values (alternative 3); somewhat more than one in four employees gives primacy to work values (1); and approximately another one-quarter of the personnel in both factories select family values (2). Expressed in the form of Kluckhohn-Strodtbeck (1961) value hierarchies, the preference ranking of values among personnel in Electric Factory takes the form

Pleasure > Work > Family

while that in Shipbuilding is:

Pleasure > Family > Work

Since Japanese work organizations are sometimes described as bastions of the work ethic, it is of some interest that in two factories in leading firms in their respective industries in Japan, employees have value orientations in which work is a means to pleasure, rather than an end in itself.

TABLE 5.7

Value Orientations of Electric and Shipbuilding Factory Personnel (%)

	Percent agreeing	
	Electric (*v. 31*)	*Shipbuilding* (*v. 8*)
Which one statement best describes your own view of your work? (unmarried people also please answer the question)		
1. Work is my whole life, more important than anything else.	27.1	27.0
2. Happy family life is more important than a company job.	21.2	28.1
3. Work is only a means to get pay to spend on the pleasures of life.	51.7	44.9
	100.0	100.0
N	(1,005)	(586)
$C/C_{max} = .12^{**}$		

** Significant at the .01 level.

Before attempting to explain these three value preferences, we want to validate the responses to this question, that is, to show that by an independent measure of work commitment, employees who give primacy to work values in fact differ in the expected direction from those who choose family or pleasure values. We constructed an index of performance,[5] which measures, among other things, an employee's objective performance in terms of company goals—contributions to the suggestion system, awards, and attendance. If the values question is validly measuring what it purports to measure, those who give primacy to work (over

[5] See Chapter 11 and Appendix B for details on the index of performance.

pleasure and family) values should be significantly higher in performance score. That this is the case is shown in Table 5.8.

Granted that the values question appears to measure what it purports to measure—intrinsic work commitment—what are the determinants of variations in values among employees? We shall test the following hypotheses:

1. *Status*. The higher one's status in the company (the greater one's status rewards), the more other members of the organization *expect* one to espouse the primacy of work values (rather than family or pleasure values). Consequently, the higher one's status, the more one in fact gives primacy to work values.
2. *Sex*. Given the shorter-term labor force participation of women, they will be more likely to favor family or pleasure values over work values.
3. *Age*. Young employees are more likely to favor pleasure values, older more likely to favor work and family values.
4. *Promotion chances*. The better one thinks one's promotion prospects are, the more likely one is to favor work values over family and pleasure values. The latter values are more functional for those who see their promotion chances in the company as poor, since they provide alternative life styles and compensatory sources of satisfaction.
5. *Number of dependents*. Married employees with dependents will be more likely than single employees or those with few dependents to

TABLE 5.8
Relationship of Values and Performance in Electric and Shipbuilding Factories (%)

	Value Orientation Primacy					
	Work		*Family*		*Pleasure*	
Performance Score	*Electric*	*Shipbuilding*	*Electric*	*Shipbuilding*	*Electric*	*Shipbuilding*
High (15–19)	37.5	40.1	15.0	27.2	15.3	23.7
Medium high (13–14)	36.8	24.5	30.7	31.0	20.0	23.3
Medium low (11–12)	18.3	19.0	26.9	23.4	33.8	28.0
Low (5–10)	7.4	16.3	27.4	18.4	30.9	25.0
	100.0	99.9	100.0	100.0	100.0	100.0
N	(269)	(147)	(212)	(158)	(515)	(236)

Electric Factory: $C/C_{max} = .41$**
Shipbuilding Factory: $C/C_{max} = .22$**

** Significant at the .01 level.

favor work or family values over pleasure values. The different familial role expectations of the two groups account for these value differences.

6. *Residence*. Employees living in company housing benefit from one kind of company paternalism. In return, they will be more likely to favor work values than those who live in private residences.

7. *Size of community of origin*. Employees from rural villages and small towns will be more likely to have work values than those from cities and major metropolises. This hypothesis is based on the assumption, frequently encountered in the literature (Abegglen 1958:101), that rural existence imposes more rigorous demands on people, and also protects them more from being "tainted" by the leisure and pleasure orientations of urbanites.

Because we are measuring the values variable at the nominal scale level, it is appropriate to begin the analysis with cross-tabulations. Table 5.9 cross-tabulates values with sex, age, and rank, for Electric Factory. One of the more surprising findings of this study is that women are much more likely to favor pleasure values than family values (last column of panel 1). Almost all these women are unmarried, but they are also mostly very young, and marriage is therefore a realistic future prospect for most of them. What is striking therefore is that, at this stage in the life cycle, these Japanese women have a differentiated "modern" value pattern. What their values will be when they are older and actually married we cannot say, but we can say that when they are in their 'teens and early twenties, their values are not those of the anticipatory, "A happy family life is more important than a company job," but rather, "Work is only a means to get pay to spend on the pleasures of life." In value terms at least, the hedonism of female Japanese factory workers is quite different from the traditional values and role model for women of their age.

Age has somewhat different effects on values among men and women. Among the youngest employees (15–17 years old), men and women rank the three value orientations in the same order: pleasure > family > work. The proportion who favor pleasure values, however, varies by sex: 64 percent of the women in contrast to 49 percent of the men. Among these 15–17 year olds, sex makes less of a difference with regard to family and work values. As employees get older, the value ranking for men shifts to work > pleasure > family, while that for women shifts to pleasure > work > family. What is striking is that for both sexes, the relative value primacy among those in their thirties or older assigns family the lowest position (see Table 5.9).

Rank in the factory also affects values, as men get older (all women are in rank-and-file positions, so they are omitted here). Table 5.9 shows that as male employees get older, although work values become primary

TABLE 5.9
Value Orientations in Electric Factory, by Age, Sex, and Rank[a] (%)

	Age					
Sex	*15–17*	*18–20*	*21–26*	*27–32*	*33+*	*All Ages*
Male:						
Pleasure	48.8	60.0	35.3	39.8	29.6	37.7
Family	34.1	10.0	22.4	19.3	9.3	17.2
Work	17.1	30.0	42.4	40.9	61.1	45.1
	100.0	100.0	100.1	100.0	100.0	100.0
N	(41)	(30)	(85)	(176)	(162)	(494)
C/C_{max} = .34**						
Female:				*27+*		
Pleasure	63.6	63.8	70.8	72.2		65.4
Family	27.1	27.2	20.8	0.0		25.0
Work	9.3	8.9	8.3	27.8		9.6
	100.0	99.9	99.9	100.0		100.0
N	(140)	(257)	(96)	(18)		(511)
C/C_{max} = .22*						
Males, by Rank				*27+*		
Kachō, kakarichō:						
Pleasure				7.3		
Family				7.3		
Work				85.4		
				100.0		
N				(41)		
Hanchō, kumichō:				*27–32*	*33+*	
Pleasure				45.5	38.1	
Family				0.0	4.8	
Work				54.5	57.1	
				100.0	100.0	
N				(11)	(42)	
C/C_{max} = .16						
Rank and File:						
Pleasure	48.8	60.0	35.3	39.5	36.6	
Family	34.1	10.0	22.4	21.0	12.2	
Work	17.1	30.0	42.4	39.5	51.2	
	100.0	100.0	100.1	100.0	100.0	
N	(41)	(30)	(85)	(162)	(82)	
C/C_{max} = .28**						

* Significant at the .05 level.
** Significant at the .01 level.
[a] Blank columns signify 0 cases. Columns with N less than 10 were combined with adjacent age group.

regardless of rank, the salience of work values increases as a function of rank. Of rank-and-file workers age 33 and over, 51 percent favor work values, in contrast to 57 percent of those 33 and over who have become first- or second-line foremen, and 85 percent of those 27 years old or over who have become kakarichō or kachō. In other words, increasing age makes for a shift toward work values among men, and this shift is even more marked when not only age but also the responsibilities and expectations of rank increase.

Because there are so few women in Shipbuilding Factory, we cannot follow the interaction of sex and other independent variables in relation to value orientations. Indeed, even in Electric Factory the extent to which we can do this is greatly limited as long as we use only cross-tabulations. When we examine the effect of one independent variable on the dependent variable with more than one or two other independent variables held constant, we quickly run out of cases. Multiple regression analysis avoids this difficulty. In the multiple regression analysis, details of which appear in Appendix D, Tables D.6 and D.7, p. 372 and p. 374, we recoded the value orientation responses as (1) pleasure, (2) family, and (3) work. Although one may grant that categories (1) and (2) represent lower values on an ordered scale, "degree of 'work is my whole life' value-orientation," there is some question whether categories (1) and (2) themselves can be clearly ordered. The results of the analysis must therefore be regarded with caution.

In Electric Factory, the most important causes of these value-orientation differences are organizational status, perceived promotion chances, and sex. Among Shipbuilding employees, organizational status and size of community of origin are the best predictors. Thus, work values are most likely to be primary, in both factories, when the employee has a higher status in the firm. In Electric Factory, work values are favored by men who perceive their promotion chances to be good, while in Shipbuilding, work values are somewhat more likely to be given primacy by those who come from smaller communities—towns and villages.

Regression analysis enables us to specify the reason that Electric Factory's female employees give primacy to pleasure values. They do so not because they are women, but because they are in low status positions in the company. Men in the same low status positions have a similar propensity to favor pleasure over work values.

When these variables are taken into account, the other hypothesized causes of variations in value-orientations—age, number of dependents, and residence—exert no significant independent influence. The hypotheses that older employees, those with more dependents, and those who live in company housing are more likely to espouse work values are all disconfirmed.

Work and Leisure: The Hard-Times-to-Affluence Syndrome

We uncovered varying perceptions among Shipbuilding workers concerning what we may call the hard-times-to-affluence syndrome. These came out in interviews; since we did not explore this area in the questionnaire, we can only present these data on the structuring of work and leisure orientations as tentative findings to be followed up in future research.

Older Shipbuilding workers make the claim—often heard in other societies too—that, in contrast to the hard times of *their* youth, and to *their* work orientation, today's younger workers, the products of postwar affluence, have grown soft and pleasure oriented. There is clearly some evidence that times *were* harder for the older workers. The Engine Division factory was bombed out during the war; the machines were damaged by typhoons, and workers had to clean the salt (from the sea water) out of the machines. In the immediate postwar years, with facilities devastated and few production orders, pay was at a minimum survival level. When there was work, it was much heavier and harder than now. In sharp contrast, the younger generation of workers grew up in a more prosperous age, enjoyed a "fun childhood," stayed in school longer, and as a result has a different attitude set than the older generation. Overtime has declined from an average of 70 hours a month (and as high as 90 hours a month) to 30 hours a month. Many young workers refuse to work the night shift or do overtime work; they want to be free after their seven hours' work (six days a week).[6] There are contrary reports on absenteeism: some claim there is more nowadays than in the past, while others argue that attendance is now better than in the past, because there are regular Sundays and holidays off.

The hard-times-to-affluence syndrome is interpreted differently by others, who do not see it as a drift toward greater laxity, laziness, and leisure orientation. These employees say that even the older workers, more affluent now, have a different attitude toward work than in the past. More positively, they claim that the reason young workers now refuse to work overtime is because they want to attend night university (*yakan nai daigaku*) or night seminars given by the city to study such subjects as computing. Both clerical and manual employees take such night courses. For this group, the frittering away of leisure hours stereotype is evidently inaccurate.

[6] A 39-year-old hanchō in the Machine Section described the younger generation of workers in these terms: "Very young workers don't want to work overtime because, as 'prosperity's children,' they want to have fun at night. Young married workers also want less overtime work because their wives work too, so they have enough money and want to meet their wives in the evening. Only after they have children and the husband is the only worker in the family is he more willing to work overtime in order to buy television, and so forth."

CONCLUSION

In this chapter we have constructed an index of job satisfaction and uncovered some of the factors that influence it in Electric and Shipbuilding factories. These factors—status in the organization, age, perceived promotion chances, sex, and cohesiveness—are not unlike those that determine job satisfaction in the West. A limited test of technological implications theory partially confirmed that theory, but job satisfaction was seen to be also a function of variables not included in the theory—especially seniority and sex. The lack of strong confirmation of the relationship between technological level of the section and job satisfaction of workers in that section may be due to an insufficient range of variation in the technological variable among the five production sections in Electric Factory, which did not fairly test the theory.

An analysis of value orientations revealed that in Electric Factory, as male employees get older, there is a shift from pleasure to work values; this shift is accentuated among males who rise in rank. Pleasure values retain primacy over both work and family values among female employees, regardless of age. It may well be that women with family values are selected out after a few years' employment, but the "liberation" of Electric Factory's women workers, of all ages, from family value-primacy is one of our most unexpected findings. From one viewpoint, a fuller indication of their liberation would be the primacy of work values over both family and pleasure values. They have not gone this far, which is perhaps not surprising in view of the nature of their work.

In both factories, the modal value preference is pleasure, not work. This suggests that the popular view of the fanatically work-oriented Japanese employee requires some modification. Japanese may work harder than members of other societies in similar jobs,[7] but this does not necessarily mean they have "work is my whole life" values.

Finally, in exploring work and pleasure value orientations we uncovered in Shipbuilding Factory interviews a variety of conceptions of a hard-times-to-affluence syndrome. Most employees agree there has been such a trend from the end of World War II to the present. But there is disagreement over how employees—particularly younger ones—are now adapting to affluence. Some see the young as preoccupied with leisure and pleasure in their non-work hours; others stress that the non-work hours are being used in a variety of instrumental, achievement-oriented ways, most especially, to increase formal education and skills.

[7] At a conference on "Japan in 1980" at Yale University in the spring of 1973 it was reported that Japanese workers in Japan work one-third faster than Taiwanese workers in Taiwan, in Japanese-owned firms using the same assembly line technology. The causes of this remain to be clarified.

CHAPTER SIX

The Reward System: Pay

Throughout much of this century, pay in Japanese companies has depended more on age and seniority than on job classification or performance. Recently there have been signs of change: "Allowance for age and length of service have declined in frequency from 1956 to 1966, while allowances for position and job classification (*shokumukyū*) . . . have increased. . . . This reflects an effort on the part of many companies to move toward compensation based on job output and away from compensation based on age and education (Abeggelen 1969:112–13)." We shall treat this change as an aspect of modernization: the shift from seniority, an analytically traditional determinant of pay, to job classification, an analytically modern determinant of pay.[1] We shall examine three elements of the wage system of each firm: wage policy, the actual determinants of pay, and employees' attitudes and values concerning the wage system. Specifically, we ask, to what extent do shifts in company wage policy affect the actual determinants of pay? To what extent are employees' attitudes and values concerning how pay should be decided traditional or modern? To what extent do these values and attitudes reflect vested interests, to what extent common values?

THE PRESENT OR SENIORITY WAGE SYSTEM

The ideology used in Japan to legitimize the seniority wage system (*nenkō chingin seido*) is that it is the most prevalent one in Japan, and the best suited to Japan's economy and culture. The seniority wage system is functionally interdependent with the lifetime employment system, whereby, at least in the larger companies, regular employees entered a firm after completing school with the expectation that they would remain in that firm until retirement, typically at age 55, and that the company would not discharge them. As the employee's family size and expenditures increase with his age, the seniority wage system provides steady increments of pay. Pay rises more sharply with seniority in Japan than in the United States. National data for men employed in Japanese manu-

[1] In applying Weber's ideal-type conceptualization of rewards—remuneration, promotion—in modern bureaucracies to the Japanese case, we have seen fit to reconceptualize slightly. On our reasons for this, see below, p. 138.

facturing firms in 1967 show the following ratio of pay at age 55 to beginning pay: 4.7 for college graduate, white collar employees; 5.2 for high school graduate, white collar employees; 4.4 for high school graduate, manual workers; and 4.7 for middle school (junior high school) graduate manual workers (Tōyō Keizai, *Tōkei geppō* 1967). In other words, 55-year-old personnel typically receive from 4.4 to 5.2 times as much pay as those entering the labor force directly from school. The seniority wage system has also been shown by an economist to be functional for young companies in lowering their labor costs during the early years of their growth (Chao 1968).

Japan has no hourly or annual pay system. Rather, there has been a transition from a daily to a monthly wage system. Sake Company in 1969–70 was at an earlier point in this transition than Electric and Shipbuilding companies: Sake still had a daily wage system, whereas the other two firms had shifted to a monthly system (Electric Company in 1958, Shipbuilding Company somewhat later). It should not, however, be inferred from this that the relationship between wage policy and seniority was more traditional in Sake Company. In fact, as we shall see, pay in Sake Company is officially based on craft or technical ability or skill. Although skill tends to be positively correlated with seniority, still, in principle there was less emphasis on seniority in Sake Company than in the other two firms, prior to the introduction of their new wage systems. In more general terms, historical evidence indicates that in the early phases of Japan's industrialization—from the late nineteenth century until roughly World War I—the primary determinant of pay was not seniority, but rather craft ability, and that a full-blown seniority pay system for manual workers is a development of the last several decades, especially in the large, leading firms (Noda 1963; Takezawa 1969:184).

Pay in Sake Company

A man employed in Sake Company finds himself in a situation in which wages are based on craft or technical ability and skill, as structured in the hierarchy from master down to middle workers.[2] Table 6.1 details these wage differences. The gap between the top and the bottom of the hierarchy in total daily wages is not large: in 1966 middle workers earned 62 percent as much as a head (¥ 880 vs. ¥ 1,430), in 1969 they earned 70 percent as much as a head (¥ 1,530 vs. ¥ 2,190). The actual gap between the top and the bottom is somewhat larger than this, since we do not know the master's wages (they are decided by the owners). The gap

[2] Our discussion of pay in Sake Company in 1966 derives from Mannari, Maki, and Seino 1967:26–28. Some parts of this chapter first appeared as "Pay and Social Structure in a Japanese Factory," *Industrial Relations* 12 (February 1973):16–32. Permission from that journal to use this material is gratefully acknowledged.

TABLE 6.1
Wages for Seasonal Workers in Sake Company, 1966 to 1969

Position	1966 Daily Wages	1966 Daily Overtime Pay	1966 Total Daily Wage[a]	1969 Total Daily Wage[a]	% Increase 1969 over 1966	$\bar{X}$ Number Days Worked/Year 1967–68
Master	Not Fixed					
Head	¥ 1,280	¥ 150	¥ 1,430	¥ 2,190	53.1	175
Subhead	1,180	150	1,330	1,990	49.6	178
Upper feed mash ferment controller	1,150	150	1,300	1,970	51.5	178
Lower feed mash ferment controller	1,050	150	1,200	1,830	52.5	172
Steamer	1,050	150	1,200	1,840	53.3	176
Instrument caretaker	900	150	1,050	1,660	58.0	163
Upper worker	820	150	970	1,560	60.8	158
Middle worker	730	150	880	1,530	73.8	157

Source: Mannari, Maki, and Seino, 1967:27, and company data for 1967–69

[a] The standard daily food allowance, paid in addition to the total daily wage, was ¥ 200 in 1966 and ¥ 370 in 1969.

between middle worker and head narrowed between 1966 and 1969 because the rate of wage increases was higher at the bottom of the hierarchy than at the top (73.8 percent vs. 53.1 percent). The absolute difference in daily wages, on the other hand, increased from ¥ 550 in 1966 to ¥ 660 in 1969.

Table 6.2 shows the distribution of annual wages in Sake Company in 1966. The modal income was ¥ 100,001–150,000, earned by 39.7 percent of the personnel, followed closely by ¥ 150,001–200,000, earned by 33.3 percent of the personnel. Thus, 73 percent of the employees earn between ¥ 100,001 and ¥ 200,000 a season. Whereas the data in Table 6.2 are self-reported income, taken from the 1967 questionnaire, and may have some underreporting, the data in Table 6.1 are standardized wages for all sake workers in the Nada region in 1966. This standard is agreed upon annually in negotiations between the Association of Owners of Nada Sake Plants and the Masters' Association in Hyogo Prefecture (*Hyogo-ken Tōji Kumiai Rengō-kai*). Since the Nada area is the center of sake making in Japan, these wage agreements set the national pattern for sake workers.

An interesting traditional aspect of pay in Sake Company is the formal practice of payment in kind. Employees are given a number of bottles

TABLE 6.2
Total Wage of Dekasegi Workers in Sake Plants, 1966

Total Amount a Year	*Number*	*Percentage*
¥ 100,000 or less	3	3.8
¥ 100,001–150,000	31	39.7
¥ 150,001–200,000	26	33.3
¥ 200,001–250,000	6	7.6
¥ 250,001–300,000	4	5.1
¥ 300,001+	2	2.6
No answer	6	7.6
Total	78	100.0

Source: Mannari, Maki, and Seino 1967:26.

of sake at the end of the season. In 1968, 40 bottles of first-grade, 1.8 liter refined sake (seishu) were given to a master, 10 bottles to an associate master, eight to a subhead, seven each to upper and lower blenders, steamers, and instrument caretaker, and six each to upper and middle workers.

Electric company

Electric Company's wage system prior to 1967 was based on the following criteria: (1) level of education; (2) length of employment in the firm, (3) age; (4) sex; and (5) superiors' evaluation of performance. It should be noted that this system was not devoid of performance criteria. With regard to education, whether one has a middle school, a high school, or a university education is functionally relevant to the kind of work one does and the skills one brings to the job. And the last factor is, at least in principle, a mechanism by which the worker's actual performance is judged, and distinctions among workers of the same age, seniority, sex, and education can be made, with consequences for pay differentials. In short, it is grossly oversimple to view the previous wage system as completely ascriptive.

Shipbuilding company

Although Shipbuilding Company introduced a "job control system" in 1970, at the time of our field work this system was operative mainly in the area of job classification; its extension to wages and promotions was still under discussion.

The total pay a Shipbuilding Company employee receives derives from four general sources: basic monthly pay, overtime pay, allowances, and bonuses. Basic pay is determined on the basis of age, seniority, education, and performance evaluation. The formal criteria of performance are "skill and ability"; in practice this means such things as the nature of a worker's job, his attendance record, his willingness to work overtime, his shop record, his safety record, and even his seniority. Age and seniority are the main determinants of basic pay. The rate of increase in pay begins to decline at age 40; by age 50 a plateau is reached, and retirement is normally at age 55.

Overtime pay is allotted for work in excess of seven hours in a given day: the mean for all X Plant personnel in the month from September 16 to October 15, 1969 was 1.7 hours per day overtime, which yielded a mean of ¥ 14,978 in overtime pay out of a mean of ¥ 68,907 for total pay that month.

Allowances take several forms. There is a positional allowance which increases for each higher supervisory and managerial rank; a standard allowance of ¥ 1,500 for all A Group personnel; a cost of living allowance, which is higher for employees in the Osaka factory than for those who work and live in a less expensive area; a dependents' allowance; and a transportation allowance up to ¥3,000 per month. The dependents' allowance has not increased for a decade; management would like to abolish it, but the union has prevented this.

Bonuses are given to all employees in July and December of each year. The mean bonus in July 1969 was ¥ 110,000, and in December 1969 ¥ 133,000; together this amounted to the equivalent of 4.6 months' basic pay. The amount of bonus an employee receives varies slightly, according to the same factors noted above in relation to performance evaluation—skill, attendance, willingness to work overtime, and so on.

Wage raises occur every April, at the rate of five or six percent. Beginning pay in Shipbuilding Company in 1969 was still higher for men than for women, even controlling for education and job status. For example, for high school graduates, aged 18, hired for clerical jobs, the beginning monthly pay was ¥ 28,440 for men and ¥ 25,270 for women.

The average annual wage increases (base-ups,[3] bonuses, etc.) during spring wage offensives in 155 major private industries have risen considerably in recent years, especially since the onset of rapid inflation. The annual increase reached 10 percent in 1965, 18 percent in 1970, and 20 percent in 1973 (Koshiro 1974:5). By this standard Shipbuilding Company's wage increases have been modest.

The practice in Shipbuilding Company generally has been to equate skill with seniority. Although certain operations in the company have been mechanized and automated, it takes time to learn how to operate the new processes. The consequence is that older, skilled workers are called upon. A 45-year-old Machine Section worker described this situa-

[3] "Base-up" is an across-the-board wage hike for all employees in a firm.

tion: "Though some new machines can automatically mark and cut a part, the workers on these machines may lack basic training on how to use the automatic machine. So *they ask me to manually mark* the center, etc. They get knowledge in polytechnical high school, but not enough practice, so they don't know how to use the basic knowledge in running the new machines."

The long-run prospects are, however, that the shift from craft to more automated forms of production will continue. An Assembly worker recognized the implications of this for skill: "The skill of the craftsmen is now declining. The model is now to have only average skill." Insofar as less skill is required, workers can attain the highest level in less time, and therefore beyond a certain number of years' seniority, there will be less reason to expect skill to continue to rise. This is one of the main factors that is likely to weaken the seniority pay system.

THE NEW WAGE SYSTEMS

Electric company

Despite the factors in favor of the seniority wage system in Japan, noted above, other developments have begun to undermine it. Recent studies have shown that the lifetime commitment pattern is less prevalent than some have asserted (Marsh and Mannari 1971, 1972; Cole 1971b). The rapid pace of innovation in industry results in the quicker obsolescence of seniority-related skills. Younger, more recently graduated workers are more attuned to the new technology. Younger Japanese workers now have a different social consciousness and are less satisfied with a wage system that is tied to age and seniority; they want a wage system based on the job and skill of the worker. The previously wide wage differentials between younger and older workers performing the same job have begun to narrow (Japan Institute of Labor 1966:7–8).

"On the occasion of the Spring Labor Offensive in 1962, the Yahata, Fuji, and Nihonkōkan Steel Companies joined in introducing [job-classification-based pay] rates. As these three companies have a great influence on the Japanese economy, their action was a noteworthy event" (Okochi, Karsh, and Levine 1973:375; see also Japan Institute of Labor 1964a: 3–4). Three years later, in 1965, Electric Company management announced that the seniority wage system is "opposed to modern rationality" and started to investigate new wage systems. In 1966 it presented a "New Job Classified Wage System" (*Atarashii shigotobetsu chingin seido*) to the union. In April of the following year, agreement was reached between management and the union, and the new system was inaugurated. *Denki-rōren* (All Japan Federation of Electric Machine Workers' Unions) was also in favor of the establishment of a wage system based on job evaluation (Japan Institute of Labor 1968:3).

The main criteria the company now uses for evaluating and classifying jobs are: (1) the knowledge and education the job requires; (2) experience and practical knowledge of the job; (3) the degree of mental burden the job imposes; (4) the degree of physical burden the job imposes; and (5) the leadership and supervision the job requires. Job classification levels are shown in Tables 6.3 through 6.5. The company has specified in writing what the job classification of each skill level in a given job process will be.

TABLE 6.3
Criteria of Evaluation in Electric Company's Job Classification System

	C—Creative Work		*O—Operative Work*
C-5	Kakarichō and associated positions; supervise professional and technical workers (C-4 employees)	O-6	Kakarichō and associated positions
C-4	Professional and technical employees; wide, general, theoretical knowledge; judgment, thinking, and mental responsibilities. For example, engineering for research and development production planning, patent administration, financial administration, industrial design.	O-5	Hanchō and associated positions. For example, metal pattern and tool designing.
		O-4	Kumichō and associated positions. Higher knowledge of electricity and machines than that learned in compulsory education; high craftsman skill; 1–4 years' experience; direct junior workers.
	R—Routine, Repetitive Work		
R-3	Jobs that require one year to master; or jobs that require six months to master, but where the physical burden is great; jobs that require judgment for negotiating with outside companies. For example, welding, testing balance of electric fans, typing.		
R-2	Jobs that require six months to master; method of work is simpler. For example, assembly line workers, ordinary repair work, soldering, general bookkeeping.		
R-1	Simple work. For example, janitors.		
	T—Training		
T-3	Third year trainee for given manual, clerical, or staff jobs.		
T-2	Second year trainee for given manual, clerical, or staff jobs.		
T-1	First year trainee for given manual, clerical, or staff jobs.		

For example, welding jobs that require a high degree of skill are classified Operative-4 (O-4), electric and gas welding jobs are classified Routine-3 (R-3), and simple spot welding Routine-2 (R-2).

Earnings Structure under the New System. Under the new job classification system, the basic components of pay as of 1969 were:

1. Monthly base pay (*getsu gaku*)

TABLE 6.4
Job Titles and Ranks Included in Each Job Classification in Electric Factory

	Classification	*Number in Each Classification*
Creative (C) Jobs		
C-5	Kakarichō (staff)	19
C-4	(41 job titles) Getting petition for manufacturing, engineering for research and development, drafting for test work, engineering for experiments and design, quality control work, art, personnel management, accounts, safety-hygiene management, purchasing, production planning, sales promotion, research planning, sales control, patent administration, system programming, calculating, translating, financial administration, editing the history of the company, transportation, planning, inspection, new model planning, industrial design, engineering research planning, developing surface treatment (of fans, for example), export planning, administration of fixed assets.	145
Operative (O) Jobs		
O-6	Kakarichō (line)	11
O-5	Hanchō; metal pattern and tool making and designing, chauffers treated as foremen.	29
O-4	(15 job titles) Kumichō; directing work, repairing, inspection, welding, driver, machine work, test manufacturing, in charge of boiler, in charge of transformer of electric oven, guard.	110
Routine (R) Jobs		
R-3	(36 job titles) Welding, compounding manganese, paint spraying, inspection, testing balance of electric fan, honing, art printing, paint mixing, driver of truck and trailer, maintaining tools, adjustment work in work processing, telephone operator, key puncher, teletype operator, nurse, typist in English and Japanese, woman in charge of dormitory, boiler man, dietician, parts supplier, cook.	189
R-2	(41 job titles) General assembly line workers, packaging, welding, preparation work, varnishing, inspection, lathe operator, winding, plating, repairing, soldering, information desk, general bookkeeping, woman for dormitory maintenance, hospital worker, boiler assistant, transportation of material, receipt and shipment of material, parts and products.	262
R-1	Janitors	1
		766[a]

[a] This figure is less than the total number of employees (1,212) because it does not include training personnel or kachō and higher ranks.

TABLE 6.5
Skill Levels within the Same Job Processes in Electric Company

Job Process	*Job-Skill Classification*	*Definition of Tasks*
Coordinating	R-3	Coordinating work that requires six months to one year to learn
	R-2	Ordinary coordinating work in work shop
Repairing	O-4	Repairing productive machinery
	R-3	Work that requires more than six months' experience, but not professional knowledge
	R-2	Ordinary repair work (e.g. change parts and coordinate new parts)
Inspecting	O-4	Inspection of finished products (TV, etc.)
	R-3	Inspection of finished products (less complex)
	R-2	Inspecting work on the assembly line
Welding	O-4	Welding that requires high skill
	R-3	Electric and gas welding
	R-2	Simple spot welding
Machining	O-4	Machining that requires high skill
	R-3	Machining that requires six months to one year to learn
	R-2	Automatic machine work that requires simple skill
Crane operating	R-3	More than 15 kgs., all day
	R-2	Less than 15 kgs.
Transformer	R-4	In charge of transformer; has license for electrical engineering
	R-3	Lower level license for electrical engineering
	R-2	No license for electrical engineering
Packing	R-3	Wood packaging for large products
	R-2	Packing using ordinary (cardboard) packing cases

2. Position or rank allowance (*yakuzuke teate*)[4]
3. Allowance for jobs with special danger, heat, or smell (*tokushu kinmu teate*)
4. Allowance for those who work more than five days a week but do not get overtime pay (*jitan teate*)

[4] The monthly allowance, over and above base pay, for each position or rank in the company is as follows:

buchō (department head)	¥ 15,000–18,000
associate buchō	¥ 16,000–17,000
kachō (section chief)	¥ 12,000–14,000
younger kachō, kachō-dairi (acting kachō)	¥ 10,000
kakarichō (subsection chief)	¥ 2,000
shunin	¥ 1,600
hanchō (second-line foreman)	¥ 1,300
kumichō (first-line foreman)	¥ 1,000

5. Allowance for those who work more than eight hours a day but do not get overtime pay (*shokumu teate*)
6. Allowance for dependents (*kazoku teate*)—wife, children and parents over 65 are eligible.

These components, the first of which is the most important, determine total regular monthly pay (excluding overtime pay and midsummer and year-end bonuses).[5] Earnings under the new pay system can be seen in the following figures from a 1967 company document:

Job Classification	*Minimum Monthly Base Pay at Age 25*	
Creative-5, Operative-6	¥ 36,000	
Operative-5	¥ 31,000	
Operative-4, Creative-4	¥ 29,000	
Routine-3	¥ 27,500	(In 1967,
Routine-2	¥ 27,500	$1.00 = ¥ 360)
Routine-1	¥ 26,000	

The most striking change is that a university graduate with only three years' seniority would receive, at age 25, a higher base pay than middle school graduates with ten years' seniority. This is so because the university graduate would be in a higher job classification (e.g., Creative-4) than the middle school graduate (e.g., Routine-3). By contrast, under the seniority system, it was not uncommon for a 25-year-old manual worker with ten years' seniority to receive higher base pay than a 25-year-old university graduate technical or staff employee, because the latter had only three years' seniority.

Advancement in Job Classification and Rank. Advancement in job classification results in an increase in monthly base pay. During the first three years' employment after school graduation (T-1 through T-3 job classification), wages increase by a standard amount. Beginning in the fourth year (R-2 job classification) individual differences in wage increases of ¥ 300–500 a month emerge, based on performance evaluation. Wage raises are also specified by educational level. For example, of those who graduated in 1965, the raise in 1967 was ¥ 3,200 per month for middle school

[5] Biannual bonuses, given to all employees, constitute an important source of pay. In 1969, the average June bonus was equal to 3.2 months' pay, and the December bonus was also 3.2 months' pay, making the total average bonus for 1969 equivalent to over six months' regular pay. The amount of bonus an individual employee gets is affected by his attendance record and performance record; the range from highest to lowest bonus is about ¥ 10,000. Ministry of Labor data indicate that nationally employees received bonuses equivalent to 3.87 months' average monthly salary in 1969 (Ono 1971:8, Table 2). Electric Company's bonus is therefore considerably above the national average. Bonuses reflect a firm's success, and as noted earlier, Electric Company is one of Japan's most successful firms.

graduates, ¥ 3,900 for high school graduates, and ¥ 4,500 for college graduates (C-4 job classification).

Advancement in rank—for example, from kumichō to hanchō—would also result in a pay increase. It should be noted that although in most instances advancement in rank entails advancement in job classification, the reverse is much less often the case. The six levels of job classification from T-1 through R-3 are all rank-and-file positions, and movement within them involves no advancement in rank. Moreover, of those in given middle-level job classifications (O-4, O-5, O-6, C-5), some hold a concurrent rank (kumichō, hanchō, kakarichō) while others do not, i.e., are still rank-and-file or staff workers.

Dependents' Allowances. Every married male employee receives a dependents' allowance of ¥ 1,800 a month for his wife and ¥ 900 for each dependent other than his wife. Allowances for dependents are administered in a completely universalistic manner: given the same number of dependents, the allowance is fixed regardless of rank or education or performance. (In Electric Factory, only one-fourth of the employees are married and those who are have small families, so allowances for dependents is a small item in the total factory payroll.[6])

Equal Pay for Men and Women. The new pay system provides for equal starting wages for men and women who are "regular" recruits—those hired directly from school. This policy went into effect in 1968. The initial pay for those hired with previous job experience is determined by the job they hold; if over 25, their pay is calculated on an individual basis in accordance with their previous job. In this respect, Electric Company's social organization is more modernized than that of Shipbuilding Company, where, as we saw, beginning pay in 1969 was still higher for men than for women, even controlling for education and job status.

In summary, all components of pay except dependents' allowance are presumably directly related to work performance (Electric Company still considers education and seniority to be functionally related to performance).

Shipbuilding company

A job control system (*shokunō kanri seido*, or shokkan seido for short) has been under consideration in Shipbuilding Company since the early 1960s. Between 1967 and the time of our field work there had been con-

[6] Among Electric Company employees, 70.7 percent have no dependents, 5.4 percent have one, 9.6 percent two, 10.2 percent three, 3.2 percent four, and 0.9 percent five or six, the largest number. The mean number of dependents for male employees is 1.37, for females 0.02, and for all employees 0.74.

ferences between management and the union with a view to putting the system into effect in 1970. Management proposed that under the new system pay would be based equally on three factors: seniority, ability or skill, and job classification. A kachō in the company headquarters Planning Section explained the new system in these terms: "In our X Plant there is now a transitional situation. . . . Young workers are dissatisfied with the present [1969] seniority-based wage system. Also, the company can't afford the higher wage costs which increasingly result from having a larger proportion of high seniority employees. The low birth rate of recent years reduces the supply of young workers."

The 1969 job classification system works as follows. In the Diesel Engine Division of X Plant 30 job titles were specified, most of which had four skill levels: junior (2), junior (1), middle, and senior. In a given job one would be at the junior (2) skill level for, say, the four and one-half months' apprenticeship period; in the junior (1) level one would have the informal status of sakite;[7] in the middle level, the informal status or skill comparable to boshin; and at the senior skill level, either the formal rank of hanchō, or skill comparable to that of hanchō. The four skill levels are applied even to the lowly job title of sweeper (sōji). A senior sweeper, for example, supervises sweepers from outside the company, who use complex chemicals to clean the insides of the machines.

The proposed job classification pay system, as stated in a company manual, is outlined in Figure 6.1. It is of very general significance, since it will be used for pay, promotions, and "ability development." The higher positions of secretary, assistant secretary, and manager are listed as equal to the rank of buchō, kachō, and kakarichō, respectively. This means that some employees in these positions will also be holders of the equivalent rank, but some will not. Not everyone can become a kakarichō or kachō, since there are simply not enough of those ranks at the higher reaches of the company pyramid to go around, but for those who do not achieve the rank, there will at least be a special title to signify that they have reached an equivalent skill and seniority level.

Consider how the system outlined in Figure 6.1 would apply to engineers. As in firms in other countries, engineers have at least two career tracks: technical and managerial. Shipbuilding Company believes a technical engineer should not be bothered with personnel or other managerial problems. Recognizing that technical engineers and managerial engineers are performing different functions, the company equalizes their status by having each move up through parallel skill grades and positions. Thus, when two engineers reach the same status—say, grade 5—they will

[7] "Sakite" is a Tokugawa period army term meaning the vanguard, or the advance guard. In its contemporary usage in Shipbuilding Company, it may be translated as "right hand" or "assistant."

FIGURE 6.1

Shipbuilding Company's New Job Classification Qualifications System

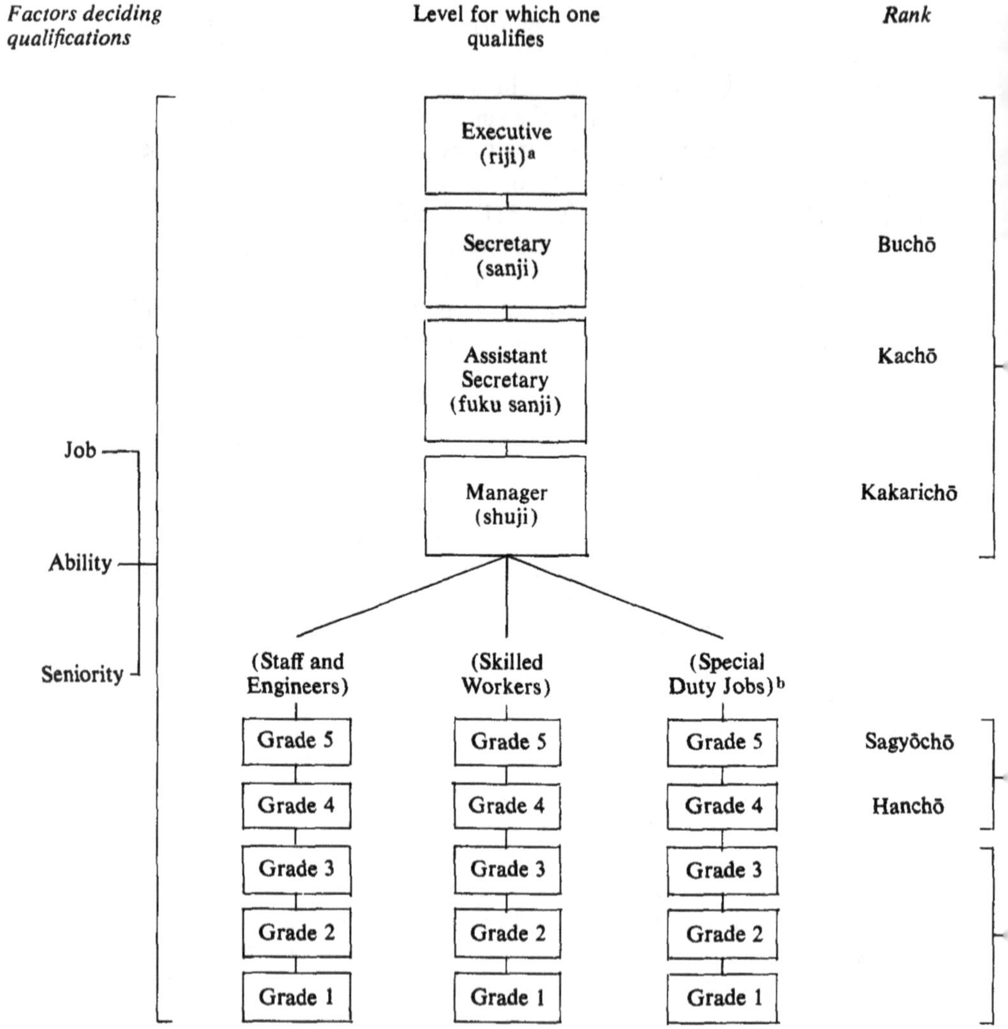

[a] Equivalent to director (*torishimariyaku*).

[b] Special duty jobs are those involving skills marketable outside the company, e.g., driver, key punch operator.

Source: adapted from Shipbuilding Company brochure.

receive equal pay, even though one is a technical engineer and the other managerial.

Decisions to move employees up through the skill grades in Figure 6.1 will be made in the annual personnel evaluations (*teiki jinji kōka*). Each employee will get an A, B, or C skill evaluation; the more A's and fewer C's he receives from his superiors, the faster he will rise through the system. Table 6.6 shows how a cohort of employees who entered the company at age 18 would move up through the grades, in accordance with the annual performance evaluations they received. Evaluations are based on two successive years; e.g., AA means A evaluation both years, AB means A the first year and B the second, and so on. The higher the grade one is in, the more the evaluation influences the minimum time one must stay in that grade before being promoted. Thus, everyone entering grade 1 at 18 would reach grade 2 by either age 20 or 21, but of those in grade 5, an AA evaluation pattern could bring promotion to *shuji* after a minimum of two years, whereas a CC evaluation would mean waiting at least eight years for this promotion. From grade 3 upward, the time intervals are couched in terms of minima, which means promotion is not automatic, and that some employees could, in principle, remain in one of these grades indefinitely. Considering the extreme cases among a cohort, those who received an A evaluation every year could reach the rank of shuji by age 31, while those who received a C evaluation every year could not reach this rank before age 50.

Although the full system of evaluations as just outlined had not gone into effect at the time of our field work, parts of it—evaluations of A, B, and C, as well as the grades of D and "other"—have been operative for some years. Though workers have not been and would not be told what jinji kōka evaluation they received, they could, of course, guess, on the basis of how long they remained in one grade.

Evaluations are clearly patterned by age. A machine Section sagyōchō told us: "These A, B, C evaluations have no relationship to length of service, but older and younger workers are *grouped separately* for these evaluations."

Table 6.7 shows the relationship between age and evaluation for all manual production workers (A Group) in X Plant in 1967. The proportion of workers who get A's rises from zero in the 18–20 age group to a peak of 39 percent in the 31–35 age group and thereafter declines to only 12 percent in the group over 50 years old. This curvilinear relationship[8] reflects both observable, objective facts of performance and company

[8] This relationship is seen more clearly if the category "other," which appears to be used mainly for the youngest age group (18–20 years old), is omitted. With only four evaluation categories—A through D—the proportions in the 18–20-year-old group are: A zero; B 27.2 percent; C 38.3 percent; D 34.6 percent.

TABLE 6.6

Proposals for Evaluation and Advancement of Workers under Job Qualifications System in Shipbuilding Company

	Criterion	*White Collar or Manual Work*					*Ability on Job*					*Special Job Responsibilities*				
	Evaluation for 2 Previous Years	AA	AB, BA	AC, BB, CA	BC, CB	CC	AA	AB, BA	AC, BB, CA	BC, CB	CC	AA	AB, BA	AC, BB, CA	BC, CB	CC
Shuji ↑ Grade 5	Minimum number of years in Grade 5 before promotion to shuji	2	3	4	6	8	2	3	4	6	8	2	3	4	6	8
	Standard age[a]	31	33	36	42	50	31	33	36	42	50	31	33	36	42	50
Grade 5 ↑ Grade 4	Minimum number of years in Grade 4 before promotion to Grade 5	3	4	5	6	8	3	4	5	6	8	3	4	5	6	8
	Standard age[a]	29	30	32	36	42	29	30	32	36	42	29	30	32	36	42
Grade 4 ↑ Grade 3	Minimum number of years in Grade 3 before promotion to Grade 4	3	3	4	6	8	3	3	4	6	8	3	3	4	6	8
	Standard age[a]	26	26	27	30	34	26	26	27	30	34	26	26	27	30	34
Grade 3 ↑ Grade 2	Number of years in Grade 2 before promotion to Grade 3	3	3	3	4	5	3	3	3	4	5	3	3	3	4	5
	Standard age[a]	23	23	23	24	26	23	23	23	24	26	23	23	23	24	26
Grade 2 ↑ Grade 1	Number of years in Grade 1 before promotion to Grade 2	2	2	2	2	3	2	2	2	2	3	2	2	2	2	3
	Standard age[a]	20	20	20	20	21	20	20	20	20	21	20	20	20	20	21

Source: Shipbuilding Company *Shokuno kanri seido manual*, November 1969, p. 9

[a] Age the person would be at time of promotion, assuming he entered the company at age 18

TABLE 6.7
Relationship between Age and Performance Evaluation, all X Plant A-Group Production Workers in Shipbuilding Company, 1967

	Performance Evaluation (%)						
Age	*A*	*B*	*C*	*D*	*Other*	*Total*	N
18–20	—	10.8	15.2	13.7	60.3	100	204
21–25	14.1	29.4	34.9	18.3	3.3	100	269
26–30	25.2	35.0	22.1	12.4	5.3	100	226
31–35	38.9	29.0	14.5	12.3	5.3	100	131
36–40	23.8	33.1	25.6	14.4	3.1	100	160
41–45	26.7	35.3	26.2	9.1	2.7	100	187
46–50	16.0	28.8	36.6	18.6	—	100	156
51 +	11.7	21.5	36.8	28.2	1.8	100	163
N	(278)	(417)	(401)	(236)	(164)		(1,496)

policy. The youngest group of workers is given a lower evaluation because of its lack of skill and experience, and the oldest group both because its skills have become somewhat obsolete in a rapidly changing technology and because its energies for manual labor have waned. The age group 31–35 has the optimum combination of skill and energy.

It is company policy for the sagyōchō to add his own evaluation to those his hanchō have given to the workers under them. The sagyōchō is free to use his judgment to "smooth out" the overall number of A's, B's, and C's. Thus, if he thinks a hanchō has given too many high marks, or too few, he can change the evaluation. Since this "editing" function takes place anew at each higher level in the hierarchy, two things can happen: first, an employee's mark may change one or more times; and second, his mark (and therefore his advancement and pay) is ultimately decided by higher level managers, and it is the attitudes of these men concerning the relative importance of the official qualifications—job, ability, seniority—that may count most heavily.

What we have described is essentially Shipbuilding Company management's wage policy. The union and other groups in the company have somewhat divergent views on the relative importance that seniority, skill, and job classification should have in determining pay. We shall return to this below, when we analyze attitudes of personnel toward the wage system.

We have examined changes in wage policy in Electric and Shipbuilding companies in some detail. We have done so not only because there is so little in English on how wage policy operates in particular Japanese firms, but also because we must be clear about what the policy changes are before we ask: to what extent have wage policy changes actually altered the relative influence of the various determinants of pay?

THE ACTUAL DETERMINANTS OF PAY

A THEORY OF REWARDS

The basic theory we shall test is that there are four main sets of independent variables that influence organizational rewards, in this case, pay.

I. *Bureaucratic Influences on Pay.* In any bureaucratic organization, as Weber and others have theorized, remuneration varies directly with status in the organization, performance (merit), and attendance. Thus, we shall test the hypotheses that pay varies directly with:

1. Rank
2. Seniority
3. Education
4. Job classification
5. Performance
6. Attendance[9]

II. *Cultural Influences on Pay: Present Extra-Organizational Statuses.* Distinctive sociocultural settings in which organizations exist modify in varying degrees the influence of strict bureaucratic elements on pay. In the Japanese case, the cultural setting is such that pay is also formally based on such present extra-organizational statuses as one's age and and number of dependents. A third extra-bureaucratic status is sex. Although officially and formally sex is not a determinant of pay in Electric Company, it does have important direct and indirect implications for pay. Most female employees work only a few years in the firm; hence they do not rise high enough in job classification and seniority to earn higher pay; partly as a result of women's lower seniority, but also because of prevailing societal norms, all supervisors and managers are males. As a result, women do not attain the higher rank that would entail higher pay. Accordingly, our text three hypotheses are:

7. Pay varies directly with age
8. Pay varies directly with number of dependents
9. Men receive higher pay than women

III. *Informed Influences on Pay: the Social Integration of the Employee into the Organization.* Pay is not solely an automatic function of such objective factors as rank, seniority, education, and job classification. In order for performance to influence pay, performance has to be evaluated by other members of the organization. When this evaluation

[9] Attendance (ratio of absences to legitimate days off) is also one item in the index of performance. Initially we regressed pay on both attendance and performance to see which has the stronger relationship with pay.

takes place, the evaluator takes into account, in varying degrees, what may in general be called the social integration of the employee into the company; specifically, his cohesiveness with other employees, how often he participates in company recreational activities, and the extent to which he exhibits paternalistic and lifetime commitment attitudes and behavior, favored by management. Thus, our next four hypotheses are that pay varies directly with:

10. Employee cohesiveness
11. Participation in company activities
12. Paternalism
13. Lifetime commitment

IV. *Previous Extra-Bureaucratic Statuses' Influence on Pay*. The first three sets of influences on pay (I–III) all refer to factors that operate contemporaneously with the decision on a person's pay. The last set of factors, on the other hand, refers to the influence on present pay of statuses prior to entering the present organization. Specifically, we refer to social origins variables—father's occupation, size of community of origin, and its distance from the location of the organization for which one presently works. Theoretically, rewards in formal organizations appear to be less influenced by this kind of extra-bureaucratic factor than by factors II and III above. The main reason for this may be that while factors II and III are presently operating, factor IV is, by definition, part of one's more or less remote past. Whatever the initial effects of social origins on prerecruitment behavior and attitudes, they become progressively attenuated as time passes. Thus, our next hypotheses are null hypotheses. Pay varies independently of:

14. Father's occupation
15. Size of community of origin
16. Distance of community of origin from location of factory

A final previous status factor is prior work experience, as measured by number of years in jobs before entering one's present firm. Except for the fact that this is an earlier rather than a present status, this factor could be considered within (I) above, as a bureaucratic influence on pay. The distinction is a subtle one. There is nothing in the theory of bureaucracy that precludes taking into account work in previous organizations as a determinant of pay. However, we argue that whatever significance the duration of previous work experience has on current pay will operate primarily through one's present rank and job classification. It is these that are adjusted to take into account previous work experience, the more so when that experience is considered relevant to one's role and status in the present firm. It would seem that the primary factor is not the number of years in other jobs, but the kind of jobs. Certain kinds would give one some edge, but others—in small firms, or offering little responsibility—would not. Also counter-balancing the positive advantage of previous jobs is the preferred pattern of coming up through the ranks in one company, rather than being recruited from another firm at a higher level of pay or rank. For all these reasons we hypothesize that when rank and job classification are held constant, pay varies independently of:

17. Number of years' previous work experience.

To summarize the theory, there is one set of strictly bureaucratic influences on pay (I) and three sets of non- or extra-bureaucratic influences (II–IV). In this sense, the strength of the variables in hypotheses 1–6 relative to that of the variables in hypotheses 7–17 is a measure of the relative bureaucratization of pay in the organization.

One further distinction within the theory will be of interest. In Weber's theory of bureaucracy, rewards—e.g., remuneration, promotion—depended on seniority or performance (merit), or both (Weber 1946:203). But the two variables that in this view are both analytically "modern bureaucratic" influences on pay need in the case of Japan to be separated into traditional and modern influences. We have seen that, at least in recent decades, Japanese organizations have decided pay primarily on the basis of seniority; more recently, there has been some tendency to assign greater importance to job classification as a determinant of pay. Hence, among analytically bureaucratic determinants of pay we shall want to consider the relative influence of seniority and job classification (and performance). Specifically, when pay is predominantly based on seniority, we shall define the pattern as traditional; the greater the influence of job classification and performance relative to seniority, the more modern the pattern will be defined as being.

Data on only a few of the variables in our theory of rewards were gathered in Sake Company. Therefore, we offer only some descriptive statements about the determinants of pay in Sake Company, and attempt to test the theory when we move to Electric and Shipbuilding factories.

Sake company

In principle, even in the past, a sake worker's pay was determined by his position (skill), and all steamers, for example, would receive the specified daily wages of a steamer, regardless of age of seniority. Thus, one Sake Company worker noted in the questionnaire: "Wages depend on the length of service [in other companies]. But it is not so here. . . . So I hope this company will consider the length of service in deciding wages." However, although formally pay has been based on job position and skill, in general one advances up the hierarchy in relationship to seniority, as

Table 6.8 shows. There is a seniority bonus starting in one's fourth year in Sake Company; according to this, one's daily wages are increased by a factor computed on the basis of the number of years' service. As a result, there is a positive relationship between seniority, position in the craft hierarchy, and pay.

TABLE 6.8
Relationship Between Mean Seniority and Mean Pay in Sake Company, 1968

Position	*X̄ Years' Seniority in Company*	*X̄ Daily Pay Exclusive of Food Allowance*	N
Head	16.4	¥ 2,341	9
Subhead	11.0	2,193	2
Upper blender	9.0	2,175	1
Lower blender	7.0	2,048	35
Steamer	8.4	2,035	7
Instrument caretaker	3.9	1,865	30
Upper worker	3.3	1,762	24
Middle worker	1.4	1,726	23
Other	5.4	1,999	19
Total	5.4	¥ 1,935	150

Source: Company records.

ELECTRIC FACTORY

We were given complete data on all components of pay for all 1,212 Electric Factory employees for the month of May 1969. Employees are identified on payroll records by a number, and since this number was used on the questionnaire each employee filled out for us, we were able to collate the two sets of data.

For 61 percent of the employees in Electric Factory, base pay was the only source of pay (other than overtime); 23 percent also received pay from a second source, typically positional allowance or dependents' allowance; 14 percent received pay from two sources other than base pay; and only two percent received pay from as many as three sources other than base pay. Expressed in terms of the percentage of one's total monthly pay that came from base pay:

Base Pay as % of Total Monthly Pay	*% of Employees*
100	61
90–99	35
80–89	4
	100

In the multiple regression analysis, the dependent variable is an employee's total monthly pay, coded as: under ¥ 20,000; ¥ 20,000–39,999; ¥ 40,000–59,999; ¥ 60,000–79,999; and ¥ 80,000 or more. The main results of the multiple regression analysis[10] are that the best predictors of the amount of pay are seniority and number of dependents in Electric Factory, and age and rank in Shipbuilding Factory. All these relationships are positive, i.e., pay increases as one accumulates more seniority (or gets older), has more dependents, and attains higher rank in the firm. It is these factors which determine pay, *not* the much larger number of other hypothesized variables—sex, job classification, attendance, performance, the degree of social integration of the employee into the firm, or his social origins.

The theoretical issue of whether pay depends more on seniority or on job classification cannot, unfortunately, be resolved on the basis of our data. This is because seniority and job classification are so strongly positively correlated among Electric Factory employees ($r = .82$) that they are measuring almost the same thing; consequently, we cannot disentangle their *relative* effects on pay.

This problem aside, these findings lend themselves to at least two substantive interpretations. One is that Electric Company's new job classification wage system is only the older seniority system in another guise. This view calls attention to the fact that the company admits that its job classification pay system will not totally eliminate the importance of seniority, even when the system is fully institutionalized. Provisions that make a certain minimum number of years in a given job classification or rank a precondition for advancement, and thus higher pay, are retained in the new wage system.

The second interpretation is less cynical. The company acknowledges that it cannot erase existing seniority-based pay differentials overnight. It has not, after all, *reduced* the pay of its present high-seniority employees. It points out that for the time being differences in base pay for the same job will continue. The job classification wage system cannot have its full impact until the older cohort of long-seniority employees has left the company. The expectation, therefore, is that over time the influence of seniority on pay will in fact decline, and that of job classification will increase.

Although we cannot at present empirically reject either of these two interpretations, we shall be able to in the future. To the extent that a multiple regression analysis of pay in Electric Factory in, say, 1979 yields the same findings as those observed for 1969, the view that the job classification system is simply the seniority system under a new guise will be confirmed. The second interpretation will be confirmed to the extent that

[10] For details, see Appendix D, Tables D.8 and D.9 p. 375 and p. 377.

the future regression analysis shows that job classification is more predictive of pay than is seniority.

Shipbuilding factory

Turning to Shipbuilding Factory, the most important determinant of pay is age, followed by rank: the older one is, and the higher one's rank, the more pay one receives. As much of the variation in pay can be explained on the basis of these two factors as when all 17 hypothesized independent variables are taken into account!

It appears that although supervisors and managers in Shipbuilding go through the motions of evaluating the performance and skill of their subordinates each year, and although the company states that pay formally depends on attendance, these factors do not exert any impact on pay independent of the major determinants of age and rank. The interview comment that in practice performance is largely equated with age and seniority is borne out by the results of the multiple regression analysis. A further implication, however, concerns the three-way relationship between age, performance, and pay. We stated earlier that the company views the relationship between age and performance as curvilinear, with the optimum combination of skill and energy coming in employees' thirties, and then declining somewhat. If pay were based on performance, therefore, this relationship would also be curvilinear. Yet in fact it is strongly positive. Insofar as our measure of performance is valid, therefore, it appears that relative to their performance (skill and energy) older workers are overpaid, and that this results from the fact that pay is much more strongly based on age (and seniority) than on performance as independently measured.

Our findings show, then, that as of 1969–70, the seniority wage system was so highly institutionalized in these two firms that we can predict more of the variance in pay than in any other dependent variable in this study (Cf. R^2's in all tables in Appendix D).

ATTITUDES TOWARD THE WAGE SYSTEM

Having examined two aspects of the wage system—company policy and the actual determinants of pay—we turn now to a third: employees' attitudes toward the wage system. The data are taken from the questionnaires.

Electric factory

Once again, more questions, and more comparable questions, were asked in Electric and Shipbuilding factories than in Sake Company, so we shall restrict the analysis to the first two firms. In the Electric Factory questionnaire we asked, "Do you think the present job classification wage system is better, or the previous wage system?" There were four pre-coded answer categories in this variable (v. 101). Two distributions of responses are shown in Table 6.9, one for the 976 employees who checked one of the four categories, including "don't know," and the second for 643 employees, omitting the 333 who checked "don't know."

TABLE 6.9
Wage System Preferences in Electric and Shipbuilding Factories

Electric Factory (v. 101)		*Percentage of Respondents*[a]	
1. The job classification system is better		33 0	50.1
2. The previous wage system is better		14.1	21 5
3. No change		18.8	28.4
4. Don't know		34 1	—
		100.0	100.0
	N	(976)	(643)

Shipbuilding Factory (v. 65)		*Percentage of Respondents*[a]	
1. The present seniority-based wage system is better.		30.6	42.4
2 The new *shokkan* system is better.		27.5	38.1
3. No difference		14.1	19.5
4. Don't know		27.8	—
	Total	100 0	100.0
	N	(553)	(399)

[a] The first column includes all respondents, the second omits those who answered "Don't know."

Since the new wage system was just beginning to be institutionalized, 34.1 percent of the employees said they did not know which of the two systems is better. Of the 643 who expressed a definite opinion, over twice as many favored the new system as the previous one (50 percent vs. 22 percent). In addition, more than one in four (28 percent) believe that there is no difference between the two systems in practice.

High status employees—as measured by job classification, total pay, education, and rank—are more likely than lower status employees to favor the job classification wage system. The strongest support for the previous, seniority-based wage system, on the other hand, comes from the middle strata of the organization—Routine-3 workers and first-line foremen. These are personnel with relatively long seniority but poor or uncertain chances for further advancement in job classification. In the past they could count on regular pay increases for seniority, which is no longer the case; thus their preference for the previous system.

From these findings it appears that the new job classification wage system has the support of the more elite segments of Electric Factory. Given this, its further institutionalization should be less problematic.

Returning to the 34 percent who said they did not know which wage system is better, we found that whereas preference for the new job classification wage system is greatest among higher status employees, it is lower status employees who are most likely to give "don't know" responses. They are more likely to give this noncommittal response than to say they favor either the new or the previous wage system.

To what extent does the overall pattern of response fit an interest group theoretical model? Table 6.10 divides employees into two interest groups:

TABLE 6.10
Wage System Preferences, by Education, Job Classification, and Seniority (%)

Wage System Preferred	*University and High School Graduates: Kachō, Creative, Operative, or Training Job Classification*		*Middle School Graduates: Routine or Training Job Classification in Six Production Sections*	
	Seniority		*Seniority*	
	0–7 years	*8+ years*	*0–7 years*	*8+ years*
Job classification system	68.9	64.2	17.4	31.6
Previous wage system	3.2	9.5	7.8	33.6
No difference	21.3	16.8	19.5	22.1
Don't know	6.6	9.5	55.3	12.7
	100.0	100.0	100.0	100.0
N	(61)	(95)	(333)	(244)

those whom the new job classification wage system will benefit most—university and high school graduates in Creative jobs or being trained for such jobs (columns 1 and 2); and those who would benefit more from the seniority system than from the new system—middle school graduates in the six production sections, all of whom are either already in Routine jobs or are likely to receive that classification after their training period is over (columns 3 and 4). Let us refer to these as the job classification interest group and the seniority interest group, respectively.

Within the job classification interest group, seniority has no impact on wage system preferences: over two-thirds of both the 0–7 years' and the 8+ years' seniority groups favor the job classification system; very few indeed prefer the previous seniority system. Among the seniority interest group, on the other hand, seniority has important influences on wage system preferences. One influence of greater seniority is that the proportion who don't know which wage system is better sharply declines (from 55 percent to 13 percent). Of those with 8+ years' seniority, virtually as many favor the job classification system as the seniority system.

This presumably indicates the legitimacy effect of the newly institutionalized job classification system, even among the seniority interest group. More to the point, however, is that among the seniority system interest group, greater seniority increases the support for the previous, seniority wage system more dramatically than for the job classification system: from eight percent to 34 percent, in contrast to from 17 percent to 32 percent. Support for the interest group hypothesis is also seen when we compare employees in both interest groups who already have 8+ years' seniority (columns 2 and 4). Among these high seniority employees, twice as many in the job classification interest group as in the seniority interest group prefer the job classification system (64 percent vs. 32 percent), while over three times as many in the seniority interest group as in the job classification interest group favor the seniority wage system (34 percent vs. 10 percent).

We thus find clear support for an interest group model, modified somewhat by the legitimacy effects of the recent institutionalization of the job classification system among all employees and the greater support for that system among more elite employees.

Shipbuilding factory

Management and the union have somewhat different views on the relative importance of seniority, skill, and job classification in determining pay. The company wants a productivity-oriented incentive wage system, while the union prefers a system of base pay plus a flat percentage addition for all staff and production personnel. Management is more favorable to job classification-based pay than is the union. The union's position is that union members are a conglomeration of machinists, assemblers, gas welders, crane operators, tamakake, and so on, and there should not be wage differences among these different job groups. Instead, the union prefers wages to be based upon ability and skill, as evaluated by supervisors, and upon seniority. The qualifications system (*shikaku seido*) outlined above (Figure 6.1) fits this viewpoint, in that the skill with which one performs a job will have more influence than the content of the job on wages (and promotions). This feature is said to be favored by the workers, who feel strongly that wages can be based on skill differences but should not be based on the classification of the job as such.

We also encountered different views among management and staff personnel. Earlier we reported the management view that in the future pay will be based equally on the three factors of seniority, skill, and job classification, which would give each factor a weight of 33 percent. In contrast, a kachō in the headquarters Planning Section saw seniority as having a weight of 50 percent even in the future. "At the present [1969] base pay [*honkyū*] is completely determined by age. In the future, we shall shift to pay being based 50 percent on age and 50 percent on job skill. At the present, base pay is the major segment of total pay, job skill a minor segment. But base pay will be increased only through the annual 'base-up,' whereas job skill-based pay will increase more, until they are about fifty-fifty as a determinant of pay." These divergent views are surface manifestations of underlying conflicts within the company, and also reflect the fact that Shipbuilding Company is in an earlier stage of the transition from seniority- to job classification-based pay than Electric Company.

Turning now to the questionnaire, the first question (variable 65) closely parallels that considered earlier in Electric Factory: "Which do you think is better, the present seniority-based wage system, or the new shokkan system, which is based on job, ability and seniority?" Responses are again presented in two distributions, one for all respondents, and the other omitting those who checked "don't know" (see Table 6.9).

Shipbuilding Company workers are much more evenly split between older and the newer pay systems than are Electrical Company personnel. In the latter company, more than twice as many preferred the job classification system as the previous system (33 percent vs. 14 percent), while in Shipbuilding Company, employees are virtually evenly split among three responses—preference for the older system, preference for the newer one, and don't know which is better.

We asked respondents in both Electric and Shipbuilding factories who prefer either a seniority-based wage system or a job classification-based wage system to explain their preference. The questionnaire item was open-ended; an empirical code was constructed for the responses, and these are compared in Table 6.11. Among Shipbuilding employees who think the job classification wage system is better, the most common reason is that "a merit system is good for motivation or morale," while the most common reason in Electric Factory is that employees think "it's more rational to have wages fit the job." Among those who favor the seniority wage system, Shipbuilding employees are more likely than those in Electric Factory to say this is because a seniority wage system "gives stability and security as a worker gets older;" Electric Factory personnel are more likely to give as their reason an alleged defectiveness in the way the job classification wage system actually works, rather than any particular virtue in the seniority wage system itself. The strongest difference between personnel in the two factories is that those in Electric Factory are, in addition to being more modern—more in favor of a job classification wage system—also more likely than those in Shipbuilding to give as their reason for preferring a job classification wage system that it is a more rational system.

Workers' remarks during our interviews in Shipbuilding Company illustrate these findings. A Machine Section worker gave his reason for liking the job classification system: "I like the shokkan seido. People are

TABLE 6.11
Reasons Employees Give for Preferring Job Control Wage System or Seniority Wage System (%)

		Shipbuilding Factory (v. 66)	*Electric Factory (v. 102)*
The job classification (job control) system is better because:			
1. A merit system encourages motivation, is good for morale		36.4	25.2
2. Wages fit the job; it's rational		11.1	39.8
3. Other (I am young, and it favors the young, etc.)		7.8	5.6
	Subtotal	55.3	70.6
	N	(134)	(227)
The [present] seniority wage system is better because:			
1. It gives stability, security to worker as he gets older and his expenses mount; recognizes the importance of experience		25.2	8.4
2. Job control system doesn't really operate according to theory; isn't fair or consistent.		5.8	16.1
3. New job control system makes for competitiveness, makes human relations worse		2.1	3.1
4. Other reasons		11.6	1.9
	Subtotal	44.7	29.5
	N	(108)	(95)
	Total	100.0	100.1
	N	(242)	(322)
$C/C_{max} = .57^{**}$			

** Significant at the .01 level.

classed by their degree of skill, in grades one through five, so each worker has to compete, and this is good. I shall be in grade three when the system is set up."

Another Machine Section worker gave a more complicated reason for favoring the job classification system: "I favor wages based on job classification because at present a worker using a big machine is more likely to make mistakes, and he gets a pay reduction for that. Since the risks of losing pay are greater on the big machines, and since those jobs are more important, those workers *should* get more pay than men on small machines." It is management policy that men who operate large machines—e.g., the large rigide and milling and boring machines—receive higher pay than those who work on lathes and other small machines. A third Machine Section worker told us that many workers don't agree with the policy: "Workers think everyone should get the same pay, regardless of size of machine. At present, all workers want to do large machine work, to get higher pay, but they can't all do that kind of work; some have to work on small machines. Therefore, workers think it is most fair if all workers get the same pay."

Another axis of attitude difference is age. According to a 20-year-old Assembly worker: "Older workers get more pay, but we marry when we're about 27 or 28 years old, and from age 27 to 40 is when we do our best work. Therefore, pay should go up faster by age 27–28." From a Machine Sections worker: "Younger workers are more favorable to the new wage system. Older men don't like personal evaluation to be the basis of pay." A Personnel Department staff man said: "There are many complaints now from younger workers, who monopolize certain skills, but are low in seniority and therefore pay." An older Assembly worker, about to retire, admitted having encountered criticisms directly from younger workers: "Young workers today tell me, 'If you get twice as much pay as I do *you should work twice as fast*.' "

An Assembly hanchō anticipated that the change in the system of pay would have differential impact on various workers: "Personnel rationalization (*jinji gōrika*) here has been worked out as a policy, but not applied much yet. It's good, but as in any change, there must be some sacrifices. *Here the high seniority group will have to make the greatest sacrifices.*" Echoing this view, but from a more personal, subjective side, a Machine Section worker said: "I favor the new wage system, but how much will the company really practice it? One problem is that if only 'elite merit' workers are advanced, while the rest have trouble existing, it will be bad. Now everybody is treated as average. Advancement of the meritorious shouldn't be too much at the expense of the mediocre workers." An Assembly worker on the verge of retirement voiced the sentiments of the older worker: "I'm satisfied with my job, but dissatisfied with management's policy of lowering (or at least not raising) wages after age 42." Implicit in this complaint is the recognition that management knows that after age 40 it becomes steadily harder to improve one's wages by changing firms. The more seniority one has the more difficult is it to transfer it in toto to a new company. Thus, the company can afford to offer minimal increases after age 40 or so because it knows it "has" the worker.

VALUES CONCERNING THE DETERMINANTS OF PAY

To elicit employees' *values* concerning how pay *should* be determined, Electric and Shipbuilding factories' questionnaires asked:

> "Look at the following six factors that determine pay. Which do you think should be most important? Please rank these in order from 1 to 6 in the empty block.

1. Education
2. Number of dependents
3. Skill and job classification
4. Ability
5. Seniority
6. Contribution to company profit."

Responses were weighted and summed in order to give an aggregate ranking of these six factors in each factory.[11] The resulting ranking appears in Table 6.12. The best way to compare the two factories is to look at the percentage of the total weighted points for each of the six factors. Ranked in this way, the only difference is that the rank of the two lowest factors in Electric Factory—education and number of dependents—is reversed in Shipbuilding Factory. But the two factories' percentages are almost identical for each factor, suggesting a more general value pattern concerning the determinants of pay. Despite the differences in industry, technology, and social characteristics of employees between Electric and Shipbuilding factories, personnel in both firms agree pay should be based on ability, skill and job classification, seniority, and contribution to company profit, in that order, with education and number of dependents being evaluated as the least important determinants of pay.

Personnel in both factories are more modern than traditional in their values concerning pay. The highest ranked factors are the individual's ability and his skill and job classification, and the factor most irrelevant to his actual job performance—number of dependents—is at or next to the bottom in rank. At the same time, seniority has by no means been repudiated: it ranks third in both factories. One curious finding is that education is ranked so low. This is less surprising when we note that education is uncorrelated with job classification in Electric Factory ($r = .04$) and only moderately correlated in Shipbuilding Factory ($r = .30$). Thus, the respondents appear to be saying, "Pay should not be based on education as such, but on a person's actual skill and job."

The factor of contribution to company profit was ranked fourth. The meaning of this factor admits of several interpretations: a measure of company identification; a measure of ideological orientation (interests of

[11] The factor a person ranked as first (most important) was given a weight of 6, the second-ranked factor a weight of 5, and so on down to the sixth (lowest-ranked) factor, which was weighted 1. This weight was then multiplied by the number of respondents who gave a given factor a given rank. For example, of the 939 in Electric Factory who answered, 471 ranked "ability" first, 232 ranked it second, 130 third, 73 fourth, 21 fifth, and 21 sixth. The total (weighted) score for ability is therefore: $(471 \times 6) + (232 \times 5) + (130 \times 4) + (73 \times 3) + (21 \times 2) + (21 \times 1) = 4{,}788$. Since from 939 to 952 employees ranked each of the six factors, the theoretical range of scores is from 0 to between 5,634 (939×6) and 5,712 (952×6) for each factor.

TABLE 6.12
Employees' Ranking of Determinants of Pay

Electric Factory	*Number of Employees who Ranked Factor*							
Factor	*First*	*Second*	*Third*	*Fourth*	*Fifth*	*Sixth*	*Sum of Weighted Scores*	*Percent*
Ability	471	232	130	73	21	21		
	(2,826)	(1,160)	(520)	(219)	(42)	(21)	4,788	24.1
Skill, job classification	217	349	192	108	58	20		
	(1,302)	(1,745)	(768)	(324)	(116)	(20)	4,275	21.5
Seniority	129	166	248	239	124	40		
	(774)	(830)	(992)	(717)	(248)	(40)	3,601	18.1
Contribution to company profit	74	91	152	172	249	208		
	(444)	(455)	(608)	(516)	(498)	(208)	2,729	13.7
Education	45	76	140	227	184	266		
	(270)	(380)	(560)	(681)	(368)	(266)	2,525	12.7
Number of dependents	16	33	79	120	303	390		
	(96)	(165)	(316)	(360)	(606)	(390)	1,933	9.7
Total	952	947	941	939	939	945	19,851	99.8

Shipbuilding Factory	*Number of Employees who Ranked Factor*							
Factor	*First*	*Second*	*Third*	*Fourth*	*Fifth*	*Sixth*	*Sum of Weighted Scores*	*Percent*
Ability	284	118	52	33	23	18		
	(1,704)	(590)	(208)	(99)	(46)	(18)	2,665	23.9
Skill, job classification	69	176	155	63	48	19		
	(414)	(880)	(620)	(189)	(96)	(19)	2,218	19.9
Seniority	74	105	128	127	66	29		
	(444)	(525)	(512)	(381)	(132)	(29)	2,023	18.3
Contribution to company profit	74	82	81	88	105	97		
	(444)	(410)	(324)	(264)	(210)	(97)	1,749	15.8
Number of dependents	11	27	58	116	161	159		
	(66)	(135)	(232)	(348)	(322)	(159)	1,262	11.2
Education	19	23	57	104	125	204		
	(114)	(115)	(228)	(312)	(250)	(204)	1,223	10.8
Total	531	531	531	531	528	526	11,140	99.9

Electric Factory vs. Shipbuilding Factory: sum of weighted scores of six factors that should determine pay: C/C_{max} = .06**

NOTE: Numbers in parentheses are weighted scores. The factors ranked first were given a weight of six, the factors ranked second a weight of five, and so on.
** Significant at the .01 level.

labor versus capital); a measure of job performance (employees who have contributed much to profits have probably distinguished themselves by their innovativeness). Therefore we cannot be sure what this factor's relatively low rank means.

A rough comparison can be made between employees' value ranking of five of the six determinants of pay and the actual importance of these factors, as measured in terms of product-moment r's (see Tables D.8 and D.9). In this comparison (Table 6.13), ability was ranked in the employee values question, but we lack data on an analogous actual determinant of pay, so this is omitted. Of the five factors compared, four have direct analogues; the fifth, "contribution to company profit" is equated with the actual determinant, *performance*. The largest discrepancy between the two rankings in Electric Factory is that number of dependents is three ranks higher as an actual determinant of pay than in terms of employees' value preferences. Job classification, on the other hand, is two ranks lower as an actual determinant than as a value preference. The other three factors differ by only one rank in the two rankings.

TABLE 6.13
Employees' Values Concerning the Determinants of Pay and the Actual Importance of these Determinants

	Rank of Factor as Actual Determinant of Pay	*Correlation with Pay*	*Rank of Factor According to Weighted Employee Values*	*Weight*
Electric Factory				
	1. Seniority	.82	1. Skill, job classification	4,275
	2. Number of dependents	.80	2. Seniority	3,601
	3. Job classification	.72	3. Contribution to company profit	2,729
	4. Performance	.56	4. Education	2,525
	5. Education	.14	5. Number of dependents	1,933
Shipbuilding Factory				
	1. Seniority	.69	1. Skill, job classification	2,218
	2. Number of dependents	.48	2. Seniority	2,023
	3. Performance	.29	3. Contribution to company profit	1,749
	4. Education	−.13	4. Number of dependents	1,262
	5. Job	−.04	5. Education	1,223

Thus, if a change in the actual determinants of pay were to take the direction preferred by employees, number of dependents would become a considerably less important cause of pay differentials, whereas job classification would become more important. The other changes that would have to occur to bring the actual determinants of pay more in line with employees' values than they were in 1969 would be for seniority to become somewhat less important, though by no means unimportant, and performance (contribution to company profit) and education to become somewhat more important. The most important thing about this finding is that it corresponds quite closely to what management's new job classification pay system is intended to accomplish over the next several years. In short, if Electric Company's new pay system works as it is intended to in the future, it will have the function of bringing the actual determinants of pay more in line with employees' relatively modernized values concerning how pay should be determined.

The results for Shipbuilding in Table 6.13 show the same overall degree of discrepancy between the rankings of actual determinants of pay and those for employees' values as in Electric Factory. But the discrepancies relate to the factors somewhat differently. The major discrepancy is that one's job is the least important of the actual determinants of pay, whereas it is the most important factor in employees' value ranking. Number of dependents is two ranks higher as an actual determinant than in employees' values. The other factors—seniority, education, and performance—differ by only one rank, or have the same rank, as actual determinants and as valued determinants.

In Shipbuilding Factory, then, if the actual determinants of pay were to shift in the direction of employees' values, the greatest change this would entail would be a large rise in the relative importance of skill and job classification, and a decline in the relative importance of number of dependents. Seniority and education would each decline slightly in relative importance as determinants of pay, and performance (contribution to company profits) would retain its present relative importance.

In short, in both factories, employees' overall values concerning pay are more modern than is the present structure of the actual determinants of pay.

We next made an interest group analysis of the ranking of the determinants of pay. Three of these—number of dependents, education, and seniority—are used for each factory in Table 6.14. A vested interests model would predict that employees whose own status is high on a given one of these factors that should determine pay are more likely than those whose status is low on the factor to rank it high as a preferred determinant of pay. Thus, employees with university education should be more likely than those with only middle school education to rank education high as a determinant of pay; those with dependents would rank number of dependents high more frequently than those with no dependents; and so forth.

Table 6.14 shows our test of this interest group hypothesis; we found it grossly wanting. Those with more dependents do not accord that factor more primacy than those with no dependents; one's education does not

TABLE 6.14
Relationship between Employees' Status and Their Value Ratings of Analogous Status Factors as Determinants of Pay (%)

Employee Ratings	*Employee Status*						
Electric Factory	*Actual Number of Dependents*						
Number of dependents should be:	*None*	*1–2*	*3+*				
Factor 1 (most important)	1.8	2.0	0.7				
Factor 6 (least important)	40.2	42.4	45.3				
	Actual Education						
Education should be:	*Middle School*	*High School*	*University*				
Factor 1	5.9	2.7	2.0				
Factor 6	28.8	26.9	28.3				
	Actual Seniority (years)						
Seniority should be:	*0–1*	*2–3*	*4–7*	*8–11*	*12–15*	*16–23*	
Factor 1	14.5	18.1	7.5	8.7	12.6	13.4	
Factor 6	6.6	3.1	5.7	4.4	2.7	2.3	
Shipbuilding Factory	*Actual Number of Dependents*						
Number of dependents should be:	*None*	*1*	*2*	*3*	*4*	*5+*	
Factor 1	1.8	0.0	3.1	1.4	2.1	5.3	
Factor 6	28.5	31.7	26.5	33.1	34.0	31.6	
	Actual Education						
Education should be:	*Middle School*	*High School*	*University*				
Factor 1	3.6	3.8	0.0				
Factor 6	41.1	39.2	23.7				
	Actual Seniority (years)						
Seniority should be:	*0–1*	*2–3*	*4–7*	*8–11*	*12–19*	*20–27*	*28+*
Factor 1	13.2	10.8	17.6	9.7	15.6	11.8	7.7
Factor 6	1.9	13.5	7.6	8.2	3.9	2.8	7.7

predict one's propensity to rank education high or low as a determinant of pay; those with more seniority do not say that factor should have more weight in determining pay than do those with less seniority. The only support for the interest group model is that in Shipbuilding Factory, the more education one has, the less likely one is to rank education lowest (sixth). Even here, however, the more highly educated are not more likely to say education should be ranked first as a determinant of pay.

The relatively poor fit between these findings and an interest group model suggests an alternative model: a common value system model (Parsons 1964). Employees are in such relatively high agreement as to the ranking of the factors that should influence pay that this common value orientation overrides personal and vested interests. To explore further whether there are structured cleavages between managers and workers, between older and younger workers, and between men and women, as to the relative importance of the six factors that "should" determine pay, we used the same weights for each of the factors that should determine pay and computed the rankings for these specific subgroups.

First, we compared managerial-supervisory personnel and rank-and-file personnel. In Electric Factory the only difference in the way the determinants of pay are ranked is that contribution to company profit is third and seniority fourth among managers, while these two ranks are reversed among rank-and-file personnel. Otherwise, the two subgroups rank factors in the same order, thereby expressing common values. In Shipbuilding, both managers-supervisors and rank-and-file personnel rank ability and job classification first and second, respectively. For managers and supervisors the other four ranks are: contribution to company profit, seniority, education, and number of dependents, while for rank-and-file personnel they are seniority, contribution to company profit, number of dependents, and education.

Second, we compared how the youngest employees (15–17 in Electric Factory, 32 years old or less in Shipbuilding[12]) and the oldest (over 38 years old in Electric, 33 years old and over in Shipbuilding) ranked the six factors. In Electric Factory these two extreme age groups yield the same rank order for the six factors, while in Shipbuilding Factory there is only one set of discrepancies. Both younger and older employees rank the first four factors in the same order as Shipbuilding employees as a whole did (Table 6.12). They differ in that younger employees rank number of dependents fifth and education sixth, while older employees reverse these two ranks.

Third, where there were enough personnel of both sexes—in Electric Factory—we compared men's and women's rankings. The sexes rank the six determinants of pay in exactly the same order.

These findings in general lend more support to the common value system model than to the vested interest model, in both Electric and Shipbuilding factories. Employees share, to a large extent, a common value system concerning how pay should be determined.[13] These common values yield a hierarchy in which ability is most important, skill and job classification second in importance, seniority third, contribution to company profit fourth, and education and number of dependents either fifth and sixth, or sixth and fifth, respectively. This pattern remains quite stable, despite the counter-influences of the interests that derive from employees'

[12] Because there are so few employees in their 'teens and early twenties in Shipbuilding Factory, the absolute age categories cannot be exactly the same as in Electric Factory.

[13] This does not mean, necessarily, that these employees have common values in other areas of factory social organization. We shall return to this issue in Chapter 9, in the discussion of conflict and integration in each factory.

differential education, seniority, number of dependents, rank, age, and sex.

The one strong evidence of vested interests appears in Table 6.10, where the job classification interest group differed in ways that would be expected from the seniority interest group. Yet these Electric Factory employees expressed common values (as did Shipbuilding employees) concerning which factors should determine pay. More specifically, both those whose vested interests lead them to favor the job classification wage system and those whose vested interests lead them to favor the seniority wage system essentially agree that one's ability should be the main determinant of pay. This at first curious finding suggests that whereas sociological theory might view a job classification wage system as emphasizing ability as a prime determinant of pay more than does a seniority wage system, in the eyes of our Electric Factory respondents this is not the case. We confirmed the fact that proponents of both wage systems are essentially equally likely to say ability should be the main, or one of the three highest-ranked of the six factors that should determine pay. Thus, 47.7 percent of the advocates of the seniority wage system and 51.4 percent of the advocates of the job classification wage system think ability should be the first determinant of pay; 23.3 percent of the former and 29.7 percent of the latter think it should be the second most important determinant; and so on. The pattern is similar in Shipbuilding Factory. In other words, employees who favor the seniority wage system evidently do *not* conceive of themselves as denying that the "modern" criterion of ability should be a determinant of pay; a seniority wage system is not seen by the employees themselves as inimical to having ability be the preeminent criterion of pay.

CONCLUSION

The seniority wage system is a major component of the paternalism–lifetime commitment model of Japanese industrial firms. Seniority has in fact been the main criterion of wage policy in the large firms, and continues to be one of, if not the, major actual determinants of pay in the three firms we studied. However, a number of qualifications must be made if we are to understand wage systems in Japan correctly.

Sake Company is the most traditional of our three firms in its social organization, yet in principle it does not have a seniority wage system. Pay is formally based on skill and position in the job hierarchy, and Sake respondents believe wages *should* be based on job differences.

Shipbuilding Company's wage system is in a state of transition from a full-blown seniority system to one that accords increased weight to job classification and skill as determinants of pay. In terms of the three elements of the wage system we have analyzed—policy, the actual determinants of pay, and employees' attitudes and values concerning pay—Shipbuilding Factory is clearly more traditional than Electric Factory. A smaller proportion of Shipbuilding Factory employees than of Electric Factory employees prefer a job classification wage system. Moreover, as actual determinants of pay, job classification and performance are more important relative to seniority (the most important determinant) in Electric Factory than in Shipbuilding Factory.

Electric Company's new job classification system has not yet basically altered the previous seniority system as it affects the actual distribution of pay. As of 1969 there had been no radical break away from seniority and toward job classification as determinants of pay. In part, this is because the previous wage system was not solely a seniority system: formal education as well as actual job performance were also considered in determining pay. Just as the previous system had these modern elements, so the new job classification system retains some traditional elements. In part this is also because in the early stages of the institutionalization of the new pay system essentially the same employees are high (or low) in both seniority and job classification.

Yet we would not go so far as those who argue that the new job classification wage systems in Japan "are little more than seniority categories under another name" (Abegglen 1969: 113). Insofar as employees' attitudes and values constitute dynamic factors with regard to the future of wage systems, both Electric and Shipbuilding companies will face pressures actually to institutionalize job classification as a major basis of pay. As a first-line foreman with long seniority in Electric Company said: "Today there are seniority elements in wages, but in two or three years this will give way to job classification. It's rational in terms of craftsmanship [to pay according to job]. I'm dissatisfied with this, but it's inevitable." In Electric Factory the new job classification system is preferred more often than the previous system. This is true for employees in general but especially for higher-status employees. In rating the importance of factors that should determine pay, employees in both Electric and Shipbuilding factories rank ability and skill or job classification above seniority. Moreover, this ranking pattern appears to reflect a common value system in that differences of rank, age, sex, etc. rarely alter the order in which these factors are ranked.

Given all these considerations, including the way in which company policy describes the new job classification system as operating in the future, we agree with Kaneko (1970) and others that seniority will over time become somewhat attenuated, relative to job classification, as a determinant of pay in Japanese industry. The "inner logic of modernization"—more specifically, such changing external functional exigencies as the labor market, the educational upgrading of workers, and the shift to automation—would seem to point in the direction of this change. Tsuda reports that the job classification wage system "is well liked in the steel and electric machinery manufacturing industries and is becoming increasingly popular in other industries" (in Okochi, Karsh, and Levine 1973:436).

CHAPTER SEVEN

The Reward System: Promotion

By "promotion" we mean the movement of employees upward (downward would be called "demotion") in the formal hierarchies of job position (Sake Company), job classification (Electric Company), and rank (Electric and Shipbuilding companies). As an employee moves up in these hierarchies, his pay increases. But advancement means also increasing authority over others, a greater voice in decision making, and greater status and honor. To be promoted is intrinsically satisfying and also means that one has measured up well in terms of achievement value standards. This chapter will analyze both the objective and the subjective aspects of promotion, as an element in each firm's reward system.

CRITERIA FOR PROMOTION

SAKE COMPANY

In general, promotion in Sake Company is based upon skill, which in a craft industry like sake brewing is to a large extent a function of number of years' experience. As one returns year after year to work in a sake plant, one gradually moves up the hierarchy of positions. Promotion is perceived as a strong inducement to bring workers back year after year. As the master of Plant Number 2 told us: "If the worker is offered a higher status [for the next year] he's more likely to come again."

Promotion is also regarded as highly dependent upon the judgment of the master. In 1966 workers reported that promotion in sake plants depends on personal and community relationships. Where the master takes charge of personnel administration, one's relationship with him is a dominant factor in promotion. A 32-year-old farmer employed in Sake Company recognized this pattern: "When I was in the fourth grade of middle school [equivalent to tenth grade in the United States] I began to work at sake plants every winter. At first I worked at [another sake plant] in Kakogawa City, Hyogo Prefecture, but I was dismayed with the hard work there and the poor personnel administration. The position of master and other employees were filled by farmers who had come from one village" (Mannari, Maki, and Seino 1967:29–30).

The height of ambition in the sake industry is, of course, to become a master. The following are the main criteria of promotion to master. First, experience in every aspect of sake brewing. Second, ability to lead workers, a criterion especially emphasized by owners. Third, the recommendation of some owner, scientist, or brewing companies' association. In the words of one who had become a master, the master of Plant Number 2: "The qualifications for becoming a master have not changed. He must be experienced in everything—malt, boiling, washing tools. Seniority—this used to be more strict. Whether he has leadership and 'controlling power' or not. He must have the recommendation of the scientist, the masters' association, and the brewing company's owners. Younger men don't make an effort to fill these qualifications. It's difficult to train our successors." Throughout there is the underlying belief expressed by this master as follows: "No matter how much mechanization, sake is always made by men."

Several decades ago, the fact that a person was invited by the owner of a sake plant to come during the summer to clean the instruments meant that he would be appointed a master for the next season (Nihon Ginkō Chōsa-Kyoku 1931:20–21). Now the pattern has changed in that the master is a year-round employee of the company, although in the summer he still spends considerable time in the rural areas, recruiting.

The stages in the promotion of masters can be illustrated from the 1966–67 interviews with the master of Number 2 Plant. Born in 1925, he started brewing work at the age of 16 in Kobe. The first winter he was a cook (*meshitaki*), the next two he did malt work, developing the bacteria in the malt and preventing other bacteria from developing in the malt room:

> Then I went to Mukden [Manchuria] to be a sake brewery worker. A friend recommended me to work there. I became head of malt making there. After the war I worked two winters in this company's Kobe plant,[1] then I moved to this factory and was an upper worker (jōbito) for one year, an instrument caretaker (dōgu mawashi) for two years, a malt maker for one year, a subhead (daishi) for two years, and a head (kashira) for eight years. In 1959 I became a master (tōji).
>
> Ever since I was young I wanted a responsible position. When I got it I tried hard to make better sake. I wanted praise from my superior. This is how I became a master. It is necessary to have the attitude of wanting to learn. But young workers now lack this attitude.

The master of Plant Number 1 had a similar story:

[1] Since closed down.

> At the age of 16, I started working in a sake plant in the lowest rank and helped cooking. Later I became a middle worker (chūbito) and worked that way for two winters. I had no responsibility except to assist various other workers. Then I became an upper worker. My chief work then was separating sake from malt. Later I became a tool man and then a malt man. I did malt work for ten years, then became a subhead and malt master, then head and then master. I worked 12 years as a head. Finally in 1951 I became a master.
>
> I had my army service in 1925 in Tokyo and in 1937 in China, but except for that I have worked in sake brewing a total of 44 winters, missing only those two winters.

The same master had this to say of the requirements for promotion to master. He had been adopted by a master, but it nevertheless took him a long time to climb the ladder to the position of master:

> I was rather old when I became master, 45 years old. Without seniority, good skills, and the management ability to organize his crew, a man can't be recommended as master in the Masters' Association in the Sasayama area, and therefore can't be hired as a master by an owner. Among men eligible to become master, it seems that recommendations and promotions are a good deal influenced by personal and community factors. The community factor is more important in the case of becoming a master in this sake company, because the man has to be a member of the Masters' Association in the Sasayama area. All Sake Company masters are members of the community craftsmen's association, but they are not recommended on the basis of their being the son of a master. My father was a master in the Nada area, but in a different sake company.

We noted that in traditional plants, where the master has a major say in promotions of lower ranked workers, personal, familial, and village ties with him are important, and that some workers complained about this system. Since the master's position has been eliminated in the new, automated plant, this complaint is less relevant there. But this does not mean that workers there have uniformly positive attitudes toward the promotion system. Mannari, Maki, and Seino (1967:19) found that some workers in the new plant who had long experience in the traditional plants criticize the promotion system in the new plant, saying: "No matter how hard and successfully we work, our work achievement is not evaluated properly in the new plant. In the traditional plant, the harder we worked, the more we were promoted."

ELECTRIC FACTORY

In Electric Factory promotion can occur in two hierarchies—rank and job classification. These are related as follows:

Rank	*Job Classification*
Division director (jigyōbuchō)	
Associate division director (fuku jigyōbuchō)	
Department head (buchō)	
Associate department head (bu jichō)	
Section chief (kachō)	
Subsection chief (kakarichō)	C-5 or O-6
Second-line foreman (hanchō)	C-4 or O-5
First-line foreman (kumichō)	O-4
Rank-and-file workers	R-3
	R-2
	R-1
	T-3
	T-2
	T-1

There is some evidence that the criterion for promotion has been changing from seniority toward merit in recent years. As a staff person in Employee Training put it: "Promotions used to be by seniority only. Now they are by examination, and even more importantly, by the evaluation of one's superior. Promotion didn't formerly work according to merit always. Nowadays it is done more and more by merit." According to a kakarichō in Number 1 Production Control: "If promotion is done on the basis of recommendations, personal bias comes in, so examinations are better."

In general, advancement at the lower levels—from T-1 to T-2 to T-3—is more automatic than at higher levels. Workers who enter in a given year can all expect to be T-3 by their third year and R-2 in their fourth year. From then on the cohort will advance with greater individual variation to R-3 and then to the higher levels. Within the Operative and Creative job classification levels, advancement is a function not so much of years' experience, as of superiors' evaluations and formal examinations. Let us describe this system.

Promotion to Kumichō. Eligibility for promotion from rank-and-file worker to first-line foreman involves being over 25 years old, with more than two years' experience as an O-4 or in the C-4-1[2] job classification. Those eligible are also judged by (1) their kumichō's monthly report on their work improvement; (2) the buchō's and kachō's personal evaluation;

[2] The second digit in Creative (C-4-1, C-5-2, etc.) is described more fully below.

(3) a written examination, given by the Employee Training Section, testing general knowledge; and (4) a written examination in electricity and mechanics. Screening is important even at this level. As a hanchō in Hardware told us: "There are many workers, but very few are qualified to become kumichō. I advise the kakarichō and kachō who would be a good person to promote to kumichō."

Promotion to Hanchō. To be eligible for the position of hanchō, one must be a kumichō and have had more than one year's experience as an O-5. A C-4 person is also eligible. A few of those who are eligible are chosen to go to company headquarters for a period of special training. The written examinations are quite rigorous: in May 1969 seven of the 35 kumichō in Electric Factory took the hanchō examination, and three passed. (Those who did not take it in 1969 were either ineligible on one or more of the above mentioned formal requirements, or were not recommended by their buchō and kachō.)

Promotion to Kakarichō. One must have been an O-6 or a C-5 for at least one year to be eligible. One must not have received punishments for at least two years, nor had a long leave of absence. Among these candidates, those in Creative jobs are judged mainly by their skill, those in Operative jobs mainly by their supervisory ability. Selection is made initially by an evaluation committee in the local factory, on the basis of one's supervisor's and the division head's evaluations, and finally, on the basis of written examinations, by the Personnel Department in company headquarters.

Promotion to kakarichō requires skills usually developed as a result of formal education—general and technical knowledge and facility in writing reports. In the line, candidates for kakarichō are already hanchō, but since kakarichō is the lowest supervisory rank for staff people (Creative or C job classification), applicants for staff kakarichō rank need not be hanchō. The written examination for kakarichō is taken at company headquarters, and covers general knowledge, mathematics, and the Japanese language. Typical questions in this examination are, "What have you done in the last year?" "What problems have you had and how have you solved them?" Those who pass this test are given special training for kakarichō duties. But many workers find their promotion blocked:

> I have been a hanchō since 1959 [now in Hardware]. I took the examination for kakarichō but failed it. It requires higher studies. Now, promotion requires intelligence concerning technology. I am all right in skill, but my general knowledge and report writing are not good. I can master knowledge of the job, but in the future it is necessary to have general knowledge to be a kakarichō. Our work still requires skill. We

try hard, but from management's point of view, school graduates are better.

College graduates who entered Electric Company before [1958] didn't have to take examinations to become kakarichō. Now—for example in 1969—400 took the company-wide kakarichō exams, and only 20 passed.

A hanchō in Number 1 Production declared: "I took the kakarichō examination two years ago, but failed it because it needs someone who is a high school graduate, or even with some college." The results may even mean demotion, as in the case of a Hardware worker who had been a kumichō (O-4) but had been demoted to a regular worker, with an R-3 job rating: "While I was kumichō my written reports were poor. Mr. N. [the factory manager] wrote critical comments on them, so I couldn't sleep nights. *I resigned as kumichō.*" This man had entered the company in 1949, after middle school. His age and seniority cohort are now mostly hanchō or kumichō in the factory. He was and is a skilled, hard worker, and both superiors and subordinates like him. But his lack of ability in report writing brought about his decline in status. Note that he resigned rather than being formally demoted. From the way he talked about his experience, it appears that he accepted the company's criteria for supervisory positions as legitimate ones.

Promotionfrom C-4 to C-5. Table 7.1 shows that within both the C-4 and the C-5 groups there are five subdivisions, designated as C-5-5, C-5-4, and so on. When first introduced in 1968 these subdivisions meant years: a person who became a C-4-1 in a given year could expect to reach C-4-2 the next year, and so on, up to C-5-5. In 1969, however, this advancement was made less automatic: the subdivisions became levels of skill, to be determined by the supervisor's evaluation of performance. For example, performance and motivation (*yaruki*) were evaluated in deciding whether a worker deserved promotion from C-4-1 to C-4-2; these two factors and leadership were the criteria for promotion from C-4-3 to C-4-4 and to C-4-5; promotion from C-4-5 to C-5-1 was to be based on a company-wide examination; and promotion from C-5-1 up to C-5-5 was based increasingly on leadership qualities. Since these were all achievement-based promotions, not everyone would advance at the same rate. Electric Company's table for these variations is shown in the last three columns of Table 7.1. C-4-5 employees are evaluated for promotion to C-5: those who pass this examination become C-5-1; those who do not will remain C-4-5.

Promotion to Kachō. Employees with at least four years' experience as line or staff kakarichō (including C-5-3 people) are eligible to take the examination for kachō. If they pass, they spend one year as probationary

TABLE 7.1
Time Requirements for Promotion at Different Skill Levels, Creative Employees in Electric Company

Job Classification	Degree of skill	Number of Years in Skill Level before Promotion: Minimum	Standard	Maximum
	5			
	4	3	5	
C-5	3	3	5	
	2	2	4	
	1	2	4	
	5			
	4	1	2	3
C-4	3	1	2	3
	2	1	2	3
	1	1	2	2

kachō; then the decision is made whether to appoint them as full-fledged kachō.

To understand how employees feel about their promotion chances it is important to know what their reference groups are. We were told that the opportunity structure is not as good in the Hardware Section as in the assembly line sections. A Hardware worker who takes the assembly line sections as his reference group had this to say: "Now it takes 15 years [in Hardware] to become a kumichō. There are more senior workers in Hardware than in the assembly line sections, so our chances of promotion are less." There is also a much smaller proportion of males in the assembly sections than in Hardware, and this is an important factor in a setting where all supervisory personnel (kumichō and above) are male.

However, a kumichō in Number 1 Production—one of the assembly line sections—took as *his* reference group not Hardware, but one of the company's new color television plants. Since the factory we studied is the company's oldest, it has more senior workers eligible for promotion than a new division. Thus, the Number 1 Production kumichō could regretfully say that he had had to spend 10 years in the position of kumichō. Although his objective chances may have been better than those of his counterparts in Hardware, subjectively he compared himself (unfavorably) with his counterparts in the newer color television plant, with its even better opportunity structure.

General Nature of the Promotion System. Electric Company's promotion system strikingly combines impersonal and personal criteria; it also combines achievement and ascriptive criteria. With regard to the latter, no matter how outstanding a worker's performance, he must, for example, be 25 years old and have at least three years' seniority in Electric Company before he is eligible for promotion to kumichō. With regard to the first set of criteria, he must exhibit certain universalistic cognitive standards, as shown in impersonal written examinations on both general and technical knowledge. But since personal relationships with those already in managerial positions (especially high positions) are functionally important, the system operates in such a way that a candidate eligible on other grounds but not satisfactory to his buchō and kachō will not be allowed to take the universalistic examination in the first place. The examinations are decisive, and therefore no one is recommended to take them—since he might pass—who would pose a problem to higher management if he were promoted. In other words, impersonal and universalistic achievement standards do operate in the promotion system, but only after certain ascriptive and personal-social relationship factors have operated as preliminary filtering mechanisms.[3]

Shipbuilding factory

We shall discuss first the general promotion system in Shipbuilding Factory and then the specific criteria for promotion to different levels.

Promotion in General. The fullest account of how the promotion system operates in Shipbuilding Company was provided by an Assembly Section kakarichō:

> First, the qualifications for promotion to hanchō, etc. are not age, but rather seniority and merit. The Machine Section kakarichō, for example, is only 32 years old, younger than several of his hanchō and sagyōchō. The second factor is the quality of job performance: not paper test results, but our observation of his long-term performance. We classify workers in a han as upper upper, upper middle, middle middle, etc. Only the upper half are considered for promotion. Third, we consider personality: do all sakite like him? Fourth, to some degree we also consider his private life. We don't interfere with it, but if he's absent a lot, we assume he has a bad private life. Fifth, the physical strength of the worker. Sixth, his ideology and way of thinking. We

[3] An extremely apt illustration of this combination of universalistic and particularistic criteria occurred at the very top of Electric Company. As in other large Japanese companies, the two highest positions are those of board chairman (*torishimariyaku kaichō*) and president (*shachō*). The founder and board chairman of Electric Company once wrote that the spectacular growth of the company was due in large part to the young people: "In particular let me say thank you to the young. Had it not been for the abundant energy of youth, the untiring efforts, the enthusiasm and power of the young, [Electric Company] could not have become the worldwide enterprise that it is today. . . . *To keep the company young I decided . . . to hand the presidency over to my younger brother . . . .*" At the time this promotion took place, the man who became board chairman was 65 years old, and his younger brother was 59 years old.

put all these factors together and rank the members of the han. The first-ranked on all these criteria are called *hanchō-daiko* [acting hanchō]. Then we consider which hanchō-daiko should become a full hanchō.

The company now favors promoting men at younger ages. In the past, by the time a man became a buchō he was usually near the 55-year-old retirement age. As of May 1969, in X Plant as a whole, there were four buchō, ranging from 44 to 49 years old. (Above buchō are higher ranks—kōjōcho, jigyōbuchō, and executives in company headquarters—for whom the 55-year-old retirement policy is waived.

When asked about the criteria for promotion, a personnel manager replied: "At present, there first has to be a vacancy. Then we consider seniority: even if the man has all the other qualifications he can't be promoted if he hasn't the seniority. But more poeple are always qualified on the basis of seniority than there are vacancies at higher levels. Therefore, we use the ratings made by supervisors, attendance, past career, etc., to make our decision."

Promotion to Boshin. Although boshin is only an informal rank—hanchō is the first formal rank to which one can be promoted—it is a meaningful first step in a worker's climb up the ladder. An Assembly sagyōchō described how promotion to boshin is decided: "The hanchō decided who should be promoted to boshin. The hanchō and the sagyōchō don't disagree on who it should be. After all, there are few workers with the right leadership qualifications for boshin. Also we have an age distribution with too many in their twenties and forties, but not enough in their thirties, which is the preferred age for boshin. We have several very skilled men in their forties, but they have a 'craftsman's mentality' and don't get along well with other workers, so they wouldn't make good boshin. So we have to promote from among the twenties age group." An Assembly hanchō recognized one of the functions of promotion to boshin, apart from selecting the most qualified person: "In General Assembly, after about three to five years' experience a worker has as much *skill* as anyone. If I make him a boshin, he'll work harder."

Promotion to Hanchō. An Assembly Section boshin outlined the stages in Shipbuilding Company's history concerning promotion to hanchō: "In the earlier period, older workers were preferred. For example, there are still some hanchō who are 45 or 50 years old and over in the tamakake-crane operator han. In the early 1960s middle school graduates became *kunrensei* (apprentices) for a three-year training period, after which they could rise to hanchō. Now, it is senior high school polytechnical graduates who have the advantage in promotion to boshin and hanchō."

From comments by several employees in their interviews, we can piece together the following criteria for promotion to hanchō. (1) The man must rank high in performance, e.g., be among the top boshin. The major responsibility for evaluation falls to the person's two immediate superiors—hanchō and sagyōchō in the case of a man being considered for promotion to hanchō, sagyōchō and kakarichō in the case of promotion to sagyōchō. The final decision about promotion is usually made by the kachō, though officially the final judgments even go up to the levels of buchō and kōjōchō. (2) The hanchō candidate must have had an outstanding attendance record during the previous three years. (3) Leadership ability is essential. This means a variety of things: knowledge of the job and ability to improve the manufacturing method in his section; agreement by everyone that the man is the right person; balanced personality. (4) Formerly a man had to be under 45 to become a hanchō; now he must be under 40. (5) Finally, for men recommended on the basis of the above criteria by their sagyōchō, there is a competitive examination, which covers knowledge of personnel management, drawing and reading blueprints, simple mathematics, safety and hygiene, and general knowledge. Since 1969, the examination has also covered basic English. To prepare for the English test, candidates study two hours a week for six months.

When we asked an Assembly worker why one of his fellow workers had been promoted to hanchō, his reply squared quite closely with the formal criteria outlined above: "He was promoted because he was industrious, had actual talent, was 31 years old, and had passed the hanchō examination."

Promotion to Sagyōchō. The criteria here are mainly extensions of those used for promotion to hanchō. For example, a worker about to retire had noticed the company's shift in policy from seniority toward leadership qualities: "The reasons for promotion to shokuchō [an earlier term for sagyōchō] have changed from skill, seniority, and whether workers obey the man, to a greater emphasis on his managerial ability."

Other Factors in Promotion. Chūto saiyōsha status and education are also considered in relation to promotion. We did not find a consensus as to whether employees recruited directly from school had preference over chūto saiyōsha for promotion. A Machine Section worker believed that: "*Yōseikō* (workers entering the company directly from school) used to have preference over chūto saiyō men for promotion. Now it's based more on merit, so this factor is less important." A Machine Section hanchō held, on the other hand, that: "Those people trained in the company are more likely than chūto saiyōsha to be promoted to hanchō. Even if two people have the same ability, the company prefers to promote from its 'cadet corps' [i.e., those hired directly after graduation and 'brought up' in the company]."

Educational background is relevant to promotion in several ways, of which the following is worth noting. An employee who had entered Shipbuilding Company as a polytechnical high school graduate found that the kunrensei—those who had entered after middle school graduation and served a three-year apprenticeship in the company—though his age, had developed higher practical technical skills by on-the-job training. A boshin in the Assembly Section, this man continued: "Therefore, the only way I could rise faster than they did was by working harder and by having better morale, more company loyalty."

Advancement from Production to Staff Work. Though not a promotion in rank, a shift from A Group (production) to B Group (staff) is an advance in status in the company. Thus, a hanchō with wide knowledge of the factory can become a staff employee; two did in December 1969. A sagyōchō may be promoted to a staff position in Engine Production Scheduling, thereby attaining a desk job away from the production floor. Two interviews illustrate this process. The first case is a Machine Section hanchō about to retire from the company after 34 years' service: "In 1943 I was promoted to kumichō in a new shop. In 1959 I was made a sagyōchō. This was my highest rank. In the last two years, with retirement approaching, I was demoted to hanchō. In 1964 I was assigned to staff work—in production process control. In 1965 I was in the A Group, but was performing B Group work. Since 1967 I have been 'treated as hanchō,'[4] doing staff work."

The company interviews those of its workers who have increased their education by going to night school, with a view to selectively promoting them from line production to staff positions. A 24-year-old staff employee in the Diesel Design Section said: "I entered the company in 1963 as a draftsman, after graduation from polytechnical high school. Since then, I completed Kinki Polytechnical University by going to night school."

Consequences of Promotion. As in any system, promotion in Shipbuilding Company is not an unmixed blessing to those who experience it. A Machine Section worker pointed out: "Workers and boshin may work up to 70 hours a month overtime, but hanchō may not work more than 30 hours overtime. Therefore, since the position allowance for hanchō is only ¥ 500 a month, to become a hanchō really means a big drop in pay." The Assembly kakarichō saw another sacrifice entailed by reaching higher rank: "If one takes a management position, he has to make more self-sacrifice. A rank-and-file worker can say, 'I don't want to move to another plant,' but a higher rank person must be always willing to move

[4] Rather than being classified as a hanchō, since that rank exists only in the production (A Group) sections.

from plant to plant, from plant to headquarters, from headquarters to a plant, etc."

THE DETERMINANTS OF JOB CLASSIFICATION

Having considered a number of factors that influence promotion, we should now deploy them in a multivariate analysis to discover more systematically which of them exerts the most influence upon the dependent variable, job classification. Unfortunately our data are inadequate for this. We only know an employee's *present* rank and job classification, not when he was promoted to that level. In principle we could obtain this information from company records, but our data on variables of presumed causal relevance to promotion are for 1968–69, not for the time prior to the most recent promotion. Hence, we do not know whether an employee's performance (attendance, suggestions, etc.) in 1968–69 was indicative of his performance *prior* to his promotion, and therefore causally relevant to it, or simply a result of his having reached a given job classification or rank. Keeping this limitation in mind, we shall present separate multivariate analyses of the determinants of three aspects of the promotion system: (1) promotion in job classification; (2) promotion in rank; and (3) employees' perceptions of their promotion chances. The first of these will be done only for Electric Factory, the other two for both Electric and Shipbuilding factories. Since we conceptualize these three aspects of promotion, along with advancement in pay, as elements of an organization's reward system, the same theory we developed to explain variations among employees in pay should also, with minor modifications, explain variations in job classification, rank, and perceived promotion chances.

When we treat job classification as a dependent variable, the independent variables are classified in terms of the same four conceptual domains, and the same theoretical reasons are adduced for the hypothesized relationships with job classification, as when pay was the dependent variable (see the preceding chapter). However, when job classification is the dependent variable, we shall use only 12 independent variables, in contrast to the 17 used in relation to pay.[5] We hypothesize that job

[5] Six of the 17 independent variables are dropped here: job classification (because it is now the dependent variable); rank (because it is somewhat redundant with job classification, and also because it tends to follow job classification rather than precede it); number of dependents, attendance, distance of community of origin from the location of the factory, and employee cohesiveness (because there is no evidence that these *independently* influence job classification, apart from the influence of variables retained in the regression). Number of previous jobs is here substituted for number of years in previous jobs. Finally, one variable is added: there was no evidence that the informational level of the section in which one works influences one's pay, but it is possible that this influences the opportunity structure and thus one's chances for promotion in terms of job classification.

classification varies positively with:

I. Bureaucratic influences
   1. Seniority
   2. Education
   3. Informational level of section (higher informational level sections have a larger proportion of high job classification positions)
   4. Performance

II. Cultural influences: present extra-organizational statuses
   5. Age
   6. Sex (men higher than women in job classification)

III. Informal influences: social integration of employee into company
   7. Participation in company recreational activities
   8. Paternalism
   9. Lifetime commitment

Finally, we hypothesize that job classification varies independently of:

IV. Previous extra-bureaucratic statuses' influences
   10. Father's occupation
   11. Size of community of origin
   12. Number of previous jobs

As before, the form of our hypotheses follows the lifetime commitment model fairly closely in that the determinants of rewards in Japanese organizations are seen as a result of both bureaucratic and cultural influences. The model implies that if employees are equal in terms of seniority, education, and performance, those who exhibit greater lifetime commitment and other indications of social integration into the firm will be more likely to be rewarded. As should be clear, we think the model exaggerates the influence of these factors. Therefore, in the specific area of rewards having to do with job classification, we are less persuaded than the proponents of the model that social integration variables will be important independent variables. The counter-hypothesis is that the influence of lifetime commitment and other social integration variables on job classification will reduce to zero when status and performance are held constant.

Multiple regression analysis revealed that the most important determinants of job classification are seniority and the informational level of the section in which one works; the more seniority one has accumulated, and the higher the informational level of one's section, the higher one's job classification.[6] Of these two independent variables, seniority is so much more important than section informational level that seniority alone

[6] See Appendix D, Table D.10, p. 379 for the details of this analysis.

predicts job classification almost as well as all 12 original independent variables together! That 65 percent $(.81)^2$ of the variance in job classification is accounted for by seniority indicates quite clearly how closely the new job classification system is tied in with seniority, at least at present.

Performance in fact has relatively little influence on job classification, independent of the influence of seniority and the section in which one works. This outcome was adumbrated in our earlier discussion based on interviews. We also find that the strong correlation between sex and job classification is greatly reduced when seniority and the other variables are controlled, indicating that women are low in job classification mostly because they are low in seniority, are young, and work in low informational sections, rather than because they are women as such.

THE DETERMINANTS OF RANK

We hypothesize that the same variables have the same types of relationships to rank as to job classification and pay. Those having positive relationships to job classification are now hypothesized to have positive relationships to rank; variables hypothesized as having no relationship to job classification are again viewed in this light.

A comment is in order as to why we predict a positive relationship between age/seniority and rank. The attainment of higher ranks in most Japanese organizations occurs vertically within the organization rather than horizontally (from other organizations). If we were asking, "What determines promotion from rank and file to first-line foreman?" our hypothesis would be that there is a curvilinear relationship between age/seniority and reaching that rank. After a worker reached a given age, say 40, if he had not been promoted to first-line foreman, it would be unlikely that he ever would be. However, our dependent variable is the entire spectrum of ranks, and this makes it appropriate that we hypothesize a positive rather than a curvilinear relationship between age/seniority and rank.

In Electric Factory, the three best predictors of rank are number of previous jobs, seniority, and education.[7] When other hypothesized variables are held constant, employees are most likely to be in higher ranks if they have more seniority and education and have worked in other firms prior to coming to Electric Company, but are nevertheless still young. This finding disconfirms our hypothesis that there is a positive relationship between age and rank: the relationship is in fact negative. Our findings partially fail to support the paternalism–lifetime commitment model in

[7] See Appendix D, Table D.11, p. 380 for the details of this analysis. See Table D.12, p. 381, for a comparison of the determinants of job classification with those of rank.

that neither paternalism nor lifetime commitment exerts more than a marginal influence on rank.

The influence of sex on rank is reduced to virtually zero when other variables, e.g., seniority, are held constant. This suggests a modern pattern in which sex *per se* is not allowed to have a strong influence on promotion in rank. However, strictly speaking, all this finding indicates is that so few women have accumulated high seniority that they have not been promoted to supervisory or managerial ranks, and that many men are also in rank-and-file positions. It does *not* necessarily mean, however, that if there were a larger proportion of women with high seniority, some of them would have been promoted. In this sense, sex is still a crucial determinant of rank, at least indirectly, in Electric Factory.

The finding that the relationship between performance and rank is also reduced to zero when seniority and other variables are controlled indicates an analytically traditional pattern. Rank does not appear to be a reward for performance, at least, not for performance as measured independently of seniority. After number of previous jobs, seniority, and education have been taken into account, the other nine independent variables add nothing further to the explanation of differences in rank.

The variables hypothesized as explanations of rank in Shipbuilding Factory are the same as those in Electric Factory. But there are two changes in the variables that will be tested: sex is omitted in Shipbuilding, because there are too few female employees, and the recruitment channel variable, for which we have data only from Shipbuilding, is added. The lifetime commitment model implies that employees recruited by school recommendation and hired directly from school are more likely to be promoted than are those recruited by more impersonal means. We predict, on the contrary, that recruitment channel will have no effect on rank, when other variables are held constant.

The multiple regression analysis (see Appendix D, Table D.13, p. 382 for details) can be summarized as follows. Hypotheses 1, 2, and 4 are confirmed: when other independent variables are held constant, rank is positively related to education, seniority, and performance. Hypotheses 7–9 are largely disconfirmed: social integration into the company has little influence on rank, independent of performance, seniority, and education. All other variables account for little or none of the variance in rank. This negative finding supports hypotheses 10–12: size of community of origin, father's occupation, and the number of previous jobs (as well as recruitment channel) exert virtually no influence on rank. This reflects partly the operation of counteracting forces that stem from these variables—some of which enhance rank while others have the opposite effect—and also the fact that as a bureaucratized organization, Shipbuilding Factory structures its promotion system so that it is formal criteria—performance, seniority, and education—rather than social origins, recruitment channel, and number of previous jobs that are more decisive in determining rank. It is functional for the organization when rank is based more on actual performance than on how one was recruited or how many previous jobs one had. Whether an employee had no previous jobs or ten, it is performance in the present firm that constitutes the proof of the pudding as far as rank is concerned.

We saw in Chapter 6 that in Shipbuilding Factory performance did not influence pay independently of age and rank. It does, however, influence rank independently of seniority and other major predictors. This suggests that an employee's performance is taken more seriously in deciding his rank (and, by inference, his promotion) than in determining his pay. The difficulty with this interpretation, of course, is that the relationship between performance and rank is cross-sectional, not longitudinal. Our interpretation is that some employees are in higher ranks than others, *ceteris paribus*, because their *past* performance was more outstanding, and there is anecdotal evidence for this from interviews. But it should be clear that our questionnaire data and multiple regression analysis cannot reject an alternative interpretation: certain employees have better performance *because* they are in higher ranks; it is attaining a higher rank that encourages one to improve one's performance, not the other way around. In the last analysis, both interpretations are probably true: certain employees are promoted because of their outstanding performance, and, having been promoted, continue to perform well (or even better).

THE DETERMINANTS OF PERCEIVED PROMOTION CHANCES

We have analyzed the factors that account for employees' job classification and rank. We now want to delve into the subjective side of the promotion system. What do employees think about this system? How do they see their own promotion chances? The questionnaire asked, "Do you have ample chance to become a manager, or supervisor?"[8]

The pre-coded responses, shown in Table 7.2, indicate that the majority of personnel in both factories see their promotion chances as relatively poor. However, Shipbuilding employees perceive their promotion chances as significantly better than do Electric Factory personnel. This difference

[8] The intended meaning of the question was, "What are your chances of becoming a supervisor or manager if you are now rank and file, or of becoming a higher level supervisor or manager if already a supervisor or manager?" Response rates indicate that 90 percent or more of each rank and job classification in fact interpreted the question in this way. The sole exception is that only six of the 21 *kachō* (29 percent) answered.

TABLE 7.2
Perceived Promotion Chances of Personnel in Electric and Shipbuilding Factories (%)

Perceived Promotion Chances	*Electric Factory (v. 109)*	*Shipbuilding Factory (v. 73)*
Ample chance	3.9	7.1
Fairly good chance	10.0	19.0
Not too good a chance	27.3	28.0
No chance at all	58.8	45.9
	100.0	100.0
N	(981)	(564)
C/C_{max} = .23**		

** Significant at the .01 level.

is, however, a result of the fact that in neither factory are any women in supervisory or managerial positions, and that the Electric Factory respondents include a much higher proportion of women than in Shipbuilding. Only one percent of the Electric Factory women, in contrast to 27 percent of the men, think they have "ample" or "fairly good" chance of promotion. Sex is a background status variable that is strongly related to perceived promotion chances. Thus, when we compare only *male* employees in the two factories, the difference is reversed; Electric Factory men see their promotion chances as significantly better than Shipbuilding men (C/C_{max} = .28, significant at the .01 level).

As we try to explain these perceptual variations, the general problem with which we shall be concerned is the degree of correspondence between objective and subjective aspects of the promotion system. An ideal test of this would be to compare employees' perceived promotion chances at one point in time with their actual promotion or non-promotion at one or more later points in time. Another test would be to compare employees' perceived promotion chances with their superiors' independent estimates of their chances. We lack the necessary data to make either of these tests. The most we can do is to test the hypothesis that the variables that explain perceived promotion chances are the same variables that explain present rank (and, by inference, previous promotions). Therefore, we shall test the same hypotheses used to explain rank—more specifically, that the best predictors of perceived promotion chances, as of rank, in Shipbuilding company are education, seniority, and performance and in Electric Factory are number of previous jobs, seniority, and education. In other words, we assume that whatever an employee's present rank, the higher he is on these variables, the better he will think his future chances of promotion are.

Before testing our theory of rewards in relation to perceived promotion chances, it is useful to see how these perceptions vary with present rank. After all, the main meaning of promotion is a move upward in rank. Table 7.3 shows the dramatic extent to which perceived promotion chances rise with rank in both factories. The very steep positive gradient of this relationship suggests that a main reason employees think their promotion chances are good is their experience with having already been promoted in the past.

TABLE 7.3
Perceived Promotion Chances, by Present Rank, in Electric and Shipbuilding Factories

Present Rank	*Promotion Would be to*	*Number Answering*	*Percentage Who Think Their Chances of Promotion are Ample or Fairly Good*
Electric Factory			
Kachō	Buchō	6	83.4
Kakarichō	Kachō	20	80.0
Hanchō	Kakarichō	17	58.9
Kumichō	Hanchō	31	25.8
Rank and file	Kumichō (line) or kakarichō (staff)	907	10.7
		C/C_{max} = .52**	
Shipbuilding Factory			
Kachō	Buchō	4	100.0
Kakarichō	Kachō	14	100.0
Sagyōchō	Kakarichō	8	87 5
Hanchō	Sagyōchō	34	82.3
Boshin	Hanchō	38	39.5
Rank and file	Boshin	340	11.7
		C/C_{max} = .62**	

** Significant at the .01 level.

In Electric Factory, the multiple regression analysis supports our hypotheses that perceived promotion chances vary positively with organizational status variables—rank and education—and with the extra-organizational cultural variable of sex, and vary independently of previous extra-bureaucratic statuses (father's occupation and size of community of origin).[9] The hypothesis that promotion chances are positively related to performance is also confirmed, though only weakly. Finally, our

[9] See Appendix D, Table D.14, p. 383 for details.

hypotheses that perceived promotion chances vary positively with the informal factors of social integration of the employee into the company—participation, paternalism, and lifetime commitment—are largely disconfirmed by the findings.

Thus, when other factors are controlled, it is men in higher ranks, with more education, and a somewhat better performance record, who are most optimistic about their promotion chances. Sex outdistances all the other variables as a determinant of perceived promotion chances, so much less likely are women than men to think their chances are relatively good. The interaction between rank and sex reenforces this pattern: it is having already been promoted that makes one optimistic about future promotions, and, as we have seen, women lack precisely this experience of having been promoted.

Turning now to Shipbuilding Factory, Table D. 15, p. 384, four variables provide the best explanation of perceived promotion chances, when other variables are held constant. These are rank, recruitment channel, section, and participation. Employees are optimistic about their promotion chances to the extent that (1) they have already been promoted in the past; (2) they were recruited by Shipbuilding Company on the basis managers prefer, namely, school recommendations, rather than by knowing someone in the company or by more impersonal means; (3) they work in either the Assembly or Design Sections, rather than in the Machine Section; and (4) they participate more often in company recreational activities. Although (2) and (4) support the lifetime commitment model, we have also seen that other hypotheses in that model are disconfirmed. By this we mean that the influence age, seniority, and lifetime commitment have on perceived promotion chances appears to operate mainly through rank. Indeed, one can predict perceived promotion chances almost as well from present rank alone, as from the four best predictors.

SUMMARY AND CONCLUSIONS

In this and the preceding chapter we have examined several aspects of the reward system—pay, job classification, rank, and perceived promotion chances—in our three firms.[10] To explain variations among employees in rewards, especially in Electric and Shipbuilding factories where our data are more complete, we have developed and tested a theory of organizational rewards. This theory asserts that the same set of independent

[10] We discuss collective bargaining, strikes, and the like in Chapter 10, in connection with conflict between labor and management, rather than here, in relation to the reward system. For excellent treatments of collective bargaining in Japan—a subject into which we did not delve deeply—see Levine 1958, Tsuda 1965, Cole 1971a, and Okochi, Karsh, and Levine 1973.

variables explains these four aspects of rewards. The causal variables in the theory were conceptually grouped in four domains: bureaucratic influences; present extra-organizational statuses; informal influences, i.e., the social integration of the employee into the firm; and previous extra-organizational statuses.

It is instructive to summarize our findings by showing how the four theoretical categories of independent variables predict across the several aspects of rewards. Let us consider the mean influence of the variables in each theoretical category upon all the reward system variables.[11] The influence of present extra-organizational statuses (sex, age, and number of dependents) upon rewards is the strongest (mean regression coefficient = .31). Present bureaucratic statuses are next most important (mean regression coefficient = .26). That is, education, seniority, rank, section informational level, and performance tend to be important determinants of virtually all reward system variables—pay, job classification, rank, and perceived promotion chances—in both factories. The third strongest category of determinants is previous extra-organizational statuses (social origins, number of previous jobs and number of years in these jobs, and recruitment channel), with a mean regression coefficient of only .14.

Variations in rewards among employees are essentially independent of the fourth theoretical category, informal influences flowing from the

[11] The unstandardized partial regression coefficients (*b*'s) on which this comparison is based appear in Figure D. 2 for Shipbuilding. For Electric Factory employees (both sexes) the *b*'s that are significant at the .01 level are as follows:

	Dependent Variable			
Independent variable	*Pay*	*Job Classification*	*Rank*	*Promotion Chances*
Age				−.05
Sex	.12			.77
Number dependents	.28			
Number previous jobs			.26	
Years in previous jobs	.09			
Father's occupation	.02			
Distance	.03			
Education	.21		.06	.29
Section		.46		.09
Seniority	.19	.66	.08	
Rank	.19			.42
Ratio of absences	.04			
Performance		.11		.03
Participation			.04	.03
Paternalism			.04	

Note that these *unstandardized* regression coefficients are not to be confused with the *standardized* regression coefficients as given in Tables D.8–D.15 in relation to the same dependent variables. The two coefficients may differ for a given independent variable, and may yield different "best independent variables." On the distinction between the two kinds of regression coefficients, see Appendix D, footnote 16, p. 415.

degree of social integration of the employee into the firm. Pay, job classification, rank, and promotion chances are virtually unaffected by whether the employee is high or low in cohesiveness, participation, paternalism, and lifetime commitment.

These findings clarify the paternalism–lifetime commitment model. They are certainly consistent with the model in showing the importance of both bureaucratic and extra-organizational cultural influences upon the various dimensions of rewards. But contrary to the implications of the model, not only does the employee's informal social integration into the firm have little independent influence on his objective rewards—pay, job classification, and rank—it also fails to exert a strong influence on his subjective rewards—perceived promotion chances—once the more important variables of bureaucratic status, age, number of dependents, and sex have been allowed to operate. In other words, there is no evidence that these informal factors of the social integration of the employee into the firm, which figure so prominently in the lifetime commitment–paternalism model of the Japanese firm, are important determinants of the reward system.

CHAPTER EIGHT

Social Integration of the Employee into the Company (1)

INTRODUCTION

We have had much to say about processes of social differentiation, in connection with formal organization, work, and the reward system. But any social system also requires processes of integration to offset the forces of differentiation. Processes of internal integration have the function of coordinating and unifying the differentiated, specialized elements of the system.

We shall analyze social integration in each firm in terms of the concepts introduced in Chapter 1: employee cohesiveness, company paternalism, employee participation in company recreational activities, company identification, conflict, interfirm mobility, and lifetime commitment. We conceive of these as analytically distinct areas of the social integration of the employee into the company.

One interpretation of the paternalism–lifetime commitment model is that all employees in a Japanese firm tend to be high in all aspects of integration into the firm. Employees rather uniformly are highly cohesive with their fellow employees, favor company paternalism, participate frequently in company recreational activities, are highly identified with the company's instrumental goals and problems, exhibit low levels of conflict as between, for example, workers and managers, do not change firms, and have high levels of support for lifetime commitment norms and values. This interpretation, which we shall call the strong form of the model, implies that employees' scores on each integration variable will cluster at the high integration end of the scale. Such a situation would produce low correlations between social integration variables.

A second interpretation, which can be called the weaker form of the model, is that employees' scores on each integration variable have relatively high variance, but there is a strong positive correlation between integration variables. That is, some employees are highly integrated, others less integrated, on each variable; but employees who are highly integrated on one variable tend to be highly integrated on all the other variables, while those low in integration on one variable will tend to be low on all.

If neither of these interpretations fits the data, there are only two other possible explanations. Either the empirical findings for our factories are suspect on methodological grounds, or the evidence can be relatively unambiguously interpreted as disconfirming the paternalism–lifetime commitment model.

Before we can deal with this general problem of the relationship among all aspects of integration, we must first measure and analyze each individual aspect: cohesiveness and paternalism in this chapter, participation, company identification, and conflict in Chapter 9, and interfirm mobility and lifetime commitment in Chapter 10. Then, at the end of Chapter 10 we can summarize the findings of all three chapters and confront the question of which of the above interpretations most plausibly fits the data.

EMPLOYEE COHESIVENESS

Sources of cohesiveness

From our interviews and observations in each firm we identified the following sources of cohesiveness among employees.

Basic Cultural Values. An American observer is continually struck by the pervasiveness of collectivity orientations in everyday interaction in Japanese society. From primary school to the university, the student is encouraged to act and think of himself primarily as a member of a group, rather than as an individual. Thus, before a person ever enters a company, he has normally developed a collectivity orientation, and upon entering the company, he transfers this to the new group setting. In this sense a company that has high solidarity among its employees can never take full credit for this, no matter how much it emphasizes group ideology in its company policies and practices.

Norms of Cooperation and Competition in Work. In all three firms interviewees explicitly recognized that cooperation and cohesiveness are functional requirements of work and of successful production. In Sake Company the master of Plant Number 1 said in 1966: "The master can't make good sake by himself. Most important is that he gets good cooperation from all workers." In Electric Factory, each kumi (the basic work group of 17 to 40 people) is given a production goal, and encouraged to compete with other kumi in a han. The various han in turn are encouraged to compete among themselves over how much time is spent on direct production work, and how much time is lost, for example, by having to do a job over. Thus, in the Hardware Components Section in June 1969, the second kumi of han 1 spent 88.6 percent of its work time on direct production, while, at the other extreme, the first kumi of han 2 spent only 68.6 percent of its time on direct production. This is said to stimulate competition between kumi (or between han), but to encourage cooperation *within* each unit. It was clearly recognized as a source of employee solidarity by the section chief of the Hardware Components Section: "The machines in this line are the responsibility of our group solidarity. This is group life. We don't want individual competition. We help each other whenever one is in trouble. The Hardware Section is moving toward making itself a group that can attain its goals completely. Otherwise, we can't get results. Changing a moulding requires mutual help. The line as a whole cooperates with each other as much as possible for quality control. I told the workers that everyone is a quality control man; they shouldn't send defective parts to the next man on the line."

At the same time, competition between kumi or han is not so universal or rigid that work groups cannot be formed and reformed for ad hoc, particular "task force" objectives. As noted before, in Hardware: "There is no sharp distinction between the first and the second han. They give each other help. So there is no difference in morale. They work together."

Norms of mutual help among employees are strong enough so that even in an assembly line setting, work is not totally individualistic. A girl in Number 1 Production reported: "We help others who are busy. It is not a shop where a person can't help others."

In Shipbuilding Factory, although the work is more individualistic in the Machine Section than in the Assembly Section, even workers in the former section recognized the need for solidary relationships: "Even in the Machine Section, we need teamwork between day and night shift workers on the same machine. Every large machine has two or more workers, one for the day, one for the night. Teamwork exists among them because they pass information back and forth about the machine. For example, there is strong teamwork among the three workers who handle the rigide machine."

Thus, one determinant of the degree of cohesiveness in given work units is the cohesiveness objectively required by the nature of the work.

Size of the Work Group. One machine oyakata in Shipbuilding Factory gave evidence that, as would be expected from theory, cohesiveness varies inversely with the size of the social unit: "A hike that is sponsored by our han is attended by all workers in the han, and also by the hanchō and boshin. But only one-third of a whole *shop* attend when it's a shop-sponsored activity." The han to which he referred had 24 members, the shop 76. The shop, it will be recalled, contained four or five han.

Expressive, Socio-Emotional Leaders. The following remarks by a boshin in Assembly reveal (1) that not all work groups have a person who plays the role of expressive, or socio-emotional leader (see Bales 1953), and (2) what the attributes of such a role player would be:

> There's no "human relations leader" to be found in any specific person in our han. I do my best in handling human relations problems, as boshin, but they are very difficult. And it's bad if a boshin gets too involved in this aspect. So I turn these problems over to the hanchō, and if he can't solve them he hands them to the sagyōchō.
>
> An example of a human relations problem is when the work group has a member who has too strong a personality. The best person to smooth out human relations problems is neither an older man (too rigid), nor a young man (too stubborn), but middle-age employees.

It should be noted that in this work context, "middle aged" means people in their thirties and forties; the fifties are already defined as "old." Our general point here is that, other things being equal, a work group would have higher employee cohesiveness to the extent that there is an "expressive leadership" role player in the group.

Boundary Maintenance and Rites of Passage. The section (ka) in Electric Factory is another unit that, although it probably has less solidarity than smaller units like the han and kumi, at times becomes the unit in terms of which solidarity mechanisms are activated. When a young staff member of Number 2 Design Section was being transferred to another section in Electric Factory, the members of the section he was leaving gave him a farewell party, even though they would continue to see him at lunch and around the factory. The party had the function of defining and maintaining the section's solidarity in the face of the loss of one of its members. There are also welcoming parties for new employees.

Leisure Behavior. Leisure behavior can also maintain and strengthen employee solidarity. At noon time, Electric Factory employees typically spent only about 10 to 15 of their 40 minutes in the factory cafeteria, eating. After leaving the cafeteria, they formed several groups for games in open spaces near the building where they worked. One popular game was a kind of Ping-Pong, played by four players (two on a side), with a low net stretched across the ground (instead of on a table), and using Ping-Pong–type paddles and a soft hollow rubber ball. As soon as a player missed the ball, he would be replaced by one of several fellow workers looking on; in this way everyone in a group of 10 or so would have his turn during the 20 or 30 minutes' play. Indoors, near the machines and work benches, another game played was *shōgi*, Japanese chess, which is somewhat less intellectual than *go*, Japanese checkers. This game of concentration is played by two, and was usually watched by two or three other workers. These two games were primarily played by male workers, in Hardware and

Production Engineering. Women talked and walked around together during recess, sometimes holding hands; they usually were not asked to join in the Ping-Pong–type game. After work in the evening, we observed several of the male college graduate employees playing tennis in the factory court; there are even factory tennis tournaments. Other after-work activities include Ping-Pong in the Training Center recreation room, and various club meetings, e.g., the English Conversation Club. Since most employees go home directly after work most days, these clubs cater more to the minority who live in the dormitory and are therefore nearby.

The participation of workers in group leisure activities is both a result of the preexisting level of cohesiveness—generated by the nature of the work task, size of the units, etc.—and also a means by which cohesiveness is maintained and even heightened. At lunch time several groups of Shipbuilding Factory workers could always be seen playing various outdoor games such as volley ball and paddle ball, as well as more sedentary, indoor games, such as shōgi.[1] We asked whether the membership of these play groups coincided with work groups, and found, according to one Machine Section worker: "At lunch workers from one Machine shop sit at the same table, but not necessarily next to workers of their own han." According to another Machine Section worker: "The volley ball and other play groups at lunch time are not the people of one shop or han. But there are generally the same people in the play group from day to day. Anyone who wants to play can. Some workers who play are practicing for the interplant Ping-Pong competition in the company."

Employees by no means limit their leisure social interaction to fellow employees, as the remarks of a 55-year-old Machine Section worker, about to retire, indicate: "Some of my best friends are merchants, not factory workers, either here or in other factories. They are former Shipbuilding Company workers who left here to set up independent stores. But some other friends of mine are workers in this company. I also visit my relatives in Shikoku. Generally, most workers go out with their fellow workers in this company, except for the New Year's holidays." Another Machine Section worker had somewhat more intimate contacts with non-work associates: "More of my friends are outside the company—former middle school classmates. I used to belong to a badminton club in the company but I quit. I belong to the YMCA and the Protestant Church. [Why are most of your friends outside the company?] Outside contacts can be more intimate. Although I have more close friends outside the company, I have one or two inside also."

[1] One interesting difference was that whereas virtually no female workers participated in the outdoor games with the males in Electric Factory, several did in Shipbuilding Company's Engine Division.

At least one worker—a 25-year-old Machine Section worker—expressed the full-blown differentiated pattern of non-solidarity with fellow workers in out-of-work contacts: "My friends are mostly outside the company, and I don't see Shipbuilding Company employees outside work. I've belonged to an NHK (Japan Broadcasting Corporation) Correspondence Course group, in general education. My friends are Osaka people who are in this course. We get together. . . . This NHK group of friends is better than seeing my workmates [outside work]."

Characteristics of Employees. In addition to these cultural and social structural sources of employee cohesiveness, there are some characteristics of employees themselves that make for variations in cohesiveness. It is said in Electric Factory, for example, that cohesiveness increases with one's seniority. The kumi-consciousness to which we referred above is said to develop only after two or three years of working together as a group. A kumichō in Number 1 Production said: "Workers who have been here more than 12 years have very strong solidarity. Maybe we used to be trained that way. In this factory people are ranked first in terms of being honest, serious, hard workers. In Osaka, there are all types of indifferent workers. Here there aren't so many differences among workers."

In Sake Company age is seen as having consequences for cohesiveness. The master of Plant Number 1 said: "The problem is to make for harmony between young and old workers." In a sense this is not simply a result of age differences among workers, since these have always existed in sake plants. Rather, it is a result of the more modern attitudes and expectations with regard to many aspects of work and the social structure of the factory that the younger workers now bring with them. One specific difference may be that whereas in the past young, apprentice sake workers took older sake craftsmen as their role models and reference group, today's young sake workers have substituted workers of their own age and qualifications in other, more modernized industries.

Shipbuilding personnel also perceive cohesiveness as positively related to the age homogeneity of the work group. Thus, a machine worker: "Most employees here are likeable, but a few aren't. *My hanchō is old, so he's not approachable by me.* I am 24 years old. He's a good man, but sometimes has a bad temper. Other hanchō are younger." When we asked another machine worker, "What makes good human relations (*ningen kankei*) in the shop?" he replied: "Most shops have one 'outsider,' nonconformist, but we don't have any. [Q. What makes someone an 'outsider?'] For example, being very different in age from all the others in the shop or han. Our hanchō and boshin are both over 50 years old, long-time veterans in the company. That makes for better human relations than where the hanchō is younger, less experienced." Note that the second person, who is 30 years old, disagrees with the first as to whether older or younger hanchō are better for solidarity. This partly reflects the difference in age between the two workers, and the greater age homogeneity of the older worker with his (older) hanchō. But both agree on the fact that age is a determinant of work group cohesiveness.

The Control of Misfits. Management is of course very concerned to have good teamwork. Consequently, supervisors observe closely how well each worker fits in and exhibits teamwork. A second-grade boshin in the Assembly Section of Shipbuilding reported: "If a worker isn't smooth in his human relations he'll be shifted to another work group." Another mechanism by which cohesiveness is maintained in the work group is that crucial social control functions are assigned to supervisors, rather than to fellow workers: "Among workers here there are never charges by one worker to another that 'You haven't pulled your share.' We all take breaks for smokes, etc. between jobs, for 10 minutes or so. *If there's a laggard worker, the hanchō would speak to him, not a fellow worker.*"

Kinship Ties. Kinship bonds between personnel were not even mentioned as a source of cohesiveness in Electric Factory or Shipbuilding Factory. In Sake Company, however, it is said that some of the older kura, which we did not study, have higher social solidarity than the three traditional kura we did study because relatives serve as masters and associate masters of the former plants. Since we do not have cohesiveness data from all the kura, we cannot test the hypothesis that kura cohesiveness varies positively with kinship ties at the top of the kura. However, only five of the 16 masters and associate masters have kinship ties with one another. If kinship were the only factor making for cohesiveness, Sake Company's kura would not be highly cohesive.

Company Housing. This factor could only be a significant source of cohesiveness in Sake Company, where all personnel in the traditional kura live in company housing during the months of their employment; indeed, all sleep and have their belongings in the same large dormitory room. Only eight percent of Electric Factory employees and nine percent of Shipbuilding live in company housing.

Constraints on Cohesiveness. Cohesiveness is clearly a variable property of work units in Japanese organizations. We have seen some of the conditions under which it varies. Thus, if the functional requirements of cooperation in work foster cohesiveness, what some interviewees call "rivalry consciousness" has the opposite effect on cohesiveness. Even in the Assembly Section of Shipbuilding Factory, where cohesiveness is especially important, a second-grade boshin told us: "The first-grade boshin and I make a good cooperative pair. But young workers have rivalry consciousness. For example, a 28-year-old worker has longer experience, but a 23-year-old worker has more knowledge of the machines, so their rivalry is too strong, and I recommended to the hanchō that they shouldn't be assigned to work together."

An Assembly hanchō echoed this statement: "All my workers are under 30 years old and have rivalry consciousness, so they don't have good teamwork. I wish I could pair young and older workers, but there aren't any older workers."

The degree of employee cohesiveness

Having reviewed what interviews and observation uncovered as the main sources of cohesiveness among employees, the next step is to measure the degree of cohesiveness in each factory. This will be based on the cohesiveness items used in each questionnaire, which are shown in Table 8.1. These items provide a somewhat more systematic basis for comparing the degree of cohesiveness across firms than the foregoing list of the sources of cohesiveness. We did not systematically collect data on each source of cohesiveness in each factory, but the same seven questions appeared in the Electric and Shipbuilding questionnaires, and two of these were also included in the Sake questionnaire.

Let us compare the marginals for each factory on these questions, moving from those that show a higher degree of cohesiveness to those that indicate less cohesiveness. The areas of greatest average cohesiveness for both Electric and Shipbuilding employees are having one or more close friends who work near them in the factory, and saying that at least half of their best friends work inside their own factory. However, of employees who have one or more close friends at work, those in Electric Factory are more likely than those in Shipbuilding to make special dates to see them outside work. Of all respondents to the question, "Do you have close friends at work with whom you make special dates to see outside the company?" Electric Factory employees are again more likely than those in Shipbuilding to say they do. When asked how troubled they would be if moved to another job in the factory, similar to the one they do now, but away from their present work mates, Electric Factory employees are once again somewhat more likely than their Shipbuilding counterparts to express cohesiveness by saying they would be "very much" or "fairly much" troubled.

In the two questions asked in all three firms—quality of perceived teamwork and whether one thinks relationships among "the people who work here are mostly warm and family-like" or "cold and mechanical"—Sake employees are much more cohesive then employees in the other two firms. Finally, a question on union participation, which was asked only in Electric Factory, indicates a high level of attendance at union meetings.

TABLE 8.1
Aspects of Employee Social Cohesiveness (%)

Variable[a]			Shipbuilding	Electric	Sake
42/65.	How would you feel if you were moved to another job in the factory, more or less like the one you now do, but away from the people who work near you? Would you feel:				
	Very troubled		18.4	18.1	
	Fairly troubled		30.7	42.0	
	Not much bothered		41.6	35.5	
	Not bothered at all		9.3	4.4	
			100.0	100.0	
		N	(581)	(1,022)	
	$C/C_{max} = .19$**				
44/67.	How many of the people who work near you (han or kakari) would you call your close friends?				
	None		20.0	21.7	
	1 or 2		32.1	37.0	
	3 or more		47.8	41.3	
			99.9	100.0	
		N	(579)	(1,016)	
	N.S.				
45/68.	(If in previous question you have one or more close friends at work) What opportunity do you have to see them outside the company?				
	Make special date (for visiting, hiking, mountain climbing, etc.)		55.9	86.6	
	If we happen to see each other we associate, *or* no association outside the company		44.1	13.4	
			100.0	100.0	
		N	(454)	(796)	
	$C/C_{max} = .27$**				
	Percent of those answering previous question (v. 44/67) who have close friends with whom they make special dates outside work (Shipbuilding = 295/579; Electric = 689/1,016)		50.9	67.8	
43/66.	Concerning your best friends, do they work in the factory or outside?				
	1. Most of them work in the factory		25.7	25.8	
	2. About half of them work inside and half outside the factory		49.6	55.2	

TABLE 8.1 (*Cont.*)

Variable[a]			*Shipbuilding*	*Electric*	*Sake*
	3. Most of them work outside the factory		24.7	19.0	
			100.0	100.0	
		N	(579)	(1,019)	
	$C/C_{max} = .10$*				
46/69/16.	Would you say that among the people who work at this factory relationships are mostly warm and family-like, or cold and mechanical?				
	1. Warm, family atmosphere		14.8	14.3	43.5
	2. Cold, mechanical atmosphere		18.6	14.9	12.9
	3. Half and half		66.6	70.8	43.5
			100.0	100.0	99.9
		N	(586)	(1,022)	(62)
	Shipbuilding vs. Electric: N.S.; Sake vs. Electric: $C/C_{max} = .26$;** Sake vs. Shipbuilding: $C/C_{max} = .31$**				
47/70/17.	How do you perceive the teamwork in your shop (if line: in your kumi; if staff: in your kakari or ka):				
	1. Very good teamwork		12.7	10.7	13.1
	2. Fairly good teamwork		44.1	48.3	67.2
	3. Not too good		35.1	36.1	18.0
	4. Not good at all		8.1	4.9	1.6
			100.0	100.0	99.9
		N	(590)	(1,018)	(61)
	Shipbuilding vs. Electric: $C/C_{max} = .11$;* Sake vs. Electric: $C/C_{max} = .14$;* Sake vs. Shipbuilding: $C/C_{max} = .21$**				
71.	Recently, how often do you go to Union meetings?				
	Regularly (within the last month)			61.4	
	Occasionally (within the past year)			25.8	
	Rarely			9.0	
	Never			3.8	
				100.0	
			N	(981)	

* Significant at the .05 level.
** Significant at the .01 level.
[a] When more than one variable number is given, the first refers to Shipbuilding Factory, the second to Electric, and the third to Sake Company. For a complete list of variables by number, see Appendix C.

The variation in measured cohesiveness from question to question is understandable if we note that the highly cohesive areas have to do with primary group cohesiveness, whereas the less cohesive areas have to do more with secondary group cohesiveness among employees. In other words, in each factory, for any given employee, there is high cohesiveness with a small number of fellow workers. But when the question refers to cohesiveness with the larger number of employees in his entire line, shop, union, or factory as a whole, the sheer number places a limit on the intensity of the interaction that is possible, and therefore, there is less cohesiveness at the level of the larger group. This same interpretation can account for why Sake personnel are so much more cohesive than those in Electric or Shipbuilding: the much smaller number of people in a given sake kura (15–35) means that it is possible for one to have primary group ties with a larger proportion of all employees.

Do these questions measure the same underlying property, employee cohesiveness? Table 8.2 shows the patterns of association among the seven variables that were questionnaire items in Table 8.1 for the two factories that have all but one of the questions in common. When we

TABLE 8.2
Interrelations among Employee Cohesiveness Variables in Electric and in Shipbuilding Factories[a]

Variable[b]		*67/44*	*69/46*	*70/47*	*65/42*	*68/45*	*71*
66/43.	Friends inside vs. outside factory	.45** .51**	.13* .26**	.11 .27**	.20** .21**	.09 .24**	.06
67/44.	Number of friends among workmates		.19** .21*	.18** .18	.20** .21	.13 .24**	.10
69/46.	Warm/cold social relationships in shop			.50** .61**	.11 .26**	.09 .15	.12
70/47.	Teamwork in factory				.16* .28**	.09 .10	.14
65/42.	Troubled if transferred away from workmates					.12 .13	.09
68/45.	See workmate friends during time off						.12
71.	Attendance at union meetings						

* Significant at the .05 level.

** Significant at the .01 level.

[a] Upper number in each cell is C/C_{max} for Electric Factory, lower is C/C_{max} for Shipbuilding.

[b] When two variable numbers are given, the first refers to Electric Factory, the second to Shipbuilding. For a complete list of variables by number, see Appendix C.

subjected the seven variables for Electric Factory to the item-analysis program (ITEMA), the variable with the lowest item-total correlation—union attendance—was rejected. This item would have had to be dropped from any index of cohesiveness that would be comparable for Electric and Shipbuilding factories anyway, since it was not used in Shipbuilding Factory. The other six items form our index of employee cohesiveness.[2] The individual's total score on this index measures the extent to which he or she would be troubled if moved away from fellow workers, has friends among workmates, has most friends inside the company rather than outside, makes dates to interact with these friends outside of work, perceives teamwork as good, and thinks the atmosphere of social relationships in the factory is warm and family-like.

THE DETERMINANTS OF EMPLOYEE COHESIVENESS

We shall test the following hypotheses concerning employee cohesiveness. We list the hypotheses first in bare outline, and introduce the rationale for each below. Employee cohesiveness varies:

I. Inversely with organizational status:[3]
   1. Informational level of the section
   2. Rank
   3. Job classification
   4. Education
   5. Seniority

II. Inversely with previous status:
   6. Number of previous jobs

III. Inversely with extra-organizational status:
   7. Age
   8. Sex
   9. Number of dependents

IV. Directly with attitudinal variables:
   10. Job satisfaction
   11. Perceived promotion chances

V. Directly with social integration:
   12. Participation in company recreational activities

[2] By the formula for estimating index reliability from the items' intercorrelations, coefficient alpha is .30 in Electric Factory and .47 in Shipbuilding. For more details on this and other indexes, see Appendix B.

[3] As elsewhere in the analysis we had to decide whether to use the status index or the individual component status variables. With regard to cohesiveness, the latter alternative was selected, first, because some component status variables were positively related to cohesiveness, others negatively, and this would weaken the relationship between the status index and cohesiveness; and second, because component status variables yielded higher zero-order r's, beta weights, and multiple R with cohesiveness than did the index of status.

13. Lifetime commitment
14. Residence in company, rather than private, housing

Let us begin by considering these hypotheses in Electric Factory.

Informational Level of Section. The anecdotal interview and observational data reviewed above suggest that one source of variation in employee cohesiveness is the section in which one works. Table 8.3 shows that Hardware, Surface, and Motor sections have somewhat higher cohesiveness than Number 1 and 2 Production sections. Hardware, as we saw earlier, holds recreational activities about five times a year for its own members. These include trips to the sea for clam digging, trips to pick grapes, and so on. The acting kachō of the Hardware Section takes the paternalistic view that young workers shouldn't be left to their own devices too much for recreation.

TABLE 8.3
Employee Cohesiveness in Electric Factory, by Section (%)

	Index of Cohesiveness Score (v. 141)				
	High	*Medium High*	*Medium Low*	*Low*	*Total*
High informational level sections[a]	5.7	30.0	57.9	6.4	100.0
Medium informational level sections[a]	7.8	43.1	43.1	6.0	100.0
Low informational level sections[a]	14.0	42.2	38.2	5.6	100.0
Hardware Components	19.7	45.1	30.8	4.4	100.0
Surface Treatment	21.6	42.4	29.7	6.3	100.0
Motor	12.3	54.3	29.6	3.8	100.0
Number 2 Production	14.3	38.8	40.1	6.8	100.0
Number 1 Production	8.0	38.7	47.6	5.7	100.0

[a] For the names of sections classified as high and medium informational level, see Chapter 3, footnote 3.

Lower levels of employee cohesiveness are found in medium informational level sections, and lowest of all in high informational level section (Table 8.3). Thus, in general the level of employee cohesiveness in a section varies inversely with the section's informational-cybernetic level. Sections whose personnel do direct production work have more cohesiveness than those whose personnel are more highly educated and do staff work. These section differences in cohesiveness probably stem partly from the kinds of employees selected for work in the section, and partly from deliberate efforts by the leadership in some sections to raise the level of employee cohesiveness. Our data do not permit us empirically to disentangle these two sources of section solidarity.

Previous Interfirm Mobility and Seniority. The paternalism model of the Japanese factory states that lifetime commitment to one firm is supported by strong moral sanctions. A possible inference from this is that employess who enter a firm with previous employment in one or more other firms (chūto saiyōsha) would be regarded with moral disapprobation by those who have always worked for the same firm. If this is the case, the social acceptance of chūto saiyō employees, especially in their early years in the new firm, should be less than that for those who enter the firm directly from school. To test this hypothesis, we compared the degree of employee cohesiveness of Electric Factory employees (and of Shipbuilding employees) who came from another firm with those who have always worked for Electric Company (or Shipbuilding Company), holding seniority constant. The results appear in Table 8.4

TABLE 8.4
Employee Cohesiveness in Electric and Shipbuilding Factories, by Previous Interfirm Mobility and Seniority (%)

Electric Factory	Seniority						
	Under 2 Years		*2–3 Years*		*4+ Years*		
Index of Employee Cohesiveness Score	*Number of Previous Firms*						
	None	*1+*	*None*	*1+*	*None*	*1+*	
Low (6–11)	29.0	27.8	30.2	27.0	27.7	22.4	
Medium (12–13)	48.2	44.4	41.8	45.9	41.9	30.6	
High (14–17)	22.8	27.8	28.0	27.0	30.3	46.9	
	100.0	100.0	100.0	99.9	99.9	99.9	
N	(193)	(36)	(189)	(37)	(346)	(98)	(899)
C/C_{max}	.06		.05		.20**		

Shipbuilding Factory	Seniority						
	Under 2 Years		*2–3 Years*		*4+ Years*		
Index of Employee Cohesiveness Score	*Number of Previous Firms*						
	None	*1+*	*None*	*1+*	*None*	*1+*	
Low (6–11)	33.3	27.8	25.0	47.1	36.9	33.8	
Medium (12–13)	38.9	44.4	50.0	29.4	32.6	38.0	
High (14–17)	27.8	27.8	25.0	23.5	30.5	28.3	
	100.0	100.0	100.0	100.0	100.0	100.1	
N	(18)	(18)	(12)	(17)	(141)	(237)	(443)
C/C_{max}	.09		.33		.08		

** Significant at the .01 level.

The first two columns are the most crucial tests of the hypothesis. If chūto saiyō employees are socially rejected because of their earlier "disloyal" interfirm mobility, this should happen in their earliest years in the new firm; their past mobility might be progressively forgotten by their fellow employees as time passes. By this criterion, the hypothesis derived from the model is disconfirmed. Among those who have been in Electric Company or Shipbuilding Company less than two years, there is no significant difference between the mobile chūto saiyōsha and the non-mobile who have not worked for any other firms. The same is true among those who have been in Electric Company two or three years. Among those who have four or more years' seniority in Electric Company, those who have always worked for the company are significantly less likely than those who have changed firms to have high employee cohesiveness. The fact that a difference in cohesiveness emerges only after four years is evidence that some factors other than the ostracism of chūto saiyōsha are producing the difference.

Among Shipbuilding employees the difference in cohesiveness between chūto saiyōsha and those who have always worked for Shipbuilding Company is non-significant under each seniority control. Though non-significant, there is a C/C_{max} of .33 for the 2–3 years' seniority group. But again this disconfirms the model because the model cannot explain why a difference in cohesiveness should emerge only after two years of employment in the firm.

In fairness it must be admitted that our findings do not conclusively disconfirm this aspect of the model. Our questionnaire did not, after all, ask whether the employees with whom one had cohesive relationships were mobile or not. It is therefore possible that chūto saiyōsha are cohesive with each other in each factory (there are 186 such employees in Electric Factory and 317 in Shipbuilding), while those who have always worked for one company are cohesive with other non-mobile employees. If this were the case employee cohesiveness would be segregated cohesiveness: the bonds between employees would be with those like themselves in regard to interfirm mobility. Unfortunately, our data make it impossible to test this.

Other Hypotheses. We also hypothesized that employee cohesiveness varies inversely with other organizational status variables: rank, job classification, and education. These hypotheses were based on the theoretical assumption that as employees rise in organizational status, their loyalties shift from fellow employees to the organization and its goals. Selective mechanisms are posited as operating to effect this result: e.g., the differential advancement in rank, job classification, and pay of the employees who exhibit primary loyalty to the organization. We also hypothesized that cohesiveness varies positively with job satisfaction and perceived promotion chances. We regard these as weak predictions, however, since there may be pressures making for an inverse relationship. The positive relationship would be expected to the extent that employees are cohesive *because* other aspects of their job and future prospects are satisfying, and this orients them more to interaction with fellow employees. The negative relationship would be expected on the assumption that employees form solidary ties with fellow employees as a compensation for otherwise unsatisfying work, i.e., low job satisfaction and poor promotion chances.

Extra-organizational statuses are seen as competing against cohesiveness with fellow employees; thus we hypothesized negative relationships between cohesiveness and age, sex, and number of dependents. That is, people who are younger, female, and without dependents lack the commitments to family, home, and the like that their opposite numbers have, and are therefore more likely to be cohesive with fellow employees, especially in non-work settings. We hypothesized that cohesiveness also varies directly with frequency of participation in company recreational activities, degree of lifetime commitment, and living in company housing.

Our findings on the basis of the multiple regression analyses, displayed in detail in Appendix D, Tables D.16 and D.17, p. 386 and p. 387, provide little support for some of these hypotheses. In both Electric and Shipbuilding factories job satisfaction and lifetime commitment are among the best predictors of cohesiveness, both being positively related to cohesiveness. But the other best predictors in Electric Factory—education and the informational level of the section—differ from those in Shipbuilding—participation and rank. The effect of the status variables in both factories is negative: cohesiveness declines somewhat as rank rises in Shipbuilding, and as education and section informational level rise in Electric Factory. Although the specific status variables differ in the two factories, higher status employees in both factories have lower cohesiveness than their lower status counterparts. Finally, our independent variables do not go very far toward explaining cohesiveness: over 80 percent of the variance in cohesiveness in each factory remains unexplained even after all our variables have been considered.

COMPANY PATERNALISM

In the area of paternalism we are concerned, on the one hand, with the extent of paternalistic benefits each company makes available to its employees—most especially company housing—and on the other hand, with the extent to which employees prefer their company and its management to have a diffuse, paternalistic relationship with them, rather than a functionally specific relationship. The basic data appear in Table 8.5.

Consider first company housing. All Sake Company brewery workers in N-City live in dormitories. In 1970, the following kinds of housing were available for the 370 full-time and year-round employees of Sake

TABLE 8.5
Measures of Paternalism in Electric Factory, Shipbuilding Factory, and Sake Company (%)

Variable[a]			*Electric*	*Shipbuilding*	*Sake*
2/102/32.	Residence				
	Company housing		7.9	9.0	45.0
	Private residence		92.1	91.0	55.0
			100.0	100.0	100.0
		N	(1,033)	(581)	(701)
	Electric vs. Shipbuilding: N.S.; Sake vs. Electric: $C/C_{max} = .76$;** Sake vs. Shipbuilding: $C/C_{max} = .83$**				
63/40/14.	In this company suppose there are two types of superiors. Which would you prefer to work under?				
	1. A man who always sticks to the work rules and never demands any unreasonable work, but on the other hand never does anything for you personally in matters not connected with work		17.0	19.0	16.9
	2. A man who sometimes demands extra work in spite of rules against it, but on the other hand looks after you personally in matters not connected with the work.		83.0	81.0	83.1
			100.0	100.0	100.0
		N	(999)	(547)	(66)
	Electric vs. Shipbuilding: N.S.; Electric vs. Sake: N.S.; Shipbuilding vs. Sake: N.S.				
97/61.	There are two companies. From your experience which would you choose?				
	1. Management thinks they are like parents to workers. Therefore they regard it as better to take care of the personal affairs of workers.		61.2	67.0	
	2. Management thinks its relationship to the worker is simply work: therefore management doesn't find it necessary to take care of workers' personal matters.		38.8	33.0	
			100.0	100.0	
		N	(989)	(573)	
	Electric vs. Shipbuilding: $C/C_{max} = .08$*				

* Significant at the .05 level.
** Significant at the .01 level.
[a] The first variable number refers to Electric Factory, the second to Shipbuilding, the third to Sake. For a complete list of variables by number, see Appendix C.

Company (as distinct from its seasonal brewery workers): in N-City there is a four-story apartment house for 24 families; in I-City there are similar apartment houses for 100 families, built between 1965 and 1970. For single men there are dormitories in both cities, each housing about 20 employees. Thus, of a total of 701 employees (370 year-round, 331 seasonal), some 312 or 45 percent live in company housing.

Shipbuilding Company provides a dormitory for unmarried workers and some apartments for married workers. Thirty percent of our respondents are unmarried; only eight percent live in the dormitory. The full distribution of Engine Division personnel according to type of residence is:

Residence in:	*N*	%	
Company dormitory	47	8	company
Company-owned apartment	5	1	
House sold to employee by company	29	5	
Rented house or apartment	293	50	private
Own my own house	207	36	
	581	100%	

Thus, even if we include those who bought their house from the company with the other company-housed employees, only 14 percent of the Engine Division are in company housing. In a metropolitan area like Osaka, any company finds it increasingly difficult to provide company housing.

Interviews with two 19-year-old Assembly workers who live in the dormitory provide data on why they live there: "I like a paternalistic type of company. The hanchō helps workers by being considerate about our health and other individual matters, when he assigns overtime work. The dormitory rules are made by dormitory self-government. The 11 o'clock curfew is not very strict. I prefer the dormitory rather than a private house because my friends are there, and I can take part in dormitory activities." "I prefer to live in the dormitory until I get married, because my parents, who live in [a nearby city] have too small a house. Also, while young it's good to have dormitory experience." Space and age constraints prevent some workers from living in the dormitory. An assembly worker of about 30 said: "I live in a private apartment. When I entered the company there was no dormitory space. Now there's space, but I think I'm too old for the dormitory."

Electric Company is also not paternalistic as far as housing is concerned: all but eight percent of its Electric Factory employees live in private residences in the small towns, farms, and cities in the area. A major reason why the company does not have to be paternalistic with regard to housing is that the factory is not located in a densely populated metropolitan area, where inexpensive housing has to be provided for personnel who have migrated from other areas of Japan, or who live beyond a reasonable commuting distance.

The second and third items in Table 8.5 both state the same dichotomous choice between functionally specific and functionally diffuse relationships, the difference being that variable 63/40/14 refers to the employee's relationship with a specific superior in the company, whereas variable 97/61 refers to his relationship to the company and management in general. Note that Electric and Shipbuilding employees are more likely to want a paternalistic relationship with a particular superior (83 percent and 81 percent) than with management and the company in general (61 percent and 67 percent), though, of course, a majority of employees favor a paternalistic type of relationship in both respects.

When asked in interviews about foremen-worker relationships, Shipbuilding workers described their foremen as varying in degree of paternalism: "The two hanchō I've had in nine years here were different. One took interest in our out-of-work activities, our marriage, etc.; the other doesn't. The present sagyōchō is good because if we ask him to change something dangerous, he does it right away, also because he has the ability to get us to do something by jokingly suggesting that we do it."

Since the majority of Electric and Shipbuilding employees want their company to have a paternalistic relationship with them, while only a small minority actually benefit from one form of that paternalism—company housing—we may infer that for most employees these preferences for paternalism derive from factors other than housing. These could be other specific benefits, on which we lack data, or Japanese cultural emphasis on a particular nexus of *giri-ninjō* and *on*.[4]

It is our impression that management in Sake Company is in fact more paternalistic than in Electric and Shipbuilding companies. Since there is no labor union in Sake Company, management sets the salaries of its office and other non-kura personnel. Employees have the right to acquire shares in the company's stock as a reward for years of service, rather than having to buy them on the stock market. As in other Japanese firms, employees receive gifts from the company. As the operator of a squeezer in the new factory told Mannari in 1966: "I received sake and a kimono yesterday [for the Bon Festival]. We receive gifts before New Year's also, and birthday presents."

[4] Cole discusses these traditional Japanese concepts as they are used by Japanese blue-collar workers in industrial firms today. "In general, *on* refers to a sense of indebtedness by a subordinate for favors bestowed by a superior. . . . The subordinate in the factory is obliged to repay . . . by hard work and loyalty the benefits received and he does so out of *giri* feelings. . . . *Ninjō*, defined as human kindness or compassion, refers to what the individual would like to do as a response to his personal emotion" (Cole 1971a:202–03).

We have already seen that the master traditionally had a very paternalistic relationship to his sake workers. The workers call him "father" or "uncle," and if they are in trouble, the master will represent their claims to the company. The master believes the workers are his charges while they are away from their farms and home villages for the winter season of sake work. He takes responsibility for their health and welfare until they return to their home villages. This is how the master of Plant Number 2 expressed it: "My responsibility is to complete the day's work, and to send the workers home after 180 days' work in good health. The village people trust me to take care of their men who work here."

What is the effect of the new, automated sake plant structure on paternalism? Even in this plant there is no union. Management decides the level of monthly pay, annual bonuses, salary increases, and housing, recreation, and welfare programs. However, although the new automated technology has not brought a labor union, it has brought a somewhat changed social organization, and this has meant that employees view the company's paternalistic practices in a more critical light. A foreman in the new factory, with long experience in the traditional plants, said in 1966: "In today's new plant, such [paternalistic] relationships between master and apprentices do not survive." Mannari, Maki, and Seino (1967:19) found that most workers in Sake Company do not like paternalistic relationships between master and workers. Workers say: "In traditional personal relationships there is no liberty, especially when we work with a head or a master who comes from the same village we do." New workers in the automated plant are dissatisfied with any signs of traditional human relationships and paternalism in a modernized factory. Many of them complain about the company's continuing aspects of paternalistic personnel administration.

Having discussed the three paternalism items in Table 8.5 the next question is, to what extent do all three items measure the same underlying characteristic, paternalism? Since all our Sake respondents live in company housing we cannot relate residence as a variable to responses to the other two questions, as we can in Electric and Shipbuilding factories. In both Electric and Shipbuilding factories, responses to the two paternalism preference questions are significantly and positively related: $C/C_{max} = .36$ in Electric Factory and .33 in Shipbuilding. Those who favor a modern, functionally specific, relationship with their superior are more likely than those who prefer a traditional, functionally diffuse, relationship to want a functionally specific relationship with company management in general. The majority of those who prefer a paternalistic superior opt for a company management that thinks it is "like parents to the workers." However, neither of these paternalistic preferences is significantly related to living in company housing, among either Electric or Shipbuilding employees.

The same results emerge from item analysis.[5] Residence was therefore dropped from the index of paternalism. This index contains the two paternalism preference items; the higher one's score on the index, the more one favors paternalism in relationships with superiors and with the company in general.

The determinants of paternalism

In an attempt to explain paternalism, we tested the following hypotheses:

1. *Seniority.* Paternalism is directly related to seniority. Employees with more seniority may have benefited more from the paternalism of their superiors and from company management in general; if so, they might express a greater preference for paternalism.
2. *Rank.* Paternalism is directly related to rank. Cole (1971a) suggests managers in Japanese firms are more in favor of paternalistic relationships than workers.
3. *Pay.* Paternalism is directly related to pay. The more pay one receives, the more one might be inclined to reciprocate in the form of attitudes that favor company paternalism.
4. *Education.* Paternalism is inversely related to education. The United States–directed reforms in the Japanese educational system following World War II have contributed to a general de-emphasis on paternalism. To the extent that this aspect of modernization has taken hold, we might expect more highly educated employees to be less paternalistic than those with less education.

We hypothesized that paternalism is directly related to aspects of the social integration of the employee into the company. Specifically:

5. *Company Housing.* Employees who live in company housing are more paternalistic than those who live in private residences. Theoretically, these two variables are mutually dependent: being more paternalistic motivates one to live in company housing, and to live in such housing is to experience directly the benefits of this form of company paternalism.
6. *Participation.* Paternalism varies directly with participation in company recreational activities. The theoretical link posited is that these two variables reenforce each other; thus we shall treat each as

[5] The ITEMA program shows that the residence variable has an item-total *r* of only .028 in Electric Factory and .006 in Shipbuilding. Those who live in company housing are not more likely than those who live in private residences to want a paternalistic superior or a paternalistic company. The reliability coefficient of the two-item index of paternalism is .41 in Electric Factory and .38 in Shipbuilding.

an independent variable in relation to the other. In relation to paternalism as the dependent variable, employees who participate more in recreational activities are reminded of the paternalistic benefits the company offers, and thereby encouraged to maintain or increase their favorable attitude toward paternalism. Frequent participation in these activities also serves to demonstrate to the employee the diffuseness of his connections with the firm—non-work as well as work connections; this diffuseness—in contrast to a functionally specific relationship—is the heart of a paternalistic orientation.

7. *Lifetime Commitment.* Paternalism is directly related to lifetime commitment. One of the props in the ideology of lifetime commitment is, after all, paternalistic relationships between company management and employees. This hypothesis states a relationship of consistency at the level of meaning between two aspects of an ideology.

8. *Size of Community of Origin.* Paternalism is inversely related to the size of one's community of origin. Modernization in general is usually more advanced in metropolitan and urban areas than in towns and villages; therefore, paternalism, as one component of traditionalism, should be more common among employees from smaller, more rural communities.

9. *Sex.* Paternalism varies independently of sex. Paternalism is a general orientation in Japanese society and culture, and there is no evidence that it is significantly sex-patterned.

The results of the correlation and regression analysis, shown in Appendix D, Tables D.18 and D.19, p. 388 and p. 389, do not go very far toward explaining paternalism.. All we can say is that, for both factories, when other variables among our set of independent variables are held constant, employees with more education are slightly less paternalistic than those with less education, and that those who participate more in company recreational activities are *slightly more* paternalistic than those who participate less. In Shipbuilding Factory, the higher one's pay, the higher one's paternalism; and in Electric Factory paternalism rises with rank and lifetime commitment.

Our hypotheses concerning paternalistic preferences appeared to be plausible. Why, then, are they mostly disconfirmed, and why do they account for so little of the variance in paternalism? It is possible that these findings are due to measurement error. Had our index of paternalism contained more than two items, and more discriminating items, it would have been more reliable, and the variance in paternalism scores might have been greater. In that event, our hypothesized independent variables might have had stronger relationships with paternalism. Only future research can ascertain the extent to which Japanese employees vary in paternalism, when it is measured in a more reliable way; and only if

Japanese employees do vary in paternalism can we test whether it is our variables, or some others, that explain why some employees prefer more paternalism than others in the structures of social relationships and in the organizations of which they are a part.

The conditions and consequences of company housing

Before leaving paternalism we shall consider further two questions about the variable excluded from the paternalism index—residence. First, what factors account for some employees living in company housing, while the majority live in private residences? Second, does residence have any independent effect on other aspects of employees' behavior and attitudes? Eight hypotheses will be tested concerning the conditions under which employees live in company rather than private housing:

1. *Education.* University-educated recruits are likely to come from distant parts of Japan, and therefore lack access to housing they can afford. We hypothesize that living in company housing is more likely the more education one has.
2. *Seniority.* The less one's seniority, the more likely that one is young and without family. Living in company housing is a temporary expedient for such employees; therefore, it should vary inversely with seniority.
3. *Rank.* High rank employees may demonstrate their commitment to the firm by living in company housing; thus we predict that living in company housing varies positively with rank.
4. *Pay.* The paternalism model suggests that living in company housing is a symbolic expression of unity with the firm, rather than being merely economically expedient. If this is true, living in company housing should vary independently of pay. Our counter-hypothesis is that as one's pay increases one is likely to use it to purchase private housing.
5. *Number of Dependents.* The number of dependents who can be accomodated in company housing is obviously a function of the space available in that housing. If company housing consists mostly of dormitories, only unmarried employees can be accomodated. On the other hand, the smaller the proportion of employees with dependents, the more a small supply of apartments and houses can meet the demand. We hypothesize that living in company housing varies independently of number of dependents.
6. *Paternalism.* Although we have already seen that living in company housing varies independently of the paternalism index score, we shall list the original hypothesis: paternalism predisposes one to live in company housing, for this is precisely one of the kinds of diffuse,

rather than functionally specific, relationships between the employee and his firm that paternalism entails.

7. *Distance of Community of Origin* from present work place. Because private housing is expensive or involves considerable commuting, employees who have migrated from more distant parts of Japan are more likely to live in company housing than are those who come from less distant communities.

8. *Number of Previous Jobs*. We hypothesize that employees who have always worked for a company are more likely to live in company housing than those who have previously worked elsewhere. This is derived from the paternalism model, which views the various components of lifetime commitment and employee integration into the firm as all of a piece.

Two of these variables emerge in both factories as important causes of living in company, rather than private, housing (See Tables D.20 and D.21, p. 390 and p. 391). These are education and the distance of community of origin from the factory. The typical occupant of company housing in both Electric and Shipbuilding factories is a university-educated employee whose community of origin is so distant from the factory as to make him dependent upon company housing. This quite pragmatic reason for living in company housing is a more convincing interpretation of the findings than the emphasis of Abegglen and some other proponents of the paternalism model on the symbolic nature of living in company housing.

When these two variables are taken into account in Electric Factory, the remaining six independent variables do not add anything further to the explanation. Residence can be explained as well on the basis of education and distance migrated as on the basis of all eight of the original independent variables.

In Shipbuilding, two other variables are also important: seniority and rank. Living in company housing is somewhat more likely if one has low seniority but somewhat higher rank.

Turning to the second question about residence, is the paternalism model correct in its implication that living in company housing has significant positive consequences for other aspects of employee behavior and attitudes—participation in company recreational activities, performance in terms of company goals, job satisfaction, cohesiveness, paternalism, and lifetime commitment? If so, are these consequences actually a result of company housing independent of other variables that influence the given aspect of behavior and attitudes? We deal with this question in various other parts of this book, in the context of each of these aspects of behavior and attitudes, but it is well that we summarize our overall findings here.

In both factories, the multiple regression analysis consistently disconfirms the paternalism model (see Appendix D, Tables D.16—D.19, D.22—D.23, and D.28—D.29 for details). Residence has, at best, only a low correlation with the variables it supposedly influences; and when the variables that do exert the most important effects on participation, performance, job satisfaction, and so on have been allowed to operate, residence consistently fails to make any independent impact on the particular independent variable.

In short, to live in company housing is to experience a form of company paternalism. But those employees who have this experience are *not* more likely than those who live in private residences to be satisfied with their jobs, cohesive with fellow employees, paternalistic, participants in company recreational activities, supporters of lifetime commitment norms and values, or higher in performance.

The paternalism model of course recognizes that firms vary in the proportion of their personnel who live in company housing. But it implies that those who do experience this form of company paternalism are thereby drawn into a greater generalized loyalty toward the firm, and become more satisfied with their general situation in the firm. Our evidence almost totally fails to support this view, and suggests that we should de-mystify the conception of the uniquely paternalistic Japanese factory, in which living in company housing has significant positive consequences for employees' participation in company recreational activities, performance in terms of company instrumental goals, job satisfaction, and other aspects of the social integration of the employee into the firm (cohesiveness, paternalism, and lifetime commitment). Our findings show that residence either has no influence on these aspects of behavior and attitudes, or that its apparent influence is in fact a result of the operation of other, correlated variables, such as education, sex, and age.

CHAPTER NINE

Social Integration of the Employee into the Company (2)

PARTICIPATION IN COMPANY RECREATIONAL ACTIVITIES

Like most large Japanese firms, Electric Company and Shipbuilding Company provide facilities and encouragement for a number of cultural, athletic, and other recreational activities. There are, for example, group outings, which an Assembly Section worker in Shipbuilding described as follows: "Once a year workers contribute a total of about ¥ 40,000, and all members of our han go to a hot springs resort on a trip." The variety of recreational activities sponsored by Electric Company can be seen in Table 9.1.

Shipbuilding Factory personnel have a larger number of *non*-company recreational alternatives, provided by the immediate Osaka metropolitan area. The following Assembly worker, who participates in rugby, sponsored by the company, but also in leisure behavior that is unrelated to the company, is not untypical: "On Wednesday we can't work overtime because we play rugby with other workers in X Plant, about 30 employees in all. On Sundays I sometimes visit my elder brother in Osaka, sometimes see a movie, and sometimes play rugby. The [company athletic] clubs are participated in mostly by the unmarried male workers."

When employees participate in company- or factory-sponsored activities, it would appear to be an indication of their company involvement. Of British workers Goldthorpe et al. state: "If . . . workers were at all closely identified with their firms and attached to them in other than an instrumental way, one might expect that a sizable proportion of them would make their firm an important focus of their out-of-work social lives" (1968, p. 90). In a Japanese context, this interpretation of participation as company involvement might be expected to draw upon another motivational source as well: company paternalism. We have seen that the majority of employees prefer a paternalistic management. This is also true for national samples of Japanese who have been asked the same paternalism questions.[1] Presumably, the employee should reciprocate the

[1] For details see Table 12.6.

TABLE 9.1
Attendance at Activities Sponsored by Cultural and Athletic Associations of Electric Company, 1968–69.[a]

	Number Attending
Athletic festival (*tai-iku matsuri*, October 20, 1968)	712[b]
Cultural festival (*bunka matsuri*)	694
Yokata (*bon*) *matsuri* (August 10, 1968)	647[c]
Farewell to 1968 dance party (December 21, 1968)	322
Free tennis contests	231
Sports tournament (*kyōgi taikai*, November 27, 1968)	208
Exchange of New Year's greetings party	193
Bowling	160
Hiking	131
Skiing	50

[a] This same list of activities was included as a checklist in the questionnaire. Hence, we know which individuals attended which activities, but our analysis will focus upon the overall level of an individual's participation, rather than on specific activities.

[b] Total attendance at this festival was 2500, including Electric Factory employees (712), their family members, and members of subsidiary firms.

[c] Total attendance at this bon festival is said to have been 4,000, when all family members and those in subsidiary firms are included.

company's paternalism by expressing his own loyalty to the company, and one way is to participate in company recreational activities.

But the distinction between employee cohesiveness and company paternalism suggests that participation may reflect employee cohesiveness in addition to, or even rather than, company involvement as a return for company paternalism. Before exploring this possibility empirically, let us examine the specific items we used to measure participation.

The measurement of participation

The Electric and Shipbuilding questionnaires contained three questions on participation (none of which was included in the Sake questionnaire). The responses, shown in Table 9.2, indicate that in general there is a somewhat higher level of participation by Electric Factory personnel than by Shipbuilding personnel. The response to the first question (variable 98/62) by Electric Factory employees shows that somewhat fewer of them are members of clubs than the company records indicate (314 vs. 363, shown in Table 9.3).[2] The second question (variable 99/63) asked for the total number of times employees took part in activities sponsored by their section, shop, han, or dormitory, or by cutlural and athletic clubs; the

[2] The discrepancy is presumably due to two factors: the 15 percent of Electric Factory employees who did not answer the questionnaire, and those who belong to more than one club (the 314 is the number of employees who belong to either or both kinds of clubs, not the number of clubs to which they belong).

TABLE 9.2
Participation of Electric and Shipbuilding Factory Employees in Recreational Activities (%)

Variable[a]			*Shipbuilding*	*Electric*
62/98.	Do you belong to a cultural or athletic club?			
	Yes		14.5	31.5
	No		85.5	68.5
			100.0	100.0
	C/C_{max} = .26**	N	(574)	(1,005)
63/99.	How many times have you participated in (a) activities sponsored by ka (sections), kakari, or shop, and (b) activities sponsored by athletic and cultural clubs [during the past year]?			
	7 or more times		18.6	16.9
	4–6 times		25.9	20.6
	0–3 times		55.5	62.5
			100.0	100.0
	C/C_{max} = .10*	N	(533)	(1,033)
64/100.	Considering all kinds of company-sponsored, section-sponsored, and cultural and athletic club-sponsored recreational activities, approximately how often did you attend during the last year?			
	Always		4.7	7.4
	Usually		35.4	61.6
	Sometimes		46.5	28.8
	Never		13.4	2.2
			100.0	100.0
	C/C_{max} = .42**	N	(568)	(1,033)

* Significant at the .05 level.
** Significant at the .01 level.
[a] The first variable number refers to Shipbuilding Factory, the second to Electric. For a complete list of variables by number, see Appendix C.

third question (variable 100/64) asked how often employees participated in all kinds of recreational activities sponsored by the company and its subunits relative to the number of times such activities were held. One might participate in certain activities only three times a year, for example, but if those events only took place three times, then the low frequency of participation would still be participation "always." We shall refer to this question as the proportion of times (as opposed to the number of times) one participated in company recreational activities. The modal employee in both factories participated fewer than four times during the previous year; fewer than 20 percent in each factory took part seven or more times (variable 99/63). As to the rate of participation (variable 100/64). the modal rate was "usually" in Electric Factory, as opposed to "sometimes" in Shipbuilding Factory.

TABLE 9.3
Number of Electric Factory Personnel who Belong to Company-Sponsored Cultural and Athelic Clubs

Number of Members in				
Cultural Clubs		*Athletic Clubs*		
Flower arrangement	76	Tennis	25	
Home economics	34	Karate	21	
Art	14	Mountain climbing	20	
Photography	14	Basketball	16	
Nature study	10	Cycling	15	
Tea ceremony	9	Softball	14	
Go	9	Ping-Pong	14	
English conversation	8	Baseball	11	
Chorus	8	Volleyball	10	
Light music	6	Judo	9	
Chess	6	Weight lifting	7	
		Archery	6	
		Track	1	
Total	194	Total	169	363

It is apparent from the degree of association (C/C_{max}) among the three measures of participation (Table 9.4) that in Shipbuilding all three have moderately strong positive relationships to each other, while in Electric Factory the main relationship is between the number of times one participates and the rate of participation. The three items were subjected to item analysis. For comparability, it was decided to retain all three items in an index of participation for each factory. The reliability of the index, estimated from the items' intercorrelations, is .40 for Electric Factory and .64 for Shipbuilding. In each factory, the higher an employee's total score on this index, the higher his or her overall participation in terms of

TABLE 9.4
Relationship between Participation Variables in Electric and in Shipbuilding Factories[a]

Variable[b]		*99/63*	*100/64*
98/62.	Belong to clubs	.10	.13*
		.55**	.37**
99/63.	Number of times participated		.50**
			.55**
100/64.	Proportion of times participated		

* Significant at the .05 level.
** Significant at the .01 level.
[a] Upper number in each cell is C/C_{max} for Electric Factory, lower is C/C_{max} for Shipbuilding.
[b] The first variable number refers to Electric, the second to Shipbuilding. For a complete list of variables by number, see Appendix C.

club memberships, frequency, and rate of participation in company recreational activities.

THE DETERMINANTS OF PARTICIPATION

We tested the following hypotheses concerning the causes of participation as measured by the index of participation.

I. Organizational status: participation varies directly with:
   1. Education
   2. Informational level of the section
   3. Rank
   4. Seniority

   Two general explanations underlie the first four hypotheses. First, having higher status in the company and receiving more of its rewards give one more of a stake in the company, and this in turn makes for more frequent participation in its recreational activities. Second, for more educated employees, who often come from more distant communities of origin, and are thereby cut off from family and local ties, participation in company activities serves a surrogate function.

II. Attitudinal and value factors; participation varies directly with:
   5. Job satisfaction
   6. Perceived promotion chances
   7. Work (rather than pleasure or family) values

III. Extra-organizational statuses:
   8. Men have more of a long-term stake in the company than women and therefore will participate more in its recreational activities
   9. Participation varies independently of age
   10. Participation varies independently of number of dependents.

   Hypotheses 9 and 10 derive from the view that in Japanese culture there is one and only one paramount organizational focus of an individual's loyalty at any given point in the life cycle. During the years one is in the labor force, this is, especially for men, the organization in which one works, *not* family and kin. If there is a conflict between loyalty to family and loyalty to firm, e.g., in regard to how one spends free time, ideally the conflict is resolved in favor of the firm

IV. Social integration of the employee into the firm: participation varies directly with:
   11. Employee cohesiveness
   12. Paternalism
   13. Residence (employees who live in company housing participate more than those who live in private residences)

14. Performance

These hypotheses are derived from the paternalism model, which implies or states that participation is positively correlated with other aspects of the intergration of the employee into the firm and loyalty to its instrumental goals.

V. Previous, extra-organizational statuses:

15. Participation is negatively related to the size of the community of origin. This is based on the assumption that participation in voluntary organizations is higher in rural than in urban areas, and that employees' early community participation experiences predispose them to greater organizational participation when they come from rural rather than urban areas.

16. Participation is directly related to the distance of the community of origin from the location of the present work place. The reasoning here is that migrating a considerable distance from place of origin to present work and residence cuts one off more from family and local ties and throws one more into company-sponsored activities as a surrogate.

17. Participation is inversely related to the number of jobs one had prior to entering the present firm. The paternalism–lifetime commitment model suggests that since by their previous interfirm mobility, employees manifested a lack of lifetime commitment to other firms, their commitment to the present firm is therefore suspect. To the extent that participation is a manifestation of loyalty to the firm, such employees should participate less in company recreational activities than employees who have always worked for the same firm. Unfortunately, from the model one might also argue that employees who had changed firms would be all the more at pains to overcompensate by participating more in recreational activities in the new firm. If both these forces are at work, the net result might be a near-zero correlation between previous mobility and participation.

These hypotheses are tested in multiple regression analyses (see Appendix D, Tables D.22 and D.23, p. 395 and p. 396). In Electric Factory we found that participation can be explained almost as well on the basis of four variables—education, sex, performance, and seniority—as on the basis of all 17 original independent variables (multiple $R = .46$ and .51, respectively). The relatively small amount of variance in participation that we can explain—$(.46)^2$ or 21 percent—can be interpreted as follows. Employees' frequency of participation is increased if they are more educated, male, oriented to performance in terms of company goals, and low in seniority.

Thus, several hypotheses are disconfirmed, either because their variables are uncorrelated with participation, or because, if they do vary with participation, the relationships are seen to be the result of *other* variables, especially education and sex. Rather than increasing participation, more seniority acts slightly to decrease it. Employees participate not so much because they are satisfied with their jobs, think their promotion chances are good, or have work (rather than family or pleasure) values, but because they are highly educated, males, relatively new to the firm, and higher in performance. Employees from smaller, rural communities do not participate more than those from cities. Participation does increase somewhat with distance migrated, but since the latter is moderately strongly correlated with education ($r = .36$), once education has been taken into account, distance migrated adds nothing further to the prediction of performance.

Types of Participation. Table 9.4 reveals that in Electric Factory the two main components of the index of participation—belonging to clubs and participation in other company-sponsored activities, open to all employees, whether club members or not—are at most only weakly correlated. The question therefore arises, does the foregoing explanation of participation apply equally well to each component of participation? It was found that it does apply to both with only these qualifications: (1) Whereas men have higher index of participation scores, it is women who are more likely to belong to cultural or athletic clubs. (2) Whereas the fewer one's dependents, the higher one's index of participation score and the more one is likely to belong to clubs, participation in company-, section- and other subunit-sponsored, recreational activities in general varies positively with number of dependents. It appears that higher organizational status conduces both to greater commitment to work values and to greater participation in company recreational activities—*other than* membership in clubs—despite the fact that employees in higher status positions have more dependents and heavier family obligations. We suggest this pattern is made possible in part by the fact that many company-sponsored recreational activities involve an employee's whole family (see Table 9.1), and to this extent the employee does not have to choose between his family obligations and the firm. At the same time, some company recreational activities are only for employees, not their families. Yet this would not pose a conflict, because, as argued earlier, for higher status male employees, the firm takes priority over the family.

Thus, apart from the fact that club membership appeals more to female employees while other kinds of participation appeal more to males and higher status employees with more dependents, in other respects the same explanation holds for both components of participation. The multivariate analysis suggests in general that participation in company-sponsored clubs *and* in non-club recreational activities has a family surrogate function for those more highly educated personnel, far from home, without dependents, who do not have extensive local ties and obligations outside of work.

In Shipbuilding Factory the sources of participation are somewhat different than in Electric Factory. To the extent that Shipbuilding employees participate frequently in company recreational activities, it is because they have somewhat better perceived promotion chances, somewhat more cohesiveness with fellow employees, somewhat more education, and have migrated somewhat farther to Osaka for employment.

One can explain almost as much of the variance in participation on the basis of promotion chances, cohesiveness, education, and distance migrated ($R = .32$) as on the basis of all 16 of the original independent variables ($R = .42$). Each of the four best predictors explains an approximately equally small portion of the variance, but even when they all operate together, they leave most of the variance in participation unexplained.

In both Electric and Shipbuilding factories, the hypotheses that participation varies independently of age and number of dependents are in one sense, supported: relationships are either non-significant or weak. But the lifetime commitment model posits a high level of participation in company recreational activities throughout the period of employment in a firm, i.e., participation does not fall off as competing demands from family increase. In our findings, the low relationship between participation and age and number of dependents is not so much based on a continued high rate of participation as on a continued moderate-to-low rate. It is not so much that young employees without dependents and older employees with dependents both have high participation, as that both have moderate-to-low participation.

COMPANY IDENTIFICATION AND CONCERNS

Reports on Japanese companies published abroad often convey the impression of a robot-like conformity and a merging of employee individuality into the company: the symbol has been the mass of employees gathered to sing the company song. It is our thesis that this is something of a caricature of actual Japanese employees' relationship to their company. To take this caricature seriously would mean believing that workers perceived no distinction between their own needs and those of the company. Yet interviews provide evidence that workers are quite able to make this distinction. We asked employees to tell us "what's good" and "what's bad" about the company. Shipbuilding personnel are sufficiently detached from total identification with their company to note frequently such problems as the adequacy of their pay: "Among the five big dockyards of the shipbuilding industry, our company's pay is the lowest. So our union isn't doing very well. In 'base-up' collective bargaining, our company waits to see what the others will do, then gives its own decision."

Most young workers quit for money reasons, not because of human relations (ningen kankei): "In general, the workers who enter the company with skill already from a previous job are more likely to leave when they get a higher pay offer." Other workers stressed the nature of the work and the hours, as well as pay: "Other chūto saiyōsha, who enter as amateurs or without the right kind of skills, are also likely to leave because they aren't accustomed to heavy industry work. . . . What's bad about this company? Wages and lack of half a day off on Saturdays are bad. Also, I have to work overtime, then take a bus home to _____ City, so it's 8:30 when I get home, too late to read."

Managers' recognition of "self-centeredness" and "egocentrism" (*rikoshugi*) on the part of workers is further evidence of less-than-total identification by workers with the company. A kakarichō in the Engine Division stated this relationship as one in which self-centeredness declines and company loyalty increases with rank: "The young workers are mostly self-centered. Hanchō are 50 percent self-centered, 50 percent management oriented. Sagyōchō are 80 to 90 percent management oriented in their thinking." A Machine Section worker also verbalized this tension between company loyalty and egocentrism: "More workers are egocentric now—even those from rural areas. For example, a 20-year-old worker who wants to see a movie will refuse overtime night work. Even I, only 28 years old, feel I represent an older generation's culture, in contrast to these young workers' egocentrism. For example, if I want to take time off to go hiking in the mountains, I ask one month in advance, but young workers do this without warning."

Facts such as these, though presented thus far only anecdotally, should occasion surprise only to those who think of the Japanese factory as one big, happy family. To obtain more systematic data on this, all three firms' questionnaires contained an open-ended question, "Recently, what concerns, worries you most? Feel free to say anything even if it's simple, and answer below." This question was designed to uncover the extent to which employees spontaneously identify with the company and its problems, in contrast to mentioning their own personal, private problems. We analyzed in the last section the extent of employee identification with the company in the form of participation in its recreational activities. But presumably, from the company's point of view at least, identification with the company's *instrumental* goals is even more important. As we built the empirical code for responses to this question, the distinction between this kind of concern and more personal, private concerns was uppermost.

Since this question came at the end of a questionnaire that asked about a variety of aspects of the company, we cannot, of course, claim that the responses are uncontaminated, i.e., purely projective, spontaneous expressions of the real degree of salience of company versus personal concerns. We can, however, compare responses across firms. Given that this question is contaminated by the context of other items in the questionnaire, it can still elicit the degree to which employees express company concerns in different firms.

Table 9.5 shows that Electric Factory employees are significantly more likely than both Shipbuilding employees ($C/C_{max} = .42$) and Sake Company employees ($C/C_{max} = .44$) to express company concerns as such; Shipbuilding and Sake employees are not significantly different in this respect. Electric Factory personnel appear to have internalized

TABLE 9.5
Spontaneous Concerns of Employees (%)

	Electric (*v. 121*)	*Shipbuilding* (*v. 86*)	*Sake* (*v. 23*)
Company concerns as such	70.7	48.1	32.5
Respondent's personal concerns in the company or in relation to the company	6.3	27.8	35.0
Personal, private concerns unrelated to the company; no explicit reference to the company	17.9	20.4	32.5
General social (societal) problems unrelated to the company	2.9	3.3	0.0
Anti-company concerns	2.1	0.4	0.0
Total	99.9	100.0	100.0
N	(379)	(270)	(40)

Electric vs. Shipbuilding: $C/C_{max} = .42^{**}$;
Electric vs. Sake: $C/C_{max} = .44^{**}$;
Shipbuilding vs. Sake: $C/C_{max} = .18$

* Significant at the .05 level.
** Significant at the .01 level.

company goals, means, and problems to a greater extent than those in the other two firms. We also found it necessary to distinguish two kinds of personal concerns—the employee's personal relationship to the company, and the employee's personal, private problems which are stated without explicit reference to the company. Moreover, in both of the larger firms—Electric and Shipbuilding—a small number of respondents expressed concern with none of the above, but instead either mentioned general societal problems or stated anti-company concerns.[3]

[3] Space limitations prevent us from quoting extensively examples of each category of response in Table 9.5. Suffice it to say that "company concerns as such" included the need to improve technology or product, human relations, alienation, and turnover; "feudalistic,

What explains these differences in spontaneous expression of concerns? We tested basically the same set of hypotheses used to attempt the explanation of other social integration variables. The expression of company concerns varies:

I. Directly with organizational status:
   1. Rank
   2. Seniority
   3. Education
   4. Pay

II. Directly with extra-organizational statuses:
   5. Age (older more likely than younger)
   6. Sex (men more likely than women)

III. Directly with attitudinal variables:
   7. Job satisfaction
   8. Perceived promotion chances

IV. Directly with performance and social integration:
   9. Performance
   10. Lifetime commitment
   11. Residence in company, rather than private, housing

V. Inversely with previous status:
   12. The fewer one's previous jobs, the more likely that one expresses company concerns

The underlying logic of these hypotheses should be familiar by now. Even in Japanese organizations, the degree of identification of the member with the organization is neither a constant nor uniformly high. The paternalism–lifetime commitment model views Japanese employees as giving paramount loyalty to their firm, but in practice employees vary in their degree of integration into—and more specifically in the present context, their identification with—the firm. If the concerns question is a measure of company loyalty and identification, the causes of this identification should lie in the same factors that conduce to other aspects of the integration of the employee into the firm. Our main proposition is that *if* integration and identification are high, if the employee expresses *company* concerns rather than personal concerns in relation to the

authoritarian social patterns"; workers' abilities and motivations to do the job; labor supply and the job classification system. The respondent's "personal concerns in relation to the company" included the burdens and dangers of the job; wages and promotion; and one's future in relation to the industry's prospects. "Personal, private concerns unrelated to the company" involved farming problems and absence from wives and children for long periods of time in Sake Company, the cost of private housing and the cost of living in general; retirement; health, sexual relations, drinking, etc. "General social problems" mentioned included the war in Vietnam, the Japanese Communist Party, and national health insurance programs. Finally, the few outright "anti-company concerns" involved such complaints as that the company mistreats its high-seniority workers.

company, or personal, private concerns,[4] it is a result of (1) specific benefits employees differentially gain from the firm (status, job satisfaction, promotion chances); (2) employees' statuses in the wider society (age and sex); (3) previous interfirm mobility; and (4) other forms of company loyalty or integration (performance, lifetime commitment, and living in company housing).

The results of the multiple regression analysis (see Appendix D, Tables D.24 and D.25, p. 397 and p. 398, for details) confirm few of the hypotheses. Only rank is significantly related to company concerns in the expected direction in both factories. In Electric Factory, one can explain company concerns almost as well on the basis of rank, age, job satisfaction, and lifetime commitment, as on the basis of all 12 original independent variables (R = .18 vs. .22). Younger employees, with higher rank and job satisfaction and less lifetime commitment are somewhat more likely to express company concerns than are older employees with lower rank and job satisfaction and greater lifetime commitment. These relationships are all, however, weak.

In Shipbuilding Factory, the conditions under which one is more likely to express company concerns are: having higher rank in the firm, higher lifetime commitment, and having had previous interfirm mobility. The last finding is the opposite of what the lifetime commitment model would predict. One can explain variations in type of concerns almost as well on the basis of these three variables as when all 11 of the hypothesized variables are taken into account. But the more striking fact is again, as in Electric Factory, that little of the variance in the concerns variable can be explained on the basis of our independent variables. Thus, Japanese employees differ in the extent to which company concerns are uppermost in their minds, but further research is needed to uncover why they differ.

ORGANIZATIONAL CONFLICT

Conflict was defined in Chapter 1, following Coser (1956:8) as "a struggle over values and claims to scarce status and resources in which the aims of the opponents are to neutralize, injure, or eliminate the rivals." Our analysis of the degree, forms, and causes of social integration of employees into the firm would be incomplete if we failed to consider social conflict.

LABOR-MANAGEMENT RELATIONS

Sake Company. In the Western world, conflict between management and labor has been institutionalized in the form of unions, collective

[4] Societal and anti-company concerns were dropped in the multiple regression analysis, so that concerns could be coded as at least an ordinal scale: from low to high values on company concerns, the remaining categories were coded (1) personal, private concerns; (2) personal concerns in the company; (3) company concerns as such. The multiple regression analysis is done for Electric and Shipbuilding factories only.

bargaining, strikes, etc. Of Japan's five leading sake firms (as measured by sales), one has a union, a second belongs to the All-Food Products Union (*Zen Shokuhin Dōmei*) and the other three—of which Sake Company is one—have no union (Nihon Keizai Shimbunsha, *Kaisha sōkan* 1970). As we have said, the closest thing to a union in Sake Company is the Masters' Association of Hyogo Prefecture, which is the only representative the workers have in wage and other negotiations with the Nada Sake Owners' Association.

We have seen that one area of disagreement, if not of overt conflict, is the level and distribution of wages. Our 1970 questionnaire data uncovered such attitudes as: the Masters' Association should be more representative of the workers; wages are too low in comparison with other industries; and considering the heavy work junior sake personnel do, there is still too great a gap between the wages of junior and senior workers. These attitudes signify conflicts between sake workers and management on the one hand, and between junior and senior workers on the other.

Electric Company. Data on the last decade of labor-management relationships in Electric Company suggest a shift from conflict to accommodation. The relative absence of conflict in Electric Factory at the time of our study was characteristic of Japan's electric machinery industry in general at the time: " . . . The electrical machinery industry is one of the industries with the lowest propensity to strike. On the other hand, particularly in the 1960's, this industry experienced high labor turnover rates among young workers (especially females). Fierce competition for labor is being waged among the mammoth electrical machinery manufacturers and between them and textile enterprises, while the drain to tertiary industries is considerable. . . . Presumably, the culminating dissatisfactions of workers due to the low degree of militancy of the union are being relieved by the high turnover of labor" (Koshiro 1969, 8.2:8).[5]

A national union of electrical workers, Denkirōren, was formed in 1953 from the merger of three anti-Communist labor groups. But real union power resides in the "enterprise union," which comprises all employees of a given electrical machinery company. Thus each of Japan's large, and some of its medium and small electrical machinery companies, has its own labor union. The history of the formation of Electric Company's union reveals that conflict was formerly at a higher level than it was in 1969.

During the mid-1950s the division director of Electric Factory had a management policy that combined paternalism and human relations; the latter had just been introduced into Japan from the United States. He was

[5] We have also benefited much from correspondence with Professor Koshiro Kazutoshi of Yokohama National University and the Japan Institute of Labor concerning the history of the Electric Company union.

at first opposed to the establishment of a union in the company. Pressure to establish a union came from both Denkirōren and some workers in Electric Company. The union was organized in 1958. Before the union was formally recognized by Electric Company, 23 members of the organizing committee carried out a sit-down strike at one of the company's other plants. Six unionists were discharged and three others suspended, but these disciplinary actions by the company were repealed when it recognized the union. At the time of union recognition, Electric Company had 6,000 employees in its several plants.

The following year, 1959, saw several strikes of a few hours' duration each; Electric Factory had one for two hours on March 20. From November 18 to 28, 1959, Electric Factory struck, along with other company plants; the dispute concerned retirement allowances and family allowances and was settled on December 5, 1959. In early April 1960, workers in all the company's plants staged a walkout for 12 hours on each of three days, demanding wage increases. Strikes were repeated later that month at Electric Factory and two other plants. On April 20, 1960 the company staged a lockout in Electric Factory. An agreement was reached on April 30.

Between late 1960 and mid-1961, there were other strikes at specific Electric Company plants which did not, however, involve Electric Factory until the spring wage offensive in 1961. That year there was a severe conflict between the company and the union: the latter demanded both a flat wage rate increase of ¥ 2,500 plus 10 percent of basic wages and a summer bonus equivalent to 3.1 months' pay. The union walked out for 24 hours on each of two days in April, along with unions in other electric firms, and individual Electric Company plants struck on other days that month. On April 25, 1961 the company locked out two of its other plants, and on the 29th, all its plants. This strike continued for 26 days, and in the end the union was compelled to scale down its demands and accept the company's offer of an average of ¥ 2,023 ,or 10 percent wage increase and 2.85 months' summer bonus. The union leaders were criticized by the rank and file because they had failed to get the original demands. The wage increase agreement and bonus agreement were signed in May and July of 1961, respectively.

During the strike period of 1958, prior to the establishment of the union, one small group of workers in Electric Factory was strongly pro-union and another small group strongly pro-management. The majority of workers were in the middle, with much swinging of sympathy from union to management and back as a result of moves by either party that the centrist workers liked or disliked. During the strikes, the Hardware Section of Electric Factory was a bastion of pro-union sympathy. Some Hardware workers threw sand into their machines to disrupt production.

At about that time the division director of Electric Factory was promoted to company headquarters. His successor, the present division director, at first adopted a policy of selecting men for supervisory positions only on the basis of their seniority and performance, regardless of their views on management-union disputes. Among those promoted were several who were very pro-union. As a result, employees who, though pro-management, had not been promoted developed an attitude of indifference toward management. This led the division director to shift to a policy of promoting mainly men who were pro-management, a policy still in effect in 1969. Workers who were very pro-union in the strike period have been systematically kept in rank-and-file positions. Some of the anti-management workers left Electric Company, but are reported to have since undergone a metamorphosis in a more conservative direction, according to a hanchō: "In the strike period, the active strikers were those who had been trained by younger Communist Party members. The leaders of the strike left the company. Now they manage subsidiary firms and have developed a management mentality. They used to be opposed to rationalization, but now they insist that it is necessary. Of the men active in the strike who stayed in the company, most have changed their mind to some extent. Only a few haven't changed their mind at all." A Hardware worker revealed the fact that the anti-management workers have not entirely given up yet: "In the past I was associated with the leftist group of workers in Hardware Section. Now I try to keep away from them, but they still try to contact me and influence me."

However, in 1969 it appeared that the majority of Hardware workers were no longer openly anti-management. A Hardware hanchō told us: "Eighty percent of the workers are satisfied with their work. A small minority, who have a leftist ideology, are not satisfied. They don't cooperate at all. We can control them by the strength of the other 80 percent. They definitely refuse overtime work. They deny the management point of view absolutely. There are several out of 36 workers like this. They include both young and older workers." A kakarichō in Hardware gave a rather similar account: "Only about 10 percent of the Hardware employees are strongly pro-union, anti-management. The other 90 percent are indifferent or pro-management. But the 10 percent are capable, with good brains, and they make propaganda among the mass of union members in the section. I don't like their agitation activities, but it is legitimate because of the union, so I can't do anything about it."

To put this supposition to a more systematic test, we compared Hardware with the other four production sections on a number of variables that tap aspects of conflict, or consequences one might infer from the existence of conflict (see Table 9.6). The proposition to be tested is that Hardware is more in conflict with the company than are Number 1 and

2 Production, Surface, and Motor sections. Therefore, it is hypothesized that Hardware will reflect this greater degree of conflict by being significantly lower than the other four sections in performance, job satisfaction, lifetime commitment and the degree to which Hardware employees believe they have influence over their superiors in making decisions about job problems.

Table 9.6 shows that Hardware is not significantly more in conflict with the company or its management than the other production sections,

TABLE 9.6

Conflict Orientation: Hardware versus Other Four Production Sections in Electric Factory

Hypothesis	C/C_{max}	*If Significant, Does Direction of Relationship Support Hypothesis?*
140. Relative to the other 4 production sections, Hardware will have lower index of performance scores.	.27**	Opposite direction
139. Relative to the other 4 production sections, Hardware will have lower index of job satisfaction scores.	.30**	Opposite direction
146. Relative to the other 4 production sections, Hardware will have lower index of lifetime commitment scores.	N.S.	
Relative to the other 4 production sections, Hardware employees will be more likely to say they have little or no influence over supervisors and managers concerning decisions about problems on the job:[a]		
55. Influence over section chief	N.S.	
56. Influence over subsection chief	N.S.	
57. Influence over second-line foreman	N.S.	
58. Influence over first-line foreman	N.S.	
Relative to the other 4 production sections, Hardware employees will be more likely:		
118. to give "I have a poor record, reputation with management" as a reason for their poor promotion chances	N.S.	
71. to attend union meetings regularly or occasionally	N.S.	
121. spontaneously to express anti-company concerns	N.S.	

** Significant at the .01 level.

[a] The questionnaire asked, "In your workshop [han or kakari in Shipbuilding, kumi or shokuba in Electric] how much say or influence do you have in practice concerning decisions about job problems, over the following people? Very much say, fairly much say, not much say, no say at all; my shop doesn't have that position." This question was developed by Tannenbaum (1968) as the instrument for constructing control graphs in organizations.

on any of the measures. Indeed, when there are significant differences between sections, Hardware is *less* in conflict than the other sections: the relationship is in the opposite direction from the hypothesis. Therefore, the proposition that Hardware personnel are more likely than other production sections to be in conflict with the company is disconfirmed.

One may, of course, question the validity of some of the measures used. One might argue that employees with a conflict orientation toward the company would not want to change firms, since they might see their situation as the same in another firm; instead, they might want to stay in Electric Company and continue to struggle within the company. In support of our findings it should be noted that there is at least a consistency between interviews and questionnaires. The latter indicate that Hardware did not have a significantly higher level of dissatisfaction and conflict than other production sections in 1969; the former indicate that while Hardware may have been the bastion of anti-company orientations and actions in the past, it no longer had this characteristic at the time of our field work.

To put these findings in perspective, we should reiterate the fact that even during the strike period, only a minority of Electric Factory workers were strongly pro-union and anti-management (or strongly pro-management and anti-union). The majority were and are in the middle.

Disagreements *within* a union may occur in Japan as well as in the West; it is clearly not always the case that the Japanese company union presents a united front in relation to management. Illustrative of this is the recent conflict between the Electric Company union and the local Electric Factory branch union over the new job classification system. As a local union steward and kakarichō in Hardware described this: "The local factory-level union does all right, but the all-company union is the agency responsible for the shift to the new job classification wage system. Because this factory is older, it has more long-seniority workers than the company's other plants. Therefore, our factory union expressed more frustration with the new wage system, which is anti-seniority. This is less a problem in the other branch unions and in the all-company union, so the latter accepted the new job classification system more readily. But the new system is against the interests of longer-seniority workers like me."

As of 1969, a number of accommodations between management and the union in Electric Company were observable. Although some of these are strange from an American viewpoint, they are not at all uncommon in Japan's enterprise unionism. The chairman of the union is a kakarichō in the Research Section of Electric Factory. As noted earlier, the enterprise unions have more power than the national Federation of Electric Machine Workers' Unions, Denkirōren: "[There is a] lack of concentration of financial and other powers in the national federation because of the powerful influence exercised by the enterprise-wide unions in the gigantic enterprises. In response to this problem efforts have been made . . . for the gradual concentration of power in the national federation" (Koshiro 1969:7).

The structure of the national and the enterprise unions has implications for the handling of conflicts. "A notable characteristic of industrial relations in the electrical machinery industry is the existence of a joint consultation system widely operating in the very large enterprises, separate from collective bargaining and unaffected by the right to strike. Productivity, wages, and other important matters relating to production plans are discussed preliminarily by joint consultation councils at the establishment or enterprise levels. Most of the matters pertaining to working conditions, except for the size of wage increases, are decided at these levels. Any matter on which no final agreement can be reached is referred to collective bargaining" (Ibid.:7–8). In Electric Factory the Labor-Employer Joint Consultation Council (*Rōshi kyōgikai*), a committee made up of 10 union officials, the factory manager, the buchō, and the associate buchō, meets every month. The division director also attends twice a year. At the beginning of the fiscal year he explains management's goals for the coming year. At the regular monthly meetings the estimated amount of overtime work for the coming month is also settled.

The questionnaire asked, "Recently, how often have you attended union meetings?" Of the four pre-coded answers, the modal pattern was frequent attendance:

	Percent
Regularly (within the past month)	61.4
Occasionally (within the past year)	25.8
Rarely	9.0
Never	3.8
	100.0
N =	(981)

Because attendance is so skewed toward high frequency, there is relatively little variation to be explained. There is some evidence that when other independent variables are held constant, participation in recreational activities and age exert at least a weak positive influence on union participation. In this sense, union participation is somewhat more common among older than among younger employees, and to some extent it is the same employees who participate both in union meetings and in company recreational activities.

Shipbuilding Factory. Whereas Electric Company has one union, with both manual and white collar employees as members, Shipbuilding

Company has one union for its A Group (manual) and another for its B Group (non-manual). Unlike Electric Factory's union meetings, which are open to all members, in Shipbuilding, union meetings are committee meetings, and only stewards—delegates from each shop—attend. Others may attend the meetings of the individual shop union, called "House meetings." At the time of our field work the main topics discussed at house meetings were the company's new job classification wage system and the annual spring base-up wage offensive, which in 1969 resulted in a 17 percent pay increase. Another union concern is to make it warmer in the factory during the winter months. At present, small stoves are located throughout the plant, but the heat is considered inadequate if one is not near a stove. One difference between the union and management in Shipbuilding is that the union is most concerned with minimum wage levels, while management stresses performance evaluation (*kōka satei*).

We did not plumb the history of unionism in Shipbuilding Company as much as in Electric Company. But the remarks of an older worker illustrate what is known of the position of unions in Japan during the 'thirties: "There was no union in the prewar period. Leftist workers were removed from work by policemen and no one heard from them again."

In the quite different postwar climate, relations between Shipbuilding Company and its unions have been relatively strike-free. An Assembly hanchō who entered the company in 1953 recalled: "We had a weekend strike here regarding base-up [wage increases], but otherwise we've had no strike experience."

Managers and rank and file

As members of imperatively coordinated organizations (Dahrendorf 1959), managers and rank-and-file employees are objectively differentiated with regard to status and authority in all the firms we studied. The implications of this structural fact are seen differently by the conflict and consensus models discussed in Chapter 1. In the conflict model, employees with authority (managers) and those without authority (rank and file) have opposing interests, and this give rise to systematically divergent behavior and attitudes on the part of the two groups, at least some of which have implications for conflict in the organization. The consensus model, on the other hand, stresses the identity of interests and the common ideology that override the authority differences between managers and rank and file. This model predicts therefore that managers and rank and file will *not* differ significantly in their support for company goals and policies or in their performance, participation, or lifetime commitment.

The results of testing these alternative propositions are shown in Table 9.7. which compares managers and rank-and-file employees in Electric and Shipbuilding factories. It should be clear that this comparison is not based on direct indicators of conflict. Lacking data on this in the questionnaire, all we can do is examine latent grounds for conflict. The logic is that if managers are significantly higher than rank and file in performance, knowledge of procedures, primacy of work values, and integration into the company (as measured by paternalism, participation, lifetime commitment, and spontaneously expressed company, rather than private, concerns), this can provide a latent basis of conflict. For example, by pressing rank-and-file personnel to improve their performance or integration into the company, managers could engender conflict if workers, for whatever reasons, think their performance or integration are already good enough. Again, if managers and rank and file differ concerning rewards—wage system preferences, reasons for these preferences, perceived promotion chances and reasons given for poor perceived promotion chances, and perceived advantages of working in the present firm relative to other firms—these differences can also be grounds for conflict between the two strata. And if managers are more satisfied with their jobs than rank-and-file, *perceive* the latter as less satisfied and cannot "understand" why they should be less satisfied, another latent source of conflict exists.

These differences do exist between managers and rank and file, in both factories, as shown in Table 9.7. Managers are more likely than rank-and-file personnel to think they have good promotion chances, say they know procedures, spontaneously express company concerns, have high performance, profess work is my whole life values as against pleasure values, and have high job satisfaction, job autonomy, and paternalistic preferences. Rank and file are more likely than managers to think their poor promotion chances are due to not having received in-company training. Some of these differences are, of course, stronger in Shipbuilding Factory, others stronger in Electric Factory. But the conflict model is supported in the sense that in most of these areas of latent sources of social conflict there are significant differences in the direction expected.[6]

On the other hand, our findings are not inconsistent with consensus models: the differences between managers and rank-and-file personnel in both factories, while significant, are in general relatively weak ($\bar{X}$

[6] One finding shows a significant difference between managers and rank and file that is in the opposite direction from that expected in the conflict model. In both factories it is managers, not rank and file, who are more likely to say their workload is "too heavy." Apart from "explaining" this as the result of brainwashing of workers by management, one is forced to the conclusion that this finding disconfirms the Marxist notion that the burdens of work fall most heavily upon the working class. Our findings are much more in line with United States research which suggests that greater authority in fact brings greater responsibility and that this is experienced in the form of a heavier workload by managers—and by those in high status occupations in general—than by workers—and those in low status occupation (Wilensky 1960).

TABLE 9.7
Bases of Potential Conflict: Differences between Managers and Rank-and-File Employees in Shipbuilding and in Electric Factories

	Area of Behavior or Attitudes[b]	Managers vs. Rank and File[a] (C/C_{max}) Shipbuilding	Electric
73/109.	Managers more likely to see their promotion chances as fairly good or ample	.63**	.50**
18/39.	Managers more likely to know procedures in case of accidents	.46**	.54**
86/121.	Managers more likely spontaneously to express company concerns (rather than personal, private concerns)	.45**	.30*
66/102.	Managers more likely to prefer a job classification wage system because "a merit system is good for morale" (rank and file more because it provides stability and recognizes experience—as in the case of the seniority wage system)	.43**	.21
65/101.	In Shipbuilding: managers more likely to prefer the seniority wage system (rank and file prefer the job-classification wage system). In Electric Factory: managers more likely to prefer job-classification wage system (rank and file more likely to think both wage systems are the same, or to say they don't know which is better)	.42**	.35**
125/140.	Managers more likely to have higher index of performance scores	.37**	.39**
52.	As their second reason for working in Shipbuilding Company, managers more likely to cite "a large company makes stable life possible" or "good management" (rank and file more likely to cite "convenient for commuting")	.35**	
51.	As their first reason for working in Shipbuilding Company, managers more likely to cite "worthwhile, responsible job" or "good management" (rank and file more likely to cite "a large company makes stable life possible")	.34**	
56.	As their second reason for ever thinking of quitting Shipbuilding Company, managers more likely to cite "complaints about management and human relations" (rank and file and manual more likely to cite "unequal treatment, no chance for promotion" or "no hope for future")	.33**	
8/31.	Managers more likely to have "work is my whole life" values (rank and file more likely to have "pleasure" values)	.28**	.41**
132/146.	Managers more likely to have higher index of lifetime commitment scores	.20**	.11

TABLE 9.7 (*Cont.*)

	Area of Behavior or Attitudes[b]	Managers vs. Rank and File[a] (C/C_{max}) Shipbuilding	Electric
124/139.	Managers more likely to have higher index of job statisfaction scores	.27**	.41**
7/30.	Managers more likely to say their work load is too heavy (rank and file to say work load is just right)	.23**	.18*
14/35.	Managers more likely to be able to use their own judgment on the job (job autonomy)	.22**	.43**
75/111.	Rank and file more likely to say their promotion chances are poor because they have not received training by the company	.16**	.14**
129/144.	Managers more likely to have higher index of paternalism scores	.15*	.18**
84/118.	Rank and file more likely to say their promotion chances are poor because they have a poor record or a poor reputation with management	.13*	.06
49/72.	Managers more likely to think few (rather than many) other firms would give them as good conditions as their present firm	.11	.19*
128/143.	Managers more likely to have higher index of participation in company recreational activities scores	.10	.14**
32.	Managers more likely to think that conveyor belt (assembly line) work is *not* "against human nature"		.14*
$\bar{X}$ C/C_{max}, 16 items measured in both factories:		.29**	.28**

* Significant at the .05 level.

** Significant at the .01 level.

[a] In both factories, managers means all personnel with the rank of first-line foreman or above (Shipbuilding N = 67, Electric N = 95); rank and file means all other personnel, i.e., manual and non-manual in non-supervisory and non-managerial positions (Shipbuilding N = 506, Electric N = 938). Comparing managers with only manual rank-and-file personnel yielded such similar C/C_{max} relationships and similar direction of relationships that it is unnecessary to report these findings separately. When a cell is blank, it indicates that that aspect of behavior or attitudes was not investigated in one of the factories.

[b] When two variable numbers are given the first refers to Shipbuilding Factory, the second to Electric. For a complete list of variables by number, see Appendix C.

C/C_{max} = .28 in Electric and .29 in Shipbuilding). If the two imperatively coordinated strata—those with and those without authority—differ in these latent sources of conflict, they do not differ to an extreme degree. Finally, considering all the aspects of latent social conflict on which we have data from the two factories, the degree of conflict would appear to be rather similar, as measured by each factory's mean C/C_{max}.

CHAPTER TEN

Social Integration of the Employee into the Company (3)

The controversy over the extent of lifetime commitment in Japanese industry was described briefly in Chapter 1. A number of studies in Japanese and in English have refuted the usual assumptions about the high level of commitment. But the stereotype persists among non-Japan specialists, and Dore's 1973 book will probably give these assumptions a new lease on life. The controversy has lacked both conceptual and methodological rigor. There has been little attempt to measure the normative and value aspects of lifetime commitment, apart from single, isolated questions in surveys. No adequate conceptual distinction has been made between actual rates of interfirm mobility on the one hand, and values and norms concerning lifetime commitment on the other. In the theory of action, "action" has role, norm, and value components (Parsons 1961, 1966). We shall view interfirm mobility rates as lifetime commitment role behavior. Thus, lifetime commitment norms (if they exist) state the conditions under which one should stay in one firm; lifetime commitment values (if they exist) give more general reasons or legitimacy for staying in the same firm.

A role is always oriented to norms and values, but the Japanese factory controversy has too often given the impression that there is a simple, one-to-one relationship between role behavior, norms, and values. We hope to show, on the contrary, that employees may stay in one firm (lifetime commitment role behavior) for reasons other than the internalization of, and conformity to, lifetime commitment norms and values.

INTERFIRM MOBILITY

In the simplest behavioral sense, lifetime commitment means that an employee stays in one firm, that is, he has no interfirm mobility. How closely does this statement describe the employees in our three firms?

In early Japanese industry, interfirm mobility, rather than lifetime commitment, was common, and the sake industry conformed to this pattern. The long-standing practice was for workers to move from plant to plant while they gained various brewing skills. It was even common to undergo one's apprenticeship in more than one plant. The labor system in traditional plants was a merit system with frequent movement of workers (Mannari, Maki, and Seino 1967:19). The scientist we interviewed in Sake Company talked about the past this way:

> Even at the end of the Tokugawa one uniform way of making sake had been established throughout Japan. So a Sasayama sake master could go to a Hokkaido sake plant, a Sendai master could come to the Kinki area, and so forth. Masters moved around.
>
> Upper and middle workers (jōbito, chūbito) also moved. They talked to their counterparts in other firms, heard that food, beds, the master, etc. are better at another plant, so they moved there the following winter. Older workers stay put more, but younger ones try other sake plants, even other industries. When making these moves, the worker didn't lose status. He either got the same status or even a higher status.

Examples of interfirm mobility in recent years are easily come by. In the 1966–67 interviews a 32-year-old farmer–sake worker said: "I moved around among various sake plants, experiencing many positions such as upper worker, sub-head, blender and head, for about 10 years" (Ibid. 1967:30). A number of Sake Company employees exhibit not only high interfirm mobility, but even high interindustry, interfirm mobility. Here are three examples: "I have almost abandoned agriculture. The last two years I worked on construction sites. Before that I had worked at the harbor." "I have completely given up agriculture. When there is no vacancy at the sake plants I can get money by driving a dump trunk. I have a plan to become a taxi driver." "I work as a plasterer in the summer and in sake plants in the winter" (Ibid. 1967:31).

Relatively frequent interfirm mobility among Sake Company employees can also be seen from more systematic data. The kachō of the General Affairs Section and the kachō of the Technical Section (the scientist) told us in 1970 that every year they have about one-third turnover of their winter sake workers. An extra bonus is given for every three years of consecutive work in the company and a silver cup for every five years of consecutive work: these practices are held to have reduced the turnover somewhat. Although some senior workers leave to become masters in other companies, in general the turnover rates in Sake Company are similar to those in any industry: inversely related to the employees' status in the company.

Table 10.1, panel 1, shows that two-thirds of kura workers in Sake Company in 1966 had worked for other sake firms before coming to Sake Company. The proportion who had thus far in their work history remained in one sake firm declined slightly, from 34.6 percent in 1966 to

27.9 percent in 1969. Even this measure of commitment may understate the actual amount of interfirm mobility, since company data tell only how many years a man has worked in other *sake* firms, not how many *non-sake* firms he has worked for. The 1966–67 data revealed that 28 percent of the employees were over 30 years old when they began to work for Sake Company. We aggregated the number of years all employees in the three traditional plants had worked in Sake Company as a percentage of the aggregate number of years they had worked in all sake breweries (panels 2–4 of Table 10.1). In both 1966 and 1969, only 56 percent of the total time employees had worked in sake breweries had been in Sake Company itself. In the aggregate, then, not only had a large proportion worked for other firms, but this previous employment comprised almost half of their total sake brewing work histories. Since Sake Company is a leading firm, the interfirm mobility of these workers represents mainly upward mobility, i.e., movement from less well-known firms to a more well-known firm.

TABLE 10.1
Previous Interfirm Mobility of Sake Company Personnel Who Work in Three Traditional Plants

		1966	*1969*
1. Number of sake firms in which one has worked			
One firm only (Sake Company)		34.6%	27.9%
More than one sake firm		65.4%	72.1%
		100.0%	100.0%
	N	(81)	(68)
2. Aggregate number of years worked in all sake brewing firms (including Sake Company)		658	810
3. Aggregate number of years worked in Sake Company		367	454
4. Sake brewing experience obtained in Sake Company (3 ÷ 2)		56%	56%
	N	(64)	(68)

We noted earlier that the company has had to employ older men, inexperienced in sake work, in order to meet its labor needs. These inexperienced older workers have high turnover rates. The 1967 questionnaire data revealed that 56 percent of the employees had had interrupted work histories, i.e., had worked some seasons in Sake Company and skipped other seasons. Thus, the available labor supply is not only older and less experienced, but also relatively unstable. If this recruitment trend continues, one would predict even higher rates of interfirm mobility in the sake industry. Thus, Mannari, Maki, and Seino (1967:20) concluded that "loyalty (*chūseishin*) of workers to the company in sake plants is not as strong as in today's big business firms."

This is borne out by our Shipbuilding and Electric factory data. Shipbuilding Company records show the following annual quit rates for various units of the company:

A Group (production workers), total company (1969)	5.0 percent
A Group regular employees in X Plant (1969)	7.3 percent
All employees (A and B Group, manual and non-manual), in Engine Division of X Plant (1968)	4.0 percent

During the year from July 1, 1968 to June 30, 1969 the quit rate in Electric Company was seven percent—11 percent for women and three percent for men. In Electric Factory during the same period, 86 out of 1,212 employees voluntarily quit, for a quit rate of seven percent, identical to that of the company as a whole.

As a summary measure, we can compare the amount of previous interfirm mobility—the proportion of employees who have worked for one or more other firms prior to their present firm—in our three firms. Shipbuilding Factory has significantly more previous interfirm mobility than Electric Factory (C/C_{max} = .53); Shipbuilding Factory males more than Electric Factory males (.51); Sake Company more than Electric Factory (.40); Sake Company males more than Electric Factory males (.46); but Shipbuilding Factory and Sake Company are not significantly different in previous interfirm mobility (.10).

ROLE BEHAVIOR VERSUS NORMS AND VALUES

An employee who voluntarily changes firms by definition violates the lifetime commitment pattern. However, the mere fact that one does *not* voluntarily change firms is not necessarily evidence that one has internalized the full-blown lifetime commitment pattern, as Abegglen and others define it. The lifetime commitment model states that there is a tightly reciprocal set of obligations between the company and the employee. The company will not discharge the employee except in the most extreme circumstances, and the employee, in return, will not quit the company for employment elsewhere. But Abegglen's model suggests to us that there are two sources of these reciprocal obligations. One is that the employee will not leave because his loyalty will be rewarded by the company, over the years, by pay increases, bonus and fringe benefits, paid vacations, promotions, retirement benefits, and, in general, by steadily advancing status in the company. Abegglen noted this first level of reciprocal obligations by asserting that lifetime commitment was most characteristic of higher status (e.g., white collar) male employees. For convenience, we shall refer to this as the status enhancement source of lifetime commitment. The second, deeper level of commitment is suggested by Abegglen's claim that "the worker, whether laborer or manager . . . is bound, *despite potential economic advantage*, to remain in the company's employ . . . [and] . . . a system of shared obligation takes the place of the economic basis of employment of workers by the firm" (Abegglen 1958:17, italics added). This appears to mean that the employee owes the company loyalty as such, and believes that he should stay in the company because it is morally right to do so, independently of how much status enhancement the firm gives him over the years. This second type will be referred to as the moral loyalty source of lifetime commitment.

Analytically, the status enhancement source of commitment can be regarded as a norm, the moral loyalty source as a value. Norms are more situationally contingent and conditional than values. As a norm among employees, perceived adequacy of status enhancement would be a condition under which one stays in the firm; perceived inadequacy a condition under which one is less likely to stay. Moral loyalty, on the other hand, is more a value than a norm because it states relatively unconditionally, "One should not change firms, because commitment and loyalty to the firm are valued attributes in themselves, 'the Japanese way,' quite apart from considerations of concrete status advantages."

Assuming that employees want to maximize their status, the question arises, can they do this more effectively by staying in one firm or by changing firms? There is no simple answer. In the decades immediately before and after the second World War rising status in large Japanese firms was generally based on a seniority system. This was defined in such a way that most rewards, such as pay and rank, were contingent on seniority in a given firm, and one forfeited all or much of the seniority one had accumulated in previous firms upon moving to another firm. In other words, there were many institutional obstacles to directly transferring one's status from one firm to another. By the late 1960s this was beginning to change, and status could be transferred somewhat more easily. Also, apparently even in the heyday of the seniority system there was a trade off between kinds of status enhancement, which continues now; individuals working for small, relatively unstable firms would seek to switch to larger, more stable firms, even though this might mean a loss of some status rewards, e.g., seniority-based pay.

Thus, we have interpreted the lifetime commitment model as stating that it is not enough if an employee voluntarily remains in one firm from the time he finishes school until he retires. In addition, his reasons or motives for staying in one firm must be that he subscribes to a particular set of values—values that enjoin staying in one firm because it is the "morally right, loyal thing to do." If we dichotomize lifetime commitment role behavior and lifetime commitment norms and values each as present or absent, we have the following fourfold table.[1]

Lifetime Commitment Role Behavior	*Lifetime Commitment Norms and Values*	
	Present	Absent
Present	I	II
Absent	III	IV

Types I and IV are "pure," i.e., commitment behavior and commitment norms and values are both present, or they are both absent. Types II and III are mixed: in Type II, lifetime commitment role behavior is present, but commitment norms and values are absent; in Type III, the reverse is true. Let us examine these four types in turn.

Type I. Lifetime commitment role behavior present, lifetime commitment norms and values also present. This is the full-blown lifetime commitment pattern: the employee remains in one firm throughout his work life because of a combination of status enhancement and moral loyalty to the company as such.

Type II. Lifetime commitment role behavior present, lifetime commitment norms and values absent. The employee remains in one firm *not* for reasons of status enhancement or moral loyalty to the company, but for other reasons: his family or regional ties limit his mobility; he is too old to move; he has high solidarity with fellow employees, both at work and outside work; his job skills are not marketable in other firms.

Type III. Lifetime commitment role behavior absent, lifetime commitment norms and values present. The employee leaves the company for reasons beyond his control. We were not able to obtain data from employees who had left the firms we studied, so we cannot say how many fit Type III. However, Table 10.2, which presents data for all Japanese manufacturing firms, has some partially relevant data. In 1967, 14 percent of the employees who left their firm did so for involuntary reasons. The most common of these involuntary reasons were (1) the worker had been hired for a fixed period, and the contract had terminated (36 percent); (2) it was to the company's convenience to discharge him (26 percent); (3) the worker was charged with a misdeed, e.g., excessive absenteeism (20 percent); (4) illness and injury (14 percent); (5) fixed retirement age (3 percent). Unfortunately, again, there is no way of knowing what

[1] Some of this was published previously as "Lifetime Commitment in Japan: Roles, Norms and Values," *American Journal of Sociology* 76 (March 1971); 795–812. Copyright by the University of Chicago. The permission of the University of Chicago Press to use this material is gratefully acknowledged. Although the discussion at this point will ignore the further distinction between norms and values, in subsequent parts of the analysis the distinction will again be introduced.

TABLE 10.2
Separated Employees, by Sex and Reason, Japanese Manufacturing Firms, 1967 (%)

	Male	Female	Total
Involuntary reasons			
Termination of contract	33	41	36
Employer's convenience	25	28	26
Worker's misdeed	23	15	20
Sickness, injury, etc.	14	14	14
Fixed retirement	5	1	3
Total	100	99	99
N	(179,800)	(130,100)	(309,900)
Voluntary reasons			
Worker's personal convenience	100	81	90
Marriage, childbirth	—	19	10
Total	100	100	100
N	(854,200)	(1,076,100)	(1,930,300)
All involuntary reasons	17	11	14
All voluntary reasons	83	89	86
Total	100	100	100
Total N	(1,034,000)	(1,206,200)	(2,240,200)

Source: Rōdō Shō, *Rōdō tōkei nempō* 1968:42, Table 27.

proportion of these employees who left for involuntary reasons had commitment norms and values. Available data often do not fit the analytical categories of sociology, and these national data on involuntary reasons for leaving a firm do not tell us whether, if these employees had not had to leave, they would have remained in a firm because of lifetime commitment norms and values, or because of other, extraneous reasons.

Type IV. Lifetime commitment role behavior absent, lifetime commitment norms and values also absent. This is the pure instrumental relationship between the employee and the firm. The employee does not believe he owes the company lifetime commitment, and whether or not he experiences what he subjectively defines as adequate status enhancement during his period of employment in the firm, he considers it legitimate—for any of numerous reasons—to shift to another firm, and, in fact, he shifts. Among the reasons are the desire to broaden the range of job skills, to get cleaner, less monotonous, more "masculine" or more "feminine" work, the desire to "stabilize one's life" by moving to a larger firm with better growth prospects, higher pay or long-range benefits, and the like. Table 10.2 shows that 86 percent of the Japanese employees who left their firms during 1967 did so for voluntary, personal reasons.

Thus, although both parties violate the tightly reciprocal norm and value expectations of lifetime commitment, employees are much more likely to do so than is the company. From 1956 to 1967, voluntary, personal reasons annually accounted for over two-thirds of the separations. The percentage who left for personal reasons increased from 68 in 1956 to 86 in 1967 (Rōdō Shō, *Rōdō hakusho* 1969:69, Table 49). We can conclude that the incidence of Type IV (leaving for voluntary reasons) is much greater than that of Type III (involuntary separation) and that the difference has been increasing.

In his important participant-observation study of two Japanese factories in which he was employed, Cole uncovered another relationship between role behavior, norms, and values: "Despite the separation rate of 17 percent a year at the auto parts plant and 23 percent a year at the diecast plant and the absence of any formal guarantee, workers and managers at both companies insisted that they had the practice of permanent employment. The explanation is that permanent employment refers to limitations on the right of employers to fire, not on the right of workers to quit" (1967, p. 299).

This is clearly a different meaning of lifetime commitment from Abegglen's. As Takezawa (1961, p. 111) puts it: "Even the conception of career employment [lifetime commitment] itself has changed, through union efforts, from management's 'handout' to worker's 'right'." Although Cole's workers violate what Abegglen calls lifetime commitment (that the employee will not leave and the company will not discharge), both they and their employers claim they have lifetime commitment. This suggests a more adequate formulation: there is a set of reciprocal obligations between the company and its employees, but the nature of the obligations is—for many, if not most Japanese employees—only in part what Abegglen claimed. Abegglen was correct in stating that on the company's side, the obligation is not to dismiss its regular employees except under the most extreme circumstances. On the worker's side, however, the obligation is to work diligently and harmoniously for as long as he is employed in that company, but not necessarily to remain in the firm permanently. Thus, although the role behavior, norms, and values Cole describes reflect only one-sided, unbalanced commitment from the point of view of those who accept the Abegglen formulation, when the obligations are differently defined they are seen as two-way and balanced. The essential difference is that the fair return the Japanese sees himself as owing his employer takes the form of what we might call present commitment, rather than lifetime commitment.

SEX AND LIFETIME COMMITMENT

Before we attempt to measure lifetime commitment, an issue must be resolved that arises from the radically different sex composition of

Electric Factory (only 49 percent male personnel) Shipbuilding Factory (96 percent male), and Sake Company's traditional kura (100 percent male). Although the lifetime commitment items in the questionnaire were answered by both sexes, there are three reasons why we shall restrict our analysis of lifetime commitment to men. First, only by doing this can we fairly test the lifetime commitment model, which is not applied to women, who, in 1966 for example, had separation rates from firms throughout Japan of 29.4 percent, as against 15.9 percent for men (Okochi, Karsh, and Levine 1973:159, Table 5.9). Hence, it is only if we find that men lack lifetime commitment that the commitment thesis will be disconfirmed. Second, given the exclusively male referent of the lifetime commitment model we are testing, three of our five questions were worded so as to refer specifically only to male employees (variables 73/138, 93/57, and 94/58 in Table 10.3 below). Third, restricting the Electric Factory analysis to men makes the findings more comparable to those for Shipbuilding Factory and Sake Company.

THE MEASUREMENT OF LIFETIME COMMITMENT

The Electric Factory questionnaire contained five items we conceptualized as lifetime commitment variables.

1. To measure previous role behavior we asked, "How many jobs did you have before coming to work in this company?" The lifetime commitment response is "no previous jobs, have always worked in this firm."
2. To measure the individual's own intended future role behavior we asked, "Do you intend to continue to work in Electric Company?" The lifetime commitment response is, "Yes, I intend to continue."
3. "Do you think male employees intend to work in Electric Company until retirement?" measures the individual's perception of how widespread lifetime commitment role behavior is among his fellow employees.
4. "What do you think about an employee who voluntarily changes a job to another company?" Five pre-coded answers were presented. The first three were; "His behavior is not that of a Japanese person," "He is disloyal (not sincere)," and "He is an unscrupulous opportunist." These are the responses we would expect from someone who regards lifetime commitment as a value. If employees define commitment as the model says they do, and if the model calls for moral loyalty to the company, independent of status enhancement ("the worker is bound, *despite potential economic advantage*, to remain in the company's employ," as Abegglen puts it), then we would expect these responses of disapprobation. The fourth response is, "I can understand his behavior," which we

interpreted as at least mild support for the norm of mobility; that is, less than complete support for lifetime commitment norms. The last alternative, "If I had the chance, I would change jobs too," is taken as an outright rejection of lifetime commitment norms and values.

5. "Do you think male employees should work (*hataraku bekida*) for the same company until retirement?" The three pre-coded answers were "should work," "it depends," and "should not work." To say, "Males should work for the same company until retirement" is to favor the value of lifetime commitment. To say "should not work for the same company" is to reject the value of lifetime commitment. To say "it depends" is to acknowledge a norm of lifetime commitment. Unlike values, norms are by definition situationally specific. Those who say "it depends" are expressing a qualified, norm-like answer: under some conditions, males should stay; under other conditions, they need not.

To these five lifetime commitment questions in the Electric Factory questionnaire we added three new ones when we constructed the Shipbuilding Factory questionnaire.

6. "Since entering Shipbuilding Company have you ever thought of quitting?" In descending order of lifetime commitment, the three pre-coded answers were: "I have not thought of quitting," "I thought of quitting, but didn't look for another job," and "I seriously looked for another job."

7. Variable 48 in Table 10.3 poses two alternative ways of thinking about company life, and asks the respondent which he thinks is better. The lifetime commitment response is, "Once entering a company the individual gets thoroughly acquainted with the jobs in that company, the company gets prosperous, the individual has achievement, and consequently the individual becomes happier." The non-lifetime commitment response is, "The system by which the person moves from a job in one company to another company, and the company gets prosperous, the individual has achievement, and consequently the individual becomes happier." The phrasing of this item was taken from an Electric Company pamphlet.

8. The last item asked, ". . . to what extent do you feel you are 'one body' (*ittai*) with the company?" The four pre-coded responses state descending degrees of lifetime commitment identification with the company, from "very much" to "no feeling at all of being one body" with the company.

The responses of Electric, Shipbuilding, and Sake employees to the role behavior variables 148/108, 95/60, and 73/138 appear in Table 10.3. As already reported, Shipbuilding employees have had significantly

TABLE 10.3
Measures of Lifetime Commitment Role Behavior and Norms and Values in Electric and Shipbuilding Factories[a] (%)

Variable[b]			*Electric*[c]	*Shipbuilding*
148/108.	Number of firms worked in prior to present firm			
	1. None (present firm is first one)		76.3	38.6
	2. One		9.6	28.3
	3. Two		3.1	20.0
	4. Three or more		11.0	13.1
			100.0	100.0
		N	(510)	(516)
	$C/C_{max} = .51^{**}$			
54.	Since entering this company have you ever thought of quitting the company?			
	1. I have not thought of quitting			24.6
	2. I thought of quitting, but didn't look for another job			59.4
	3. I seriously looked for another job			16.0
				100.0
	N.S.		N	(589)
95/60.	Do you intend to continue to work in this company (exclude inevitable reasons for quitting such as retirement)?			
	1. I intend to quit		16.2	11.2
	2. I intend to continue to work here		83.8	88.8
			100.0	100.0
		N	(451)	(560)
	$C/C_{max} = .11^{**}$			
73/138.	Do you think male employees intend to work (here) until retirement?			
	1. I think all of them intend to work (here) until retirement		3.2	4.1
	2. I think the majority intend to work (here) until retirement		55.2	58.2
	3. I think some intend to work (here) until retirement		36.1	31.6
	4. I think almost none intend to work (here) until retirement		5.6	6.1
			100.1	100.1
	N.S.	N	(504)	(591)
93/57.	What do you think about an employee who voluntarily leaves a job in one company for a job in another company?			
	1. His behavior is not Japanese		2.8	1.9
	2. He is not sincere		8.3	5.1
	3. He is an unscrupulous opportunist		8.1	4.2
	4. I can understand his behavior		70.6	73.0
	5. If I had the chance, I would change jobs too		10.4	15.8
			100.2	100.0
		N	(472)	(544)
	$C/C_{max} = .15^{**}$			

TABLE 10.3 (*Cont.*)

Variable[b]		*Electric*[c]	*Shipbuilding*
94/58.	Do you think male employees should work for the same company until retirement?		
	1. They should	13.3	38.9
	2. They should not	2.4	8.2
	3. Depends on the time and place	84.3	52.9
		100.0	100.0
	N	(504)	(565)
	$C/C_{max} = .45$**		
48.	In company life there are two ways of thinking. From your experience which do you think is better? Please give your real feelings.[d]		
	1. The system by which the person moves from a job in one company to another company, and the company gets prosperous, the individual has achievement, and consequently the individual becomes happier.		28.8
	2. Once entering a company the individual gets thoroughly acquainted with the jobs in that company, the company gets prosperous, the individual has achievement, and consequently the individual becomes happier		71.2
			100.0
	N		(562)
137.	Through your service to Shipbuilding Company, to what extent do you feel you are "one body" (*ittai*) with the company?		
	1. Very much a feeling of being one body		8.9
	2. Fairly much a feeling of being one body		44.3
	3. Not too much a feeling of being one body		36.4
	4. No feeling at all of being one body		10.4
			100.0
	N		(583)

** Significant at the .01 level.

[a] Variables 54, 48, and 137 were asked only in Shipbuilding Factory.

[b] When two variable numbers are given, the first refers to Electric Factory, the second to Shipbuilding. For a complete list of variables by number, see Appendix C.

[c] Male employees only.

[d] Variable 48 is the only one in this table that was included in the Sake Company questionnaire. Of the 50 Sake respondents who answered, 9.1 percent checked alternative (1), 90.9 percent alternative (2).

more interfirm mobility than Electric Factory employees (variable 148/108). Shipbuilding personnel are significantly, but only somewhat, more likely ($C/C_{max} = .11$) than Electric Factory personnel to say they intend to continue working in their present firm (variable 95/60). These two factories are not significantly different in the proportion of male employees who are perceived to be committed to the present firm until retirement (variable 73/138). Considering these three role behavior questions together, we note that both Shipbuilding and Electric factory personnel are more likely to say they themselves are committed to their present firm than to perceive the majority of their fellow employees as committed (variable 95/60 vs. variable 73/138). About one-third of those who say they want to continue evidently regard themselves as in the minority, in the sense that they perceive the majority as not committed to the present firm until retirement.

Turning now to the norm and value questions concerning lifetime commitment, three interfirm comparisons are possible in Table 10.3. Only 19 percent in Electric Factory and 11 percent in Shipbuilding expressed moral disapprobation with regard to an employee who voluntarily changes firms (variable 93/57), i.e., think such behavior is disloyal, opportunistic, or not Japanese. The great majority say they can understand the behavior of the person who voluntarily changes firms. Another minority of 10 percent in Electric and 16 percent in Shipbuilding went even further and said, "If I had the opportunity, I'd change jobs too," a response that represents a repudiation of lifetime commitment values. Electric Factory males are significantly, but weakly, more committed in this aspect of values than are Shipbuilding employees (C/C_{max} = .15).

On variable 94/58, "Should males work for the same company until retirement?" an even smaller minority in Electric Factory agreed outright with this commitment value: 13 percent. Eighty-four percent gave the more norm-like answer, expressing some support for interfirm mobility; "It depends;" only two percent went to the extreme of rejecting the value of lifetime commitment ("Males should not stay in one firm"). Thus, the most common response by far is that under some conditions ("it depends") the employee may change firms. Lifetime commitment in Electric Factory, while relatively well supported in role behavior terms—both past behavior and future intentions—is relatively little supported in absolute value terms. Shipbuilding personnel express significantly more lifetime commitment to this value question than do Electric Factory personnel (C/C_{max} = .45). Over one-third say, "Male employees should work for the same company until retirement," and only 53 percent give the more norm-like answer, "It depends."

Sake Company employees are overwhelmingly committed to the staying-in-one-firm value orientation; only nine percent believe it is better for the firm and the individual to move from company to company (variable 48). We have here an example of the somewhat independent variation between role behavior and norms and values. Sake workers are more committed to one firm in the normative and value sense, but less committed to one firm in the role behavioral sense (actual interfirm movement) than are employees in the more modernized Shipbuilding Factory. Although almost all Sake employees ideally prefer to stay in one firm, the majority of them have not, do not, and apparently cannot live up to this ideal preference. Sake Company, therefore, has the lifetime commitment pattern as a value, but not as a fact of role behavior. Although the majority in Shipbuilding also favor the staying-in-one-firm ideology, Sake personnel are significantly higher on this measure of lifetime commitment ($C/C_{max} = .18$).

The remaining two questions in Table 10.3 were answered only by Shipbuilding personnel. In variable 54, one out of four respondents give the pure lifetime commitment response; "I have never thought of quitting," but 59 percent have thought of quitting, though they did not go so far as to look for another job; 16 percent have seriously looked for another job. Finally, 53 percent of Shipbuilding employees "very much" or "fairly much" feel they are one body with the company, but 36 percent feel this "not too much," and 10 percent do not feel themselves at all one body with the company (variable 137).

To what extent do responses to these questions in each factory co-vary? Only to the extent that an individual who gives a lifetime commitment response to one question also gives a lifetime commitment response to other questions is it feasible to construct an index of lifetime commitment. Table 10.4 presents the matrix of associations—eight variables for Shipbuilding and five for Electric Factory. With the exception of the number of previous jobs (variable 148/108), all the variables are significantly and moderately strongly positively related in both factories. For example, employees who think males should stay in one company are more likely than those who think it depends or that they should not stay to perceive most males as having decided to stay in the company, to feel they are one body with the company, to favor the staying-in-one-company ideology, to have not thought of quitting, to intend to continue working in the same firm, and to express moral disapproval of the employee who voluntarily changes firms. Those whose role behavior or norms and values deviate from lifetime commitment in one variable, similarly, tend to deviate in other variables as well.

The fact that seven of the 11 associations between previous interfirm mobility (variable 148/108) and other lifetime commitment variables are not significantly related in Electric Factory or Shipbuilding Factory, and that Sake personnel are committed to the ideology of staying in one company but not committed in their actual role behavior has far-reaching implications for the lifetime commitment model. It disconfirms the model's expectation of an inverse relationship between previous mobility and lifetime commitment to the present firm. Employees who have always

TABLE 10.4
Degree of Association between Lifetime Commitment Variables in Electric and in Shipbuilding Factories[a]

Variable[b]		*73/138*	*137*	*48*	*54*	*95/60*	*93/57*	*148/108*
94/58.	Should males work for one company?	.37** .41**	 .43**	 .45**	 .48**	.28** .42**	.45** .48**	.28** .23**
73/138.	How many males intend to stay?		.48**	.30**	.40**	.34** .33**	.26** .38**	.19 .18
137.	Feel "one body" with company			.35**	.42**	.41**	.47**	.23**
48.	Staying/moving as best for company and individual				.26**	.30**	.34**	.14
54.	Ever thought of quitting					.27**	.44**	.22*
95/60.	I intend to continue/quit						.40** .44**	.10 .07
93/57.	Attitude toward person leaving firm							.17 .21
148/108.	Number of previous jobs							

* Significant at the .05 level.
** Significant at the .01 level.
[a] Upper number in cell is C/C_{max} for Electric Factory males, lower is C/C_{max} for Shipbuilding Factory; if there is only one cell entry it is for Shipbuilding, since variables 54, 48, and 137 were asked only in Shipbuilding Factory.
[b] When two variable numbers are given, the first applies to Electric Factory, the second to Shipbuilding. For a complete list of variables by number, see Appendix C.

worked for the same firm are not more committed to their present firm than those who have worked elsewhere. We can only speculate, not test, why this is so. One possibility, of course, is measurement error: employees who have changed firms in the past may be somewhat cautious in answering questions about staying with or leaving their present firm. But if the finding is valid, it might reflect the fact that many with previous employment elsewhere have moved up simply by entering Electric Company or Shipbuilding Company, since they are large, stable, leading firms; this could increase their commitment. Among those who have always worked for Electric or Shipbuilding Company, on the other hand, there is no direct comparative basis of judgment; some may remain in the company more from necessity than choice and may have a "grass is greener elsewhere" orientation. Such employees may lack strong lifetime commitment norms and values, even though they exhibit lifetime commitment role behavior.

These counterpressures may result in the observed lack of relationship between past interfirm mobility and present support for commitment norms and values.

To return to our distinction between role behavior and norm and value aspects of lifetime commitment, four variables in Table 10.4 refer mainly to role behavior, while the other four refer mainly to norms and values. The mean C/C_{max} for all relationships between each role behavior variable and each norm and value variable, as shown in Table 10.5, is .29 for Electric Factory, .35 for Shipbuilding, and .33 for both factories combined. This finding on the one hand confirms our assertion that there is considerable independent variation between role behavior and norms and values in the area of lifetime commitment: the mean strength of the relationships between variables is .33, moderate at best, rather than strong. On the other hand, most of the relationships are statistically significant, and this suggests the variables are at least to some extent measuring a common phenomenon.

The lifetime commitment model asserts a higher degree of covariation between role and norm and value aspects of lifetime commitment than our findings show. At the same time, since the model states that lifetime commitment involves both role behavior and norms and values, our attempt to measure the concept of lifetime commitment would be seriously defective if we did not include both aspects in our index of lifetime commitment.

TABLE 10.5
Relationships between Role and Norm/Value Aspects of Lifetime Commitment in Electric a
in Shipbuilding Factories[a]

Lifetime Commitment Variables that Refer Primarily to Norms and Values	*Lifetime Commitment Variables that Refer Primarily to Role Behavior*			
	148/108. Number previous jobs	*54. Ever thought of quitting*	*95/60. I intend to continue/quit*	*73/138. M decision*
93/57. Attitude toward	.17		40**	.26**
person leaving firm	.21	.44**	.44**	.38**
94/58. Should males	.28**		.28**	.37**
work for one company?	.23**	.48*	.42**	.41**
48. Staying/moving as best	14	.26**	.30**	.30**
137. Feel "one body" with company	.23**	.42**	.41**	.48**

* Significant at the .05 level.
** Significant at the .01 level.
[a] Upper number in each cell is C/C_{max} for Electric Factory males, lower is C/C_{max} for Shipbuild When only one number appears in a cell, it refers to Shipbuilding respondents When two variable num are given, the first applies to Electric Factory, the second to Shipbuilding.

The index of lifetime commitment we constructed contains variables 95/60, 73/138, 93/57, and 94/58 (see Table 10.4).[2] The mean lifetime commitment index score—summed for each individual over the four items—is slightly higher in Shipbuilding Factory than in Electric Factory (8.7 vs. 8.5). This difference is significant, though weak ($C/C_{max} = .13$).

THE DETERMINANTS OF LIFETIME COMMITMENT

Since our data refer mostly to one point in time, we cannot in any strict sense treat our variables as a recursive system. We cannot, for example, say unambiguously that there is only a one-way causal influence *from* job satisfaction *to* lifetime commitment. Nevertheless, as a heuristic device, we shall consider the following model.

As shown in Figure 10.1:

FIGURE 10.1

Causal Influences on Lifetime Commitment

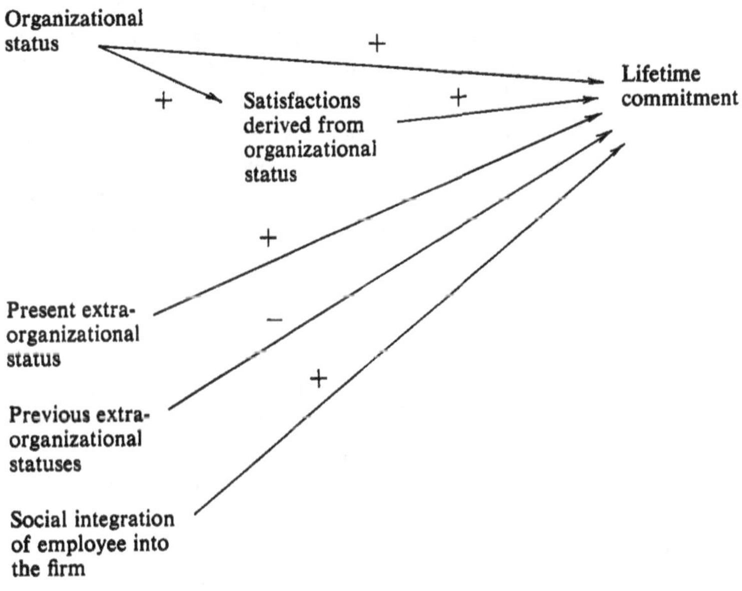

[2] The ITEMA program found that variable 148/108, number of previous jobs, had the lowest item-total correlation with the other four items in Table 10.4 for Electric Factory, and with the other seven items for Shipbuilding Factory. It was accordingly dropped from the index. The reliability of the index, alpha, is .51 for Electric Factory males and .59 for Shipbuilding Factory males. We also constructed a separate index of lifetime commitment for Shipbuilding personnel, using seven variables in Table 10.4 (but not variable 148/108); this has a reliability of .69. See Appendix B for further details on these indexes.

I. Lifetime commitment varies directly with organizational status

II. Lifetime commitment varies directly with satisfactions derived from aspects of organizational status

1. Job satisfaction
2. Perceived job autonomy
3. Perceived promotion changes
4. Perceived relative advantages of firms

Organizational status thus has both direct and indirect (through the above four kinds of satisfactions) effects on lifetime commitment.

III. Lifetime commitment varies directly with present extra-organizational status.

5. Age (Sex is not considered because we have already restricted the analysis to male employees.)

IV. Lifetime commitment varies inversely with previous extra-organizational statuses:

6. Lifetime commitment is inversely related to the size of one's community of origin. This is based on the assumption that smaller, rural communities are more traditional in social structure than larger, urban communities. Since lifetime commitment is an analytically traditional pattern, it is more likely that those from more traditional communities would develop it.

7. Lifetime commitment varies inversely with the number of previous jobs. This hypothesis derives from the paternalism–lifetime commitment model, which assumes that lifetime commitment norms and values are all of a piece with lifetime commitment role behavior, and that therefore those who have always worked for the same firm should manifest lifetime commitment values and norms more than those who have changed firms.

V. Lifetime commitment varies directly with other aspects of the social integration of the employee into the organization:

8. The higher one's cohesiveness with other employees, the greater one's lifetime commitment

9. The more one favors a paternalistic company, the greater one's lifetime commitment

10. The more frequent one's participation in company recreational activities, the greater one's lifetime commitment

11. Employees who live in company housing have higher lifetime commitment than those who live in private housing.

Before testing these hypotheses it should be noted that interviews, observation, and company records in Shipbuilding provided data on a number of factors that maintain or reduce lifetime commitment. Some of these factors and their consequences are similar to those that operate in

Electric Factory; others are unique to Shipbuilding Factory. Space limitations prevent us from detailing the factors that have positive or negative consequences for lifetime commitment; the reader is referred to an earlier analysis (Marsh and Mannari 1972).

ELECTRIC FACTORY

Although the index of lifetime commitment combines norm and value measures, we saw earlier that the kind of lifetime commitment that exists is more a norm, that is, contingent, than a value. Commitment is conditional, not ultimate or absolute. But if it is conditional, on what does it depend? Let us consider each of the explanatory hypotheses in turn.

Status and Lifetime Commitment. The hypothesis that the higher one's status in the company, the higher one's lifetime commitment, is only weakly confirmed ($r = .11$). Male employees of higher status profess slightly more lifetime commitment than those of lower status.

It is possible that while organizational status in a relatively static sense is measured by a combination of all six status variables, the more dynamic concept of status enhancement is not. Status enhancement refers to benefits that increase over time for the employee in a firm. In principle and in fact pay and job classification definitely rise with increasing seniority. In principle rank does too. Seniority by definition increases over time; however the benefits the employee realizes are not seniority itself, but its by-products—pay, job classification, and rank. Education does not increase over time: typically one's educational status is the same at retirement as it was on entering the firm. The same is true in most cases of the informational level of the section in which one works. Thus, we constructed a subindex, status enhancement, consisting of only pay, job classification, and rank, with an alpha reliability coefficient of .75.

The product-moment correlation between the original six-item status index and the new status enhancement index among Electric Factory males is .90, indicating that the two are measuring virtually the same thing.[3] Lifetime commitment varies only weakly with the status enhancement index scores ($r = .15$), just as it did in relation to the status index ($r = .11$). We therefore accept, but without much confidence, the first hypothesis derived from Abegglen, that lifetime commitment norms and values increase as a positive function of status enhancement.

The low relationship between status and lifetime commitment might have been produced, of course, by a uniformly high level of support for lifetime commitment norms and values by all employees, regardless of status. This finding, while not supporting the status enhancement hypothesis, would at least have supported the empirical generalization that Japanese males in large, successful firms are characterized by a high level

[3] See Appendix B for further details on the status enhancement index.

of lifetime commitment. It should be clear from our earlier discussion that this is not the case. The actual mean lifetime commitment score in Electric Factory, 8.5, as in Shipbuilding, falls midway between the lowest (4) and highest (12) score a person can have on the index. Lifetime commitment values of moral loyalty are not widespread among Electric Factory employees. To explore this further, employees who said they intend to continue to work for Electric Company were asked, in an open-ended question, for their reasons. These included security and other economic reasons (38.4 percent); nature of the work—I like the work, it's suitable for me, worthwhile—(32.6 percent); I live nearby (6.3 percent); I am too old to change jobs (5.8 percent); other reasons—just started work here, no other openings, all companies are the same—(16.8 percent).[4]

What we were looking for in this question was the spontaneous expression of moral loyalty to the company as the reason for lifetime commitment. Statements on the order of, "I intend to stay here because it is right, morally proper that I do so" are conspicuous by their absence. The most common answers are couched in terms of security and other economic reasons and the worthwhile nature of one's work. (This last is a clue that job satisfaction is an important source of lifetime commitment, and we shall explore this below. But job satisfaction is clearly not a distinctively Japanese reason for commitment to a firm.) The "security and other economic reasons" response might be regarded as a reflection of status enhancement providing that such a reason is given more often by higher status than by lower status employees. For it is those higher in seniority, pay, job classification, etc. who in fact have more security and higher economic benefits. Table 10.6 shows how the two most common reasons for staying in Electric Factory vary with status.

The results are in the opposite direction from the status enhancement hypothesis: it is lower rather than higher status employees who are signi-

TABLE 10.6

Males' Reasons for Continuing in Electric Company until Retirement, by Status (%)

	Index of Organizational Status Score	
Reason for Continuing (*v. 96*)	*Low* (*6–11*)	*High* (*12–18*)
Security and other economic reasons	66.1	45.6
Intrinsic work satisfactions	33.9	54.4
	100.0	100.0
N	(56)	(79)
C/C_{max} = .28*		

* Significant at the .05 level.

[4] Of the 378 males who want to stay in Electric Company, 189 did not give a reason; the N for these responses is the remaining 189.

ficantly more likely to give security and other economic reasons for staying in the company ($C/C_{max} = .28$). If these results do not support the distinctive Japanese culture emphasis, they do support a very common proposition in the sociology of work, often confirmed in studies in the West: intrinsic work satisfactions are expressed more by higher status employees, security and economic reasons more by lower status employees, even though the former are, of course, taking higher security and economic rewards for granted (Lawler 1971:37–59). Table 10.6 shows that the ratio of economic reasons to intrinsic "work itself" reasons for staying in the company is 2:1 among low status employees but less than 1:1 among high status employees.

Of the other reasons for staying, both the 6.3 percent who say it is because of the proximity of the factory to their residence and the 5.8 percent who stay because they are "too old to move" are giving reasons that are extraneous to lifetime commitment norms and values. They exemplify Type II of our fourfold table: lifetime commitment role behavior present, lifetime commitment norms and values absent. We conclude that hypothesis 1 is disconfirmed: lifetime commitment varies almost independently of status in the company.

Job Satisfaction and Lifetime Commitment. The second hypothesis is that lifetime commitment varies directly with job satisfaction. We have already seen some evidence that the worthwhile nature of the work is a spontaneous reason for commitment. This links directly to the index of job satisfaction, which is significantly positively related ($r = .25$) to lifetime commitment.

Perceived Job Autonomy. The hypothesis that lifetime commitment varies positively with employees' perceived job autonomy is not supported; the two vary independently ($r = -.07$).

Promotion Chances and Lifetime Commitment. Another factor hypothesized to have a positive influence on lifetime commitment is an employee's perceived promotion chances. Other things being equal, perceiving that one has good promotion prospects in the company might provide a stimulus toward lifetime commitment. However, we found only a weak relationship ($r = .12$) between promotion chances and lifetime commitment.

Relative Advantages of Firm. The questionnaire asked, "Do you think there are many firms which would give you the advantages you have in Electric Factory?" Our hypothesis is that perceiving that there are few other firms that equal, let alone outdo, one's own firm as a place to work is at least a negative source of lifetime commitment. If, by moving, one cannot improve one's conditions of work, why move? That the relationship between perceived relative advantages of firms and lifetime commitment works in this direction seems to us more plausible than the direction postulated in the lifetime commitment model. There, perception of the relative advantages of firms is irrelevant to lifetime commitment. Given that one has lifetime commitment, it makes little difference, according to the model, whether one perceives few or many other firms as offering the same or better conditions. The crucial fact is simply that they are *other* firms, not one's own; one's commitment means that for better or for worse one is committed to one's *own* firm. The situation is analogous to the ideal-typical monogamous marriage: there may be many other attractive women in the world, but one's wife is one's wife, and regardless of the range of hypothetical alternatives, commitment is made to one's present mate. We are not arguing that this is the way Japanese employees in fact view other firms; only that this is the way the ideal-typical lifetime committed employee would view them.

Of the three pre-coded answers to this question, 37.8 percent of the respondents chose "few firms" (with as good conditions as this company), 21.9 percent thought "many firms" have better conditions, and 41.2 percent said they did not know. As an ordinal scale variable, this is coded as 1 = many other firms; 2 = don't know; and 3 = few other firms. The higher the value on this variable, then, the more the individual sees Electric Company as offering unique advantages, relative to other firms. Before discussing this variable's relationship to lifetime commitment, we pause to note the characteristics of employees who perceive Electric Factory as offering unique advantages. The multiple regression analysis, presented in detail in Appendix D, Table D.26, p. 399 shows that this perception is most likely to be found among employees who are satisfied with their jobs, grew up relatively close to Electric Factory and therefore lack the broadening experience that coming from a more distant part of Japan may bring, and have cohesive relationships with their fellow employees.

We continue now with the analysis of variables that influence lifetime commitment. Our hypothesis that employees who think few other firms offer the advantages of Electric Company will have greater lifetime commitment than those who don't know the relative advantages of firms, or who think many other firms offer better conditions is confirmed, but there is only a weak relationship ($r = .16$).

Age. The hypothesis that lifetime commitment varies directly with age is disconfirmed. The two vary independently of each other ($r = .10$).

Size of Community of Origin. The hypothesis that the smaller and more rural one's community of origin, the greater one's lifetime commitment is disconfirmed: there is no relationship ($r = -.01$).

Number of Previous Jobs. Contrary to the hypothesis that the greater one's previous interfirm mobility, the less one's lifetime commitment (in the normative and value sense), empirically there is no relationship between the two variables ($r = .06$).

Other Social Integration Variables. Lifetime commitment is at best only weakly correlated with the other aspects of social solidarity and worker-company integration: the correlation between employee cohesiveness and lifetime commitment is .23, that between paternalism and commitment .05, and that between participation in company recreational activities and commitment .07. The hypotheses that lifetime commitment and these other aspects of social integration in the company are mutually dependent are confirmed only in the case of cohesiveness. Employees who have higher lifetime commitment have this for reasons other than their paternalistic preferences and their participation in company activities.

The hypothesis that living in company (rather than private) housing increases one's propensity to have a lifetime commitment to the firm is also disconfirmed; the two variables vary independently of each other ($r = .00$).

Although, as we have seen, most of the independent variables have at best weak relationships with lifetime commitment, we included them all in a multiple regression analysis (see Appendix D, Table D.28, p. 402), in order to check for interaction effects. The upshot of this analysis is that one can predict lifetime commitment virtually as well on the basis of job satisfaction, cohesiveness, and job autonomy, as on the basis of all 12 of the original independent variables. *Job satisfaction is the major source of lifetime commitment.* Insofar as Electric Factory males have high commitment to these norms and values of life long loyalty to the firm, it is more because they derive satisfaction from their job, think they have autonomy in their job, and have cohesive relationships with other employees, than because of their status, perceived promotion chances, view of the relative advantages of firms, age, participation in company recreational activities, paternalism, residence, previous interfirm mobility, or size of community of origin. At the same time, most of the variance in lifetime commitment is left unexplained by our indepedent variables.

SHIPBUILDING FACTORY

Before testing the same hypotheses concerning the causes of lifetime commitment in Shipbuilding Factory, let us first consider the perceived relative advantages of firms variable. Only 3.6 percent of Shipbuilding employees think, "There are few firms with as good conditions as this company;" the great majority (74.9 percent) think, "There are many firms with better conditions than this company;" 21.6 percent said, "don't know." Shipbuilding Factory employees are therefore significantly more likely than Electric Factory employees to deny that their firm offers unique advantages (C/C_{max} = .68). Nor is this simply a result of the differential sex composition: Shipbuilding males are also significantly more likely than Electric Factory males to think, "Many other firms offer better conditions than this firm" (C/C_{max} = .59).

Perceived relative advantages of firms was regressed on the same 10 independent variables used in Electric Factory, (omitting, as usual, sex). The results, shown more fully in Appendix D, Table D.27, p. 400, are that job satisfaction and distance migrated predict perceived relative advantages almost as well as all 10 variables together. It is employees with higher job satisfaction and from nearby communities of origin who are most likely to think their firm offers unique advantages. This is consistent with the findings for Electric Factory, and suggests that perceiving one's present firm as offering unique advantages may reflect a kind of innocence that comes from working in the same area where one grew up. Those who have migrated might be assumed to have a broader knowledge of firms, which slightly reduces their sense of working in a firm with unique advantages. This interpretation should be tested in further research.

We need here report only the overall findings of the regression analysis we used to test the hypothesized causes of lifetime commitment in Shipbuilding; it is displayed more fully in Appendix D, Table D.29, p. 403. In Shipbuilding Factory the characteristics that most conduce to a higher level of lifetime commitment are: having more job satisfaction, being older, being more cohesive with fellow employees, and perceiving few other firms as offering as good conditions as Shipbuilding Company. Status in the company appears to influence lifetime commitment only because higher status employees tend to have more job satisfaction, are older, and are more cohesive. Status does not exert any influence on lifetime commitment independent of the influence of these variables. This finding again undercuts the lifetime commitment model by showing that status advantages in a firm are not a main source of lifetime commitment norms and values.

SUMMARY: THE SOCIAL INTEGRATION OF THE EMPLOYEE INTO THE COMPANY

In this and the preceding two chapters we have analyzed several aspects of the social integration of the employee into the firm: employee cohesiveness, paternalism, residence, participation in company recreational activities, company identification and concerns, organizational conflict, interfirm mobility and lifetime commitment. What are our most important findings?

1. In Table 10.7, which presents a correlation matrix for the social integration variables,[5] the generally extremely weak correlations indicate that the analytically differentiated integration variables vary independently of each other to a considerable extent. Employees who are integrated into the company on one dimension—e.g., paternalism—are as likely to be unintegrated as integrated on another dimension—e.g., lifetime commitment. As noted at the beginning of Chapter 8, the strong form of the lifetime commitment model would expect low zero-order correlations between integration variables because of low variance on each variable, i.e., because all employees tend to be high in all aspects of integration into the firm. Is this what our findings reflect? High integration on each variable was operationally defined as follows: expressing company concerns, living in company housing, not having worked for any other firms, having a paternalism score of 4, and having a score in the upper third of the range on the other three indexes—14 or higher on cohesiveness, 7 or higher on participation, and 10 or higher on lifetime commitment. By these criteria, the two factories have the following percentages of highly integrated personnel:

Electric Factory		*Shipbuilding Factory*	
No previous jobs	79%	Paternalism	59%
Company concerns	71	Company concerns	50
Paternalism	56	No previous jobs	39
Cohesiveness	30	Lifetime commitment	31
Lifetime commitment	17	Cohesiveness	28
Participation	9	Company housing	13
Company housing	8	Participation	12

The evidence provides little support for the strong form of the lifetime commitment model: the low correlations among integration variables are not due to a clustering of all, or even most, employees at the high end of each integration scale.

2. The weaker form of the lifetime commitment model would suggest that although employees' scores on each integration variable have relatively high variance, there is a strong positive correlation between integration variables. Table 10.7 also fails to support this interpretation, since all the correlations are weak. However, we do not argue that the weaker form of the lifetime commitment model should be completely

[5] Because our measurement of organizational conflict is less adequate than that for other social integration variables, we omit the conflict variable from the correlation matrix.

TABLE 10.7

Pearsonian Correlation Matrix for Social Integration Variables[a]

		Range of scores[b]	*Mean* ($\bar{X}$)	*Standard deviation* (*s*)	*128/86*	*2/102*	*148/108*	*141/127*	*143/128*	*144/129*	*146/132*
128/86.	Concerns	1–4	3.456	0.890	1.00	−.02	.05	−.01	.02	.11	−.04
			2.288	0.794	1.00	.05	.10	.06	.05	.01	.20
2/102.	Residence	1–2	1.920	0.270		1.00	−.03	.13	−.29	.04	−.08
			1.865	0.341		1.00	.16	−.10	−.05	.02	.07
148/108.	Number of previous jobs	1–4	1.391	1.301			1.00	.13	.04	.06	.11
			2.077	1.052			1.00	−.07	−.21	.05	−.01
141/127.	Cohesiveness	6–18	12.491	1.772				1.00	.03	.07	.19
			12.201	2.196				1.00	.17	.07	.29
143/128.	Participation	3–9	4.647	1.241					1.00	.07	.11
			4.412	1.606					1.00	.07	.03
144/129.	Paternalism	2–4	3.439	0.690						1.00	.08
			3.485	0.678						1.00	.11
146/132.	Lifetime commitment	4–12	8.465	1.452							1.00
			8.726	1.570							1.00

[a] For each variable, the upper number refers to Electric Factory personnel, the lower to Shipbuilding Factory personnel. Data on all variables except lifetime commitment refer to both male and female employees; lifetime commitment is based on males only. The first variable number refers to Electric Factory, the second to Shipbuilding.

[b] High integration = high score on all variables except residence and number of previous jobs, where the ordering is reversed and high integration = low score.

rejected on the basis of our data. Measurement error and other methodological problems could account for the low observed correlations. For example, correlations tend to be restricted when the variables are measured only in terms of a narrow range of variation, as is the case with our measures of residence, paternalism, number of previous jobs, and concerns. Correlations also tend to be low when the variables have differently shaped distributions. Thus, in Shipbuilding, the low r between number of previous jobs and cohesiveness could result from the fact that cohesiveness has a normal, bell-shaped distribution, while number of previous jobs has a descending, monotonic distribution. For these reasons, our first conclusion is that our data disconfirm the strong form of the paternalism–lifetime commitment model: the integration of the Japanese employee into the firm is by no means as total as the model asserts. Our second conclusion is that although our data do not support the weaker form of the model, neither do they clearly disconfirm it. All we can say is that the absence of strong positive correlations between integration variables may reflect problems of measurement rather than a true absence of the predicted positive relationships.

3. Somewhat different independent variables account for each aspect of social integration; there is no one master explanation which holds for all areas of integration. Thus, the best predictors of employee cohesiveness, when other variables are held constant, are job satisfaction and lifetime commitment (in both factories), plus (in Electric Factory) education and the informational level of one's section and (in Shipbuilding Factory) participation and rank. On the other hand, living in company housing (rather than private housing) is a result of being more educated and migrating from a more distant part of Japan. Rank and lifetime commitment are among the best predictors of whether one spontaneously expresses company concerns or private concerns. And job satisfaction and cohesiveness are among the major sources of support for lifetime commitment norms and values.

4. A majority of employees in all three firms prefer diffuse, paternalistic relationships with superiors and with the company to functionally specific relationships. However, only a small minority of personnel in Electric or Shipbuilding factories live in company housing, and those who do are not more favorable to paternalistic relationships than are those who live in private residences. Our data also disconfirm the implication of the paternalism model that living in company housing has significant positive consequences, independent of the influence of other variables, for employees' participation in company recreational activities, performance in terms of company instrumental goals, and job satisfaction.

5. In Chapter 1 we asked whether employees participate in company recreational activities mainly as an expression of their loyalty to the company, their loyalty to their fellow employees, some other factors, or a combination of all of these. To test the relative influence of company loyalty (as measured by performance, paternalism, and lifetime commitment) and employee cohesiveness, we regressed participation on these four variables in each factory (table not shown). When the other three variables are held constant, only performance (beta = .22) and lifetime commitment (beta = .07) explain any significant amount of the variance in participation in Electric Factory; the betas for paternalism and cohesiveness are .04 and −.02, respectively. In Shipbuilding Factory, only cohesiveness (beta = .17) and performance (.14) explain any significant amount of the variance in participation; the betas for lifetime commitment and paternalism are −.07 and .05, respectively. All four independent variables together explain only six percent of the variance in participation in Electric Factory and five percent in Shipbuilding.

This suggests that employees' participation reflects company loyalty somewhat more than employee cohesiveness in Electric Factory, while it reflects a relatively equally small amount of each in Shipbuilding. More importantly, Tables D.22 and D.23, pp. 395 and 396, show that the best predictors of participation are for the most part variables other than those that reflect company loyalty or employee cohesiveness: education, sex, and seniority (in addition to performance) in Electric Factory; perceived promotion chances, education, and distance of community of origin from the factory (in addition to cohesiveness) in Shipbuilding. In short, employees do not seem to participate in company activities either because of company loyalty or because of cohesive ties with fellow employees. Nor is there any indication that some employees participate mainly for reasons of company loyalty, while others participate primarily because of employee cohesiveness. Instead, insofar as we can explain participation at all, it appears to be due to other factors: it serves as a family and local community surrogate for university educated employees, new to a firm, who have migrated from distant communities of origin.

6. Electric Factory employees are the most likely, and Sake employees the least likely, spontaneously to express company concerns (identification with company instrumental goals and problems). But our variables do not go very far toward explaining variations among employees within each factory in this regard.

7. As members of imperatively coordinated organizations, managers and rank-and-file employees in each of our firms are objectively differentiated with regard to status and authority. Conflict models predict

that this will give rise to structured conflict between the two strata, while consensus models emphasize the identity of interests and common ideology that override the authority differences between managers and rank and file. Our data provide some support for both models; on balance, however, our measures did not uncover a high degree of structured conflict at this time in any of the firms studied.

8. To explain variations in lifetime commitment among males in large Japanese factories we have tried to follow through some basic conceptual distinctions between lifetime commitment role behavior (degree of actual interfirm mobility) and lifetime commitment norms and values. The reciprocity between the Japanese employee and his company exists, but its symmetry is both less than, and different from, what the lifetime commitment model claims. Employees who exhibit lifetime commitment role behavior (i.e., stay in one firm) often do so not on the basis of lifetime commitment values, but for a variety of reasons extraneous to these values: job satisfaction, cohesiveness, security and other economic reasons, lack of opportunity to move to another firm, family and local ties, and age.

We disagree with those who argue that there is a distinctively Japanese pattern of lifetime commitment. Insofar as Japanese employees in large firms do remain in one firm, this is due mainly to factors other than loyalty to the company as such. Japanese employees' motives for staying in one firm are essentially the same reasons that tie Western employees to a firm.

CHAPTER ELEVEN

Performance in Japanese Firms

Most studies relevant to the lifetime commitment model have been concerned only with the nature of the social organization of large Japanese firms, what Dore (1973) calls their employment systems. Chapters 3 through 10 of our study have accordingly been devoted to a reexamination of this topic. At least one proponent of the lifetime commitment model, however, has gone beyond this, and advanced an intriguing proposition concerning the relationship between the social organization of firms and their performance. The central purpose of this, our penultimate chapter, is empirically to test Abegglen's assertion that "it is as a consequence of having developed a different, Japanese approach to organization that Japan has accomplished the industrial success that it has" (Abegglen 1969:100).[1]

Organizational performance is defined as the degree to which the organization achieves its goals (Price 1968). An organizational goal is defined, following Etzioni (1964:6), as "a desired state of affairs which the organization attempts to realize." Organizations with high performance are effective (they realize their goals) and efficient (they realize their goals with minimum cost). The ultimate goals of economic organizations are productivity and sales, and if they are part of a capitalist economy, profits. To achieve these, however, organizations must meet ancillary or subsidiary goals: a low rate of absenteeism, suggestions from employees, which improve efficiency, quality control, and the like. We refer to all these as aspects of performance, and shall analyze them in the following order: (1) absenteeism, (2) the suggestion system, (3) performance at the individual level, (4) labor productivity at the level of the section or plant; and (5) factory-level differences in performance.

At appropriate points in this analysis, we shall consider several explanations of performance: Abegglen's, a social differentiation theory of performance, human relations theory, Japanese employees' views on how to improve productivity, a Japanese managerial theory of productivity, and our own theory.

[1] See Chapter 2 for basic data on trends in performance in each of our firms and factories.

ABSENTEEISM

In both Electric Company and Shipbuilding Company the number of days' vacation to which an employee is entitled increases as a function of his seniority. For example, Shipbuilding employees are allowed nine days' vacation in their first year, 14 days in the second through fifth years, and from the sixth year on, one extra day for every additional year worked, up to a maximum of 20 days. However, the concept of not coming to work during these "vacation days" has not yet been fully institutionalized in Japan. In the words of a Machine Section worker in Shipbuilding Factory: "I have had no absences in nine years of work here. I have accumulated 35 days of paid vacation. I took six days off, but about half of the other 29 will be cancelled by mid-March if I don't take them. In the past, I've taken vacations only when a relative got married, or my wife was sick, but not when I was sick. So I'll probably forfeit these days of paid vacation again."

Our measure of absenteeism is, accordingly, somewhat different from those used in the West. Instead of assuming employees do not work in the firm on their vacation days, and asking on how many other work days during a year they do not come to work, we measure absenteeism by the variable we shall call the ratio of absences, i.e., the number of days an employee was absent in relation to the number of vacation days to which he was entitled, during the previous year. This variable is coded (1) absent one or more days *more* than number of vacation days, (2) worked 1–3 days of vacation, (3) worked 4–7 days of vacation, (4) worked 8 or more days of vacation. The modal pattern in both Electric and Shipbuilding factories, as shown in the first panel of Table 11.3 is to work eight or more days of one's vacation. Only a small minority—six percent in Electric Factory and 12 percent in Shipbuilding—were absent one or more days in excess of their vacation during the year prior to our field work.

This ratio of absences variable is one component of our index of performance; we shall have more to say about it below.

THE SUGGESTION SYSTEM

A second specific aspect of performance is the suggestion system (*teian seido*). In Electric Factory each worker is encouraged by his foreman to think of a suggestion each month. A theme is often stressed in a given month, e.g., how can we reduce costs of production? Forms available for writing one's suggestion ask for the person's name, a name for the suggestion, a description of the present method, and a description of idea for a new, better method.

Electric Factory has developed an elaborate set of criteria and weights for grading suggestions. Table 11.1 presents an abbreviated view of the system for evaluating suggestions.

The amount of award money given for the suggestion is determined by the total number of weighted points, shown in Table 11.2.

A Hardware kumichō told us that in its actual operation, the suggestion system includes several elements not explicitly contained in its ideal formulation: "The voice of the workers is reflected in their suggestions. However, the number of suggestions and the morale of workers are not always highly correlated. There is 'section consciousness' and rivalry between sections to outdo each other in making suggestions. Therefore supervisors aid their workers in writing and presenting suggestions." A hanchō in Number 1 Production echoed this view: "I have to give hints to my workers for suggestions. If they are left alone, they don't think up good suggestions." Thus, one female worker, when asked what her suggestion had been, replied that she couldn't remember, and admitted that her hanchō had actually submitted a suggestion in her name. Another worker reported: "I gave four suggestions last year. But actually my hanchō and kumichō wrote two of them." There is, of course, not only nothing illegitimate about this, but in fact this kind of cooperation is functional for the organization. A vague formulation has less value than a well-written suggestion. Since workers often lack the skill to put a suggestion in its most usable form, the assistance of a supervisor is distinctly of value. Thus, when a Hardware worker submitted a suggestion for a new air spring for use in removing pressed products he received only a seventh grade award (¥ 500). His kumichō commented to us: "He did not write it in a good way. It should have been more than a seventh grade award. Workers want to make suggestions that will make their work load lighter, and increase productivity."

The overall importance of the suggestion system is indicated by the fact that during the year from July 1968 to June 1969 Electric Factory, by making use of employees' suggestions, realized a savings of ¥ 65,400 per worker (¥ 5,450 per worker per month). During this period employees submitted 4,442 suggestions—3.6 per employee. Foremen encourage each worker to make one suggestion a month, 12 a year, but this goal is only 3.6/12 or 30 percent realized. Another difference between ideal and actual patterns is that although in principle an employee can receive as much as ¥ 30,000 for a "special grade suggestion," no one in 1968–69 had received more than ¥ 3,000 as a total amount of awards for all suggestions submitted.

We uncovered relatively little on the suggestion system from interviews in Shipbuilding Company. We asked an Assembly Section worker what

TABLE 11.1
Criteria for Grading Suggestions in Electric Factory

Criterion	*Number of Points Allowed up to Maximum Weight*					*Maximum Weight*
Amount of cost reduction[b]	0–6 Under ¥ 50,000	7–11 ¥ 50,000	. . . 28–30[a] ¥ 1.0 million	31–34 ¥ 1.5 million	35 ¥ 2.5 million	35
Scope of application	1 Almost no application	2 Very limited	3 Limited	4 Large	5 Overall application	5
Difficulty of application	1 Impossible	2 Hard to execute	3 Requires changes in other parts	4 Possible to execute but requires changes in other parts	5 Easy to execute	5
Morale, Q.C., sales increase. service	0–1 No improvement	2–4 Little	5–12 Some	13–19 Considerable	20–25 Improves a great deal	25
Extent of originality	0–4 Imitative	5–9 Little	10–13 Some	14–17 Very original	18–20 Very considerable originality	20
Extent to which the suggestion can eliminate hard work	0–2 Not at all	3–4 Not very much	5–6 Some	7–8 Considerable	9–10 Very much	10
					Total	100

[a] Ellipses indicate omission of some categories, done to simplify the table.

[b] Five extra points are given for a cost reduction of five million yen or more. This means that theoretically an employee can receive 40 points for this criterion, and, should he or she receive the maximum on all the other criteria, a total of 105 points.

TABLE 11.2
Scale of Awards for Suggestions in Electric Factory

Grade of Suggestion	*Points*	*Amount of Award*
Special	96+	¥ 30,000
First	95–90	10,000
Second	89–80	5,000
Third	79–70	3,000
Fourth	69–62	2,000
Fifth	61–55	1,000
Sixth	54–48	700
Seventh	47–40	500
Eigth	39–30	300
Good work (*kasaku*)	Less than 30	200
For trying hard (*doryoku*)	Less than 30	200
Idea only	Less than 30	200

the sources of suggestions from workers were: "Those workers with the most toilsome jobs try to think of ways of easing them, so they give suggestions." An oyakata in the Machine Section told us the nature of his suggestion: "I tried to shorten the official time on the work schedule allotted for finishing a job. I did this by suggesting improvements on the machine's method of bite." This person described the changes he had suggested with the real interest, even passion, of a young middle school graduate worker with a strong achievement motive. He was a paragon of the worker who contributes to the suggestion system, as opposed to the worker who merely goes through the paces of his job.

The Design Section is the source of a number of suggestions. One of the staff members of this section, a 1962 polytechnical university graduate, described the suggestion system from his angle:

> Workers' award suggestions come to us in Design to be evaluated. Our own suggestions are bigger, and have to be evaluated by kakarichō and senior engineering and staff personnel.
>
> I received two awards for suggestions in 1969. One was for a plan for a remote control system for our company's diesel engines. This thesis was a joint research project, and we received a fifth grade award for it. I also got a second grade award for a suggestion on "The Turbocharger: The Effect of the Press Function." My awards were for draft changes, and didn't involve laboratory work—I could do it right here in the Design Section without going over to the plant laboratory. Laboratory work would have required a budget request and approval. My draft changes for the remote control system have been incorporated.

The stimulus for my suggestions came from two sources: customer demand and the fact that I understood the mechanism of our company's diesel engine. I had to understand that mechanism before I could develop a remote control system for it.

I received a higher award for my turbocharger suggestion than for my remote control system suggestion because the former could be manufactured right in the company, whereas the latter had to be built outside our company.

Since employees' behavior and attitudes in regard to the suggestion system are measured as one component of our index of performance, we shall defer the attempt to explain differences in this area until later in this chapter. We have described two aspects of performance in Electric and Shipbuilding factories—absenteeism and the suggestion system. We turn next to a measure of performance that combines an employee's attitudes and behavior concerning absenteeism and suggestions, as well as a third aspect of performance, fulfilling company production goals.

PERFORMANCE AT THE INDIVIDUAL LEVEL

In Chapter 5 we set up the conceptual domain of job satisfaction, defined as the satisfaction the employee sees his job as giving, or not giving him. We now distinguish a second conceptual domain related to the job—performance, defined as those contributions the employee makes or wants to make to the company that are related to the achievement of company goals.

The five variables used to measure individual performance, with the responses of Electric and Shipbuilding Factory personnel, are shown in Table 11.3. With regard to the first variable, attendance or the ratio of absences, Electric Factory employees work significantly more of their vacation days than do Shipbuilding employees ($C/C_{max} = .21$). As noted above, employees in both factories are entitled to more days' paid vacation as their seniority increases. High seniority employees in this sense can eat their cake and have it too: they can be absent more days than low seniority employees, and yet, because they are allowed more days off, still be credited with working on more of their days off than employees with less seniority. However, the fact that Electric Factory employees work on more of their vacation days than do Shipbuilding employees cannot be due to their having more seniority, since in fact Shipbuilding employees have significantly more seniority ($C/C_{max} = .45$)—and therefore more vacation days—than Electric Factory employees. Electric Factory employees thus work on a larger proportion of their (fewer) vacation days. One possible explanation of this is that attendance may be negatively

TABLE 11.3
Measures of Performance for Electric and Shipbuilding Factory Employees[a] (%)

Variable[b]		Electric Factory	Shipbuilding Factory
17/103.	Attendance. Number of days absent in relation to the number of legitimate vacation days during the previous year.[c]		
	1. Worked 8 or more days of vacation	41.8	43.1
	2. Worked 4–7 days of vacation	30.3	18.7
	3. Worked 1–3 days of vacation	21.6	26.0
	4. Absent one or more days more than number of legitimate vacation days	6.3	12.2
		100.0	100.0
	N	(1,033)	(566)
	$C/C_{max} = .21^{**}$		
33/10.	How often do you think about making suggestions?		
	Always	18.5	17.9
	Sometimes	60.0	67.6
	Rarely	19.3	11.0
	Not at all	2.1	3.5
		99.9	100.0
	N	(1,030)	(593)
	$C/C_{max} = .17^{**}$		
18/11.	How many times have you submitted suggestions during the last 12 months?[c]		
	0–2 times	31.8	66.8
	3–4 times	31.8	22.2
	5–11 times	31.2	8.9
	12 or more times	5.2	2.1
		100.0	100.0
	N	(1,033)	(585)
	$C/C_{max} = .47^{**}$		
19/12.	During the last year, how much money in awards for suggestions did you receive in all?[c]		
	¥ 0	65.3	31.6
	¥ 499 or less	20.9	14.2
	¥ 500 or more	13.8	54.2
		100.0	100.0
	N	(1,033)	(572)
	$C/C_{max} = .56^{**}$		
28/5.	Are you anxious to fulfill each day's production goal?		
	1. Very anxious	22.9	21.6
	2. Fairly anxious	42.3	48.1
	3. Not very anxious	31.4	26.4
	4. Not anxious at all	3.4	3.9
		100.0	100.0
	N.S. N	(1,027)	(594)

** Significant at the .01 level.

[a] None of these performance questions was included in the Sake Company questionnaire.

[b] The first variable number refers to Electric Factory, the second to Shipbuilding. For a complete list of variables by number, see Appendix C.

[c] This question was used only in the questionnaire in Shipbuilding Company; the data for Electric Company were taken from company records.

related to the length of the work week. Electric Factory's higher attendance may be due to the fact that it has a five-day week, while Shipbuilding Factory has a six-day week.[2]

Variables 33/10, 18/11, and 19/12 in Table 11.3 concern the suggestion system. Although Electric Factory employees claim they think about making suggestions somewhat less often than those in Shipbuilding (C/C_{max} = .17), they are much more likely than those in Shipbuilding to have made five or more suggestions during the preceding year (C/C_{max} = .47). Yet the amount of award money for suggestions is significantly higher in Shipbuilding Factory (C/C_{max} = .56). Only two percent of all Electric Factory employees received ¥ 2,000 or more in awards, in contrast to 14 percent of those in Shipbuilding; 16 percent of the latter as opposed to three percent of the former had won awards of ¥ 1,000–1,999.

These differences in award money are probably related to the technological and economic scale and complexity of the two factories. A suggestion is more likely to involve larger and more complex tools, machinery, and products in Shipbuilding than in Electric Factory. Although there are more suggestions made in Electric Factory, the typical suggestion involves relatively minor and small-scale adjustments, and thus brings a smaller money award than in Shipbuilding Company. The pattern appears to be one of an inverse relationship between the proportion of workers who are actively involved in the suggestion system and the scale of their typical contribution.

The fifth performance variable in Table 11.3 shows no significant difference between Electric and Shipbuilding factory personnel in the extent to which they are anxious to fulfill each day's production goal. The modal response in both factories is "fairly anxious," and approximately two-thirds of the employees in each factory say they are either "very anxious" or "fairly anxious" to fulfill production goals.[3]

Table 11.4 presents a matrix of relationships between all pairs of these performance variables in each factory. All relationships are positive: all but two are statistically significant, and in strength they range from weak (C/C_{max} = .12) to strong (.78).

Since variables 17/103, 18/133, and 19/134 in Table 11.4 measure actual or objective performance, in contrast to variables 33/10 and 28/5, which

[2] Thus far, a literature search has not yielded any evidence on the relationship between length of the work week and attendance. We are greatly indebted to Professor William Form for helping us in this search.

[3] To our mild surprise, we even encountered the phenomenon of the rate-breaker, and the attendant social interaction between the rate-breaker and his fellow workers, in Shipbuilding Factory. A 55-year-old worker, about to retire after 22 years' service, said: "I'm seen by my fellow workers as a rate-breaker because I try to finish a blueprint job in eight hours, and then go home. My fellow workers want to stretch out the same amount of work into the evening and get overtime pay, so I have a low evaluation among them."

TABLE 11.4
Relationships among Performance Variables in Electric and in Shipbuilding Factories[a]

Variable[b]		33/10	18/133	19/134	28/5
17/103.	Attendance	.34** .21*	.30** .20	.19** .21**	.34** .23**
33/10.	Think of suggestions		.30** .50**	.16** .42**	.57** .43**
18/133.	Number of suggestions			.44** .78**	.21** .21
19/134.	Awards for suggestions				.12* .25**
28/5.	Production goal				

* Significant at the .05 level.
** Significant at the .01 level.
[a] Upper number in each cell is C/C_{max} for Electric Factory employees, lower is C/C_{max} for Shipbuilding Factory employees.
[b] The first variable number refers to Electric Factory, the second to Shipbuilding. For a complete list of variables by number, see Appendix C.

measure more subjective aspects of performance, we computed the mean association ($\bar{X}$ C/C_{max}) between (1) all pairs of objective performance variables, (2) the subjective performance variables, and (3) between all objective and all subjective performance variables. The mean association among objective performance variables is .31 in Electric Factory and .40 in Shipbuilding. The association between the two subjective performance variables is .57 in Electric Factory and .43 in Shipbuilding. The mean association among all objective performance variables and the subjective performance variables is .25 in Electric Factory and .30 in Shipbuilding. When objective peformance variables are correlated with each other, or when subjective performance variables are correlated with each other, there are somewhat stronger relationships than when objective variables are correlated with subjective variables. Thus, we may tentatively conclude that our performance cluster contains two somewhat distinct subclusters of variables, an objective and a subjective performance cluster. There is considerable independent variation between employees' claims to be concerned about suggestions and production goals on the one hand, and their objective performance on the other.

The five performance variables were subjected to ITEMA analysis, and taken together form our index of performance. The alpha coefficient of reliability is .57 in Electric Factory and .57 in Shipbuilding Factory.[4]

[4] Another variable (70/47), measured by responses to the questionnaire item, "What do you think about the teamwork in your shop?" had conceptual links with both employee

Earlier we distinguished conceptually between job satisfaction and performance and on the basis of item-analysis constructed an index of job satisfaction and an index of performance. Since there are significant relationships between some of the six job satisfaction variables on the one hand and some of the five performance variables on the other, the question arises whether there are statistical grounds, in addition to conceptual ones, for maintaining this distinction. To test this we subjected all 11 of these satisfaction and performance variables simultaneously to ITEMA analysis. Among Shipbuilding employees, in the first five trials the variables eliminated for having the lowest item-total correlation were consistently performance variables. This was also the case in Electric Factory, with one exception.[5] In this sense we may conclude that the variables we had conceptually grouped together belong together statistically as well. The job satisfaction index and the performance index are measuring different phenomena.

Before attempting to explain performance, we should note that there is one bit of evidence for the validity of our index of performance. That the index in fact measures performance is indicated when we compare the labor productivity of the five production sections in Electric Factory (see Table 11.6). Labor productivity, which is described more fully below, is a company measure, constructed entirely independently of our measure of performance. Basically, it measures the number of units produced per worker per month. Our data show that employees in the sections with higher labor productivity have significantly higher index of performance scores ($C/C_{max} = .40$) than those in sections with lower labor productivity.

THE DETERMINANTS OF PERFORMANCE

To explain performance at the individual level, as well as other aspects of performance considered later in this chapter, we begin with Abegglen's theory and then entertain alternative theories.

---

cohesiveness and performance. On the basis of item-analysis, it was decided that the teamwork variable belonged more with the index of employee cohesiveness. For details on how the coding of each performance variable was standardized, and other details on index construction, see Appendix B.

[5] In the first six trials for Electric Factory personnel the variables eliminated for having the lowest item-total correlation were all performance variables, with the exception of, "How often do you want to be moved to another job in the factory?" (v.40/13), which had been conceptualized as a job satisfaction variable. This variable is something of a maverick in that it has the lowest item-total r of the job satisfaction variables with the job satisfaction index (.464), and the lowest item-total r of any of the performance variables with the performance index (.050). However, it is more of a job satisfaction variable than a performance variable because it has a higher item-total r with the former than with the latter index, and because it has stronger relationships with each of the other five job satisfaction variables ($\bar{X}\ C/C_{max} = .47$) than with each of the five performance variables ($\bar{X}\ C/C = .14$).

Abegglen on social organization and performance

Many students of modernization, complex organizations, and industrial sociology outside Japan who cite Abegglen claim that he asserts that Japanese firms, with a social organization basically different from their counterparts in the West, nevertheless achieve as high levels of performance as do firms in the West. Two points need to be made concerning this radically relativistic assertion. The first is that although Abegglen is certainly not alone in claiming Japanese firms have a paternalistic-lifetime commitment type of social organization, most proponents of this view see this type of organization as resulting in less effective performance. For example, Kurihara (1971) reports that Japan's Economic Planning Agency estimated that the productivity of labor for the Japanese economy as a whole would decrease despite an upward trend in general technology, for the 1960–70 period, in contrast to the 1950–60 period. Kurihara notes: "One possible answer is that there are many determinants of labor productivity other than technological progress, especially in the Japanese economy with its traditional permanent employment practice and the seniority wage payment system *obstructing the mobility of labor from less productive to more productive industries or sectors*" (1971:82–3, italics added). In other words, Kurihara makes the non-relativistic statement that low rates of interfirm mobility make for low labor productivity—even in Japan.

The second point is that Abegglen's own position on the relationship between social organization and performance has changed over the years. In *The Japanese Factory* (1958) Abegglen actually argued that the Japanese paternalistic type of factory organization results in considerably lower performance than the more modern (Western) type:

> . . . in comparing their plants with American factories producing similar items, few Japanese executives would venture a productivity proportion as high as 50 percent of the comparable American unit. . . . One plant, using American processes and machines to produce a product under American patents and thus identical in factory setup to the American firms, was reported to produce at a rate of about 60 to 70 percent of the American company. This was the highest percentage reported by any plant observed. . . . *A good part of the problem in Japan may be attributable not so much to technical factors as to the effects of the social organization on the productivity system* (1958:110–11, italics added).

Abegglen specified the aspects of Japanese factory organization that depress productivity: lifetime commitment (". . . The national economy would be aided by an increased efficiency in factory output resulting from increased labor mobility . . ." 1958:16); recruitment ("The most effective placement and the most productive use of personnel undoubtedly require rather careful fitting together of position and person. . . . In these respects the Japanese firm is considerably handicapped by its present practice . . ." 1958:115); the reward system ("There can be no question that promotions so heavily, indeed almost exclusively, governed by age and seniority are costly to efficiency. . . . From the point of view of efficiency it makes no sense whatsoever that two men doing identical work side by side should, by reason of size of family, age, or some other consideration irrelevant to their work output, receive wholly different [pay]" 1958:116); authority and responsibility (there is a "customary avoidance of fixing on particular persons the responsibility for errors or failure in production. . . . Authority is not well defined and responsibility is not easily assigned" 1958:118).

These quotations from Abegglen's 1958 book should come as some surprise to those who know it only from its citations in the literature. At that time Abegglen was a relativist and an anti-convergence spokesman only in the sense that he believed that in any society, if the social organization of factory production is to be effective, it must be adapted to that society's distinctive sociocultural traditions and present social structure. But he did not deny the validity of invariant cross-societal propositions about the relationships between social organization and performance, and in that sense he was not a relativist.

In 1958 Abegglen could validly state that industrial performance was higher in the West than in Japan. Japan was still in a period of sluggish postwar recovery. Events since 1958, however, provide increasing evidence of Japanese firms overtaking Western firms in industrial performance.[6] To be sure, we lack controlled comparisons of productivity in specific Japanese and Western firms and industries. But in an overall sense it is increasingly doubtful that Japanese industry lags in performance. If the "performance gap" between Japanese and Western industry is narrowing, what, according to Abegglen, has been happening to the social organization of Japanese firms? In 1966 Abegglen collected data for a 10-year comparison (1956–1966) of 25 large Japanese firms in a variety of industries. By the logic of his position in 1958, and certainly by the logic of ours in this study, one would expect a trend from paternalistic toward more modern forms of social organization. Since the above quotations acknowledge that performance is positively related to the degree of modernization of social organization, one would expect the increase in performance to be associated with an increase in the modernization of factory social organization.

[6] See pp. 4–5, where we show the increasing number of Japanese firms in the top 200 manufacturing and mining firms in the world.

In part this is what Abegglen's 1966 data do show. There was a decline from 1956 to 1966 in the proportion of employees who live in company housing (Abegglen 1969:111); a "move toward compensation based on job and output and away from compensation based on age and education" (Ibid.:113). Most interestingly, although Abegglen generally found lifetime commitment to be still the normal pattern, one of the 25 firms "deviated from normal practice and recruited a number of middle management personnel from the outside labor market [instead of directly from school]. This was, by a good margin, *the fastest growing company in this group*" (Ibid.:116, italics added).

Yet the basic position Abegglen appears to take in this study is that (1) the increased performance in Japanese firms is due primarily to greater capital investment and more advanced technology (Ibid.:104), since (2) the social organization of these 25 firms is for the most part unchanged, i.e., as "pre-modern" and paternalistic as a decade earlier. Thus, "Japanese methods of organization, when different from those in the United States, are not necessarily less effective or less rational. . . . There is little indication that these companies are moving toward the American, or Western, model in the ways in which they recruit, reward, and punish personnel. . . . There has been . . . a general tendency to assume that Japanese companies are changing in their methods of organization. These data do not encourage that view" (Ibid.:118).

We conclude that performance, according to Abegglen, is maximized when the following conditions are met: (1) technology is modernized, and (2) factory social organization is adapted to the traditions and present social and cultural structure of the society of which it is a part; i.e., it may be relatively paternalistic or relatively modern, as long as it is adapted to the wider society. On the question of the degree of modernization in Japanese firms and in the wider society Abegglen appears to have changed his position somewhat between 1958 and 1969. In 1958 (Chapter 6) he speaks of factory social organization as lagging behind the newer urban social patterns. In 1969 he claims that Japanese factories are still paternalistic and pre-modern and yet are adapted to Japanese society, which appears to mean that the factory no longer lags behind the wider society. Our position is that there is no significant lag, not because both factory and society are still pre-modern in social organization, but because both factory and society are relatively modernized, and are becoming more modernized at about the same rate. We argue throughout this book that Abegglen has underestimated the trends toward modernization in Japanese firms; we shall argue in the next chapter that he also underestimates the trends toward modernization in Japanese society and culture in general.

Hypotheses to be tested

Of the hypotheses concerning the determinants of individual performance (as measured by the index of performance), we list first those in which we make the same prediction as Abegglen; then we list those in which we make different predictions.

1. *Organizational status.* Because higher status employees are expected to perform better and are rewarded better, we hypothesize a positive relationship between status and performance.
2. *Sex.* Men are higher in performance than women. This is more a result of the fact that men have higher organizational status and job satisfaction than of sex as such.
3. *Age.* On the one hand, given rapid changes in technology, younger employees should be higher in performance than older ones. On the other hand, the experience that comes with age has some positive consequences for performance; moreover, older workers are more likely to have responsibilities and a work rather than pleasure orientation. Given these cross-pressures, the best hypothesis is a weak relationship between age and performance.
4. *Size of community of origin.* It is often said that employees from rural areas have been socialized into an ethic of hard work to a greater extent than those from urban and metropolitan areas. Therefore, we hypothesize that performance varies inversely with size of community of origin.
5. *Job satisfaction.* The main causal flow, we hypothesize, is from organizational status to job satisfaction to performance. High job satisfaction results from high status, and high performance results from both high status and high satisfaction.
6. *Knowledge of procedures.* One condition of high performance is knowledge of company procedures concerning work. The same individuals who are motivated to perform well, we hypothesize, are motivated to learn company procedures. Therefore, there is a positive relationship between knowledge of procedures and performance.
7. *Perceived promotion chances.* Expectation theory would suggest that the better one thinks one's promotion chances are, the more incentive for high performance; therefore, we predict a positive relationship between perceived promotion chances and performance.
8. *Values.* We hypothesize that employees who have a "work is my whole life" value orientation will have higher performance than those who give primacy to family or pleasure values.

Up to this point, Abegglen and we—and probably most other students of complex organizations—would make the same predictions concerning the relationships between social organization and performance. With regard to the following independent variables, however, we think there are important alternative hypotheses to those of Abegglen. The variables are:

1. Wage system preferences of employees (seniority versus job classification)
2. Several aspects of the social integration of the employee into the firm versus the social differentiation of the employee from the firm:
   a. Cohesiveness
   b. Paternalism
   c. Residence (company versus private housing)
   d. Participation in company recreational activities
   e. Concerns (company's instrumental goals and problems versus personal, private concerns)
   f. Previous interfirm mobility
   g. Lifetime commitment

As we interpret Abegglen, his theory is that in Japan performance is maximized when employees prefer seniority-based rather than job classification-based wage systems; and there is the maximum degree of integration of the employee into the company (minimum differentiation) on variables 2a–2g. Thus, performance, according to Abegglen, is maximized when employees are high in cohesiveness, paternalistic preferences, participation, company concerns, and lifetime commitment; live in company housing; and are low in interfirm mobility. We shall refer to this theory as the integration theory of performance.

Consider this in a broader context. As societies become more differentiated internally, new integrative problems arise. The theoretical issue is, to what extent can work organizations be foci of integration at successively higher levels of differentiation? In societies with little differerentiation, integrative problems are handled mainly in terms of kinship, local community, and religion. Specialized, large-scale organizations do not exist, and therefore cannot be foci of integration, though small work organizations may have this function. When large organizations do emerge, there is little theoretical closure as to (1) the extent and conditions under which the social integration of the individual is institutionalized primarily within these organizations, or outside of them; and (2) the relationship between degree of integration of the individual into the work organization and performance. Let us consider these in turn.

Some theorists have argued that work organizations in highly differentiated societies are not the main loci of the individual's integration. Weber's ideal-type analysis of modern bureaucratic organization has more to say about coordination than about social integration. Coordination is achieved on the basis of a well-defined hierarchy of authority and a system of rules that covers the rights and duties of positional incumbents. The social integration of the employee, on the other hand, is apparently neither the manifest nor the latent function of the bureaucratic organization itself. Interpersonal relations between employees (as between them and the public, their customers, etc.) are impersonal ("*sine ire et studio*"). The organization "separates the bureau from the private domicile of the official, and, in general, bureaucracy segregates official activity as something distinct from the sphere of private life" (Weber 1946:197). The implication is that to the extent individuals are socially integrated, it is in terms of structures outside their bureaucratic work organizations: voluntary associations, selective kinship ties, the mass media, identification with elites, national political and ideological symbols and the like (Smelser 1968).

The empirical support for this position comes from studies that show that employee cohesiveness and other aspects of integration into the firm are low in bureaucratic organizations with routine, standardized tasks, e.g., auto plants (Blauner 1964, Fullan 1970) and clerical agencies (Crozier 1964). Other theory and research, however, suggest that an individual's social integration may be realized *within* his large-scale work organization. One alternative in early industrialization, according to Smelser, is that villages may be "built around" paternalistic industrial enterprises, thereby enabling such traditional integrative structures as kin and community ties to be maintained. Blauner gives this as the reason why textile workers in the southern United States are more highly integrated than would be expected on the basis of their technology and work. Many analysts, of course, see this as a passing phase. Durkheim, however, argued that even in highly differentiated societies (high organic solidarity) the corporation and the occupational association can be the main positive sources of integration. Recent research has found that this varies with type of technology and work. Integration within the firm is more likely with continuous process technology (e.g., petrochemical industry) than with mass production assembly line technology (Blauner 1964, Fullan 1970).

The lack of closure with regard to the conditions under which social integration is institutionalized primarily within large-scale work organizations, or outside of them, is bad enough. There is even more ambiguity on the second question, the relationship between integration and performance. Weber may be interpreted as stating that the relationship is negative, since performance is maximized in a modern bureaucratic type of organization, and as we have already noted, one of the characteristics of such organizations according to Weber is the segregation of official activity from private life, i.e., the low integration of the employee within the organization. Most other analysts either do not address themselves to the question of organizational performance (Durkheim), or if they do, assert a basically variable, indeterminate relationship between integration and performance. Eisenstadt, for example, argues that at each higher level of differentiation, *if* the "new integrative problems" are solved, performance will be successful ("sustained growth"); if not, the system will retrogress or break down—performance will be unsuccessful. But the locus of this integration may be "bureaucratic organizations" (Durkheim's view), or other, non-work structures (Weber's view). In other words, successful performance requires both differentiation and integration, but the *type* of integration—within the large-scale work organization or outside of it—has no determinate relationship to performance.

What is the upshot of all this for our own hypotheses? What may be called the Durkheimian view may be formulated as: the large-scale organization is the main focus of social integration, and performance varies positively with the social integration of the employee into the organization. Although this goes beyond Durkheim, it is at least not inconsistent with what he might have said. However, it should be clear that this Durkheimian proposition makes the same prediction as the paternalism–lifetime commitment model; hence, this version is already represented in what we have called the integration theory of performance.

The opposite prediction can be inferred from Weber: performance varies inversely with the social integration of the employee into the firm (though it varies directly with the coordination of his activities within the firm). More specifically, to return to our hypotheses, organizational performance is maximized to the extent that employees (1) prefer a job classification-based wage system; and (2) are maximally differentiated from the firm on variables 2a–2g. That is, performance is maximized when employees' statuses, roles, and perceptions make a sharp differentiation between work and non-work: functionally specific relationships are preferred, housing is private, leisure participation is not formally organized in terms of the firm, and so on. This can be referred to as the differentiation theory of performance.

As we see it, there are theoretical difficulties in both integration theory and differentiation theory. We have aired the former at some length in Chapters 8–10. The basic difficulty we find with differentiation theory as applied specifically to performance is that it is unclear why employees should maximize their performance in terms of organizational goals if they never participate in company recreational activities, if their preoccupation with personal concerns obliterates company concerns, and so forth. For this reason, we shall adopt a position that is a logical alternative to both integration and differentiation theories, namely, the theory that the main determinants of performance are other than the integration or differentiation of the employee in relation to the firm. We shall predict that when each aspect of performance is regressed on the relevant set of independent variables, the best predictors of performance, when other variables are held constant, are organizational status variables, and that performance will vary independently of wage preferences and the degree of integration of the employee into the firm. Put another way, we find it more plausible that employees (or aggregated organizational units) high in performance will include those both high and low in integration into the firm, than that they will include disproportionately either those highly integrated into the firm or highly differentiated from the firm.

We are under no illusion that our data provide a fully adequate test of these hypotheses (any more than do Abegglen's). What is ideally required are data on a large number of firms from Japan and the United States, (and other societies) in which the firms are matched cross-societally on all relevant non-social organizational variables that make for variations in performance, and in which the Japanese firms systematically vary from the American (or other) firms along the above dimensions of social organization. Controlling for industry, size, and technology and assuming comparable cross-societal measures of performance of firms, one could then ascertain the independent effect of social organization on performance. This is a major task for future research; by contrast, our empirical tests in the remainder of this chapter are at best a very modest first step.

Findings

The multiple regression analyses that test our hypotheses appear in Appendix D, Tables D.30—D.32, pp. 404, 406. The results confirm our hypotheses that performance is a positive function of status (hypothesis 1), of sex—men higher than women—(hypothesis 2), and of knowledge of procedures (hypothesis 6). We also found support for hypothesis 3, that performance is positively related to age, in Electric Factory, but could eliminated age as a predictor of performance because of its redundancy with status. Our findings largely disconfirm the hypotheses that size of community of origin (hypothesis 4), job satisfaction (5), perceived promotion chances (7), work values (8), and five aspects of the social integration of the worker into the company—cohesiveness, participation, number of previous jobs, paternalism, and lifetime commitment—exert more than minimal influence on performance, when other variables are controlled. In Shipbuilding, however, lifetime commitment is a significant cause of performance.

We have achieved a parsimonious explanation of performance. One can predict employee performance as well on the basis of knowing sex (male), status in the firm (high), knowledge of procedures (high), and, in Shipbuilding, lifetime commitment (high) as on the basis of the larger set of independent variables from other conceptual domains.

For three of the four best predictors of performance in the two factories—sex, status in the company, and knowledge of procedures—we make the same prediction as Abegglen. When we turn to five of the hypotheses for which we make different predictions—cohesiveness, paternalism, participation, number of previous jobs, and lifetime commitment—each of which is tested in two factories, we have 10 tests. Using the beta weight in the first run in Tables D.30 and D.32 as the criterion, eight of these tests confirm our predictions that the social integration of the employee into the firm is not significantly related to performance, while two (lifetime commitment in Shipbuilding and participation in Electric Factory) confirm Abegglen's prediction that the relation is positive. The beta weights in the two confirmations of Abegglen are, however, low, and the confirmations are therefore only weak.

Thus, the best predictors of performance do not allow us to reject either Abegglen's theory or our own, because our hypotheses are identical. But where we make alternative predictions about performance, our hypotheses fit the facts better than do either Abegglen's integration theory or the differentiation theory of performance.

Human Relations Theory and Performance. Human relations theory (Likert 1967) and the integration theory we have derived from the paternalism–lifetime commitment model both argue, on somewhat different grounds, that employees' performance in terms of company instrumental goals is maximized to the extent that they have (1) high cohesiveness with fellow employees, (2) frequent participation in company recreational activities, and (3) high job satisfaction. Some versions of human relations theory posit a causal chain in which cohesiveness and participation influence job satisfaction, and these three variables in turn influence performance. When we asked a Machine Section worker in Shipbuilding Factory how he thought productivity could be increased, he stated an essentially human relations explanation: "[One method of increasing productivity is] to have workers know each other outside of work. Going hiking together, for example, leads to higher morale, and that leads to higher productivity. In our han, even older workers come hiking."

A causal model for this human relations theory of performance is tested in Appendix D, Tables D.33 and D.34, p. 408, and Figure D.1, p. 407 for Electric and Shipbuilding factories. We found that in only one case is even *one* of the three human relations variables moderately strongly related to performance, either at the zero-order level, or when the other two variables are held constant. This variable is job satisfaction in Electric Factory. But even this gives only cold comfort to human relations theory, for we have previously shown that job satisfaction in Electric Factory is more a result of organizational status than of cohesiveness or participation (see Table D.3, p. 369). This means that the only successful predictor from human relations theory is itself a function of a variable—status—that is more often associated with Weberian and with neo-structural[7] theories of organization than with human relations theories. In this sense, we conclude that neo-structural theory provides a better explanation of performance than does human relations theory; (see also Carey 1967).

One other indication of the weakness of human relations theory is that in each factory the multiple *R* between our three best predictors and performance is higher than that between the three human relations variables and performance (.61 vs. .45 in Electric Factory, and .48 vs. .26 in Shipbuilding; compare Tables D.30 and D.32, p. 404 and p. 406, with Tables D.33 and D.34, p. 408.

LABOR PRODUCTIVITY

What is probably the most important single aspect of performance is labor productivity, defined as the output per worker per month. Attendance and suggestions are, in important respects, only means to an end—high productivity. Productivity is to a greater extent an organizational goal in itself. It is obvious why management in any economic organization is anxious to increase labor productivity; there are also reasons that are more distinctive to particular organizations. Electric Company's management impressed upon us that their concern with labor productivity was motivated by the fact that they had given wage increases averaging 19 percent in 1967, 22 percent in 1968, and 24 percent in 1969. The company's mean wage increased from ¥ 15,500 in 1960 to ¥ 35,000 in 1969. Management argued that their goal in both 1968 and 1969 of a 30 percent increase in productivity was necessary in order to keep pace with increases in wages and in capital investment.

[7] Etzioni (1964:41–44, 47–49) has a succinct discussion of neo-structural theory. Classical theories of administration stress formal structure, while human relations theory stresses informal structure. Neo-structural theories attempt a more integrated analysis of the formal and informal aspects of organizational structure. Thus, they stress formal rewards—status—as well as informal rewards—cohesiveness, participation. Neo-structural theories also emphasize objective status and power differentials and the conflicts of interest that derive therefrom. In this area as well as in the conflict between personal needs and organizational requirements, neo-structural theory is contrasted with human relations theory by its recognition that these facts are to some extent inevitable in organizations, and cannot be eliminated by "better communications."

We shall first state the facts about labor productivity in each factory, and then try to explain them. We shall consider how the Japanese themselves in these organizations explain productivity, especially what can be called the managerial theory of performance; Abegglen's explanation; and our own explanation. In reviewing these theories we shall be at pains to show how each views the relative causal influence of technology and social organization on performance.

Sake company's kura

Productivity per worker in the sake industry has increased over time. One reason for this is common to other industries as well: technological modernization. One of the sake companies studied by Mannari, Maki, and Seino in 1966–67, a separate firm located in Oimatsu, had about the same number of employees at that time as in 1886. Yet it produced 2,500 koku of sake a year in 1966, 2.5 times greater output than in 1886, as a result of its partial modernization of facilities. The same relationship is found within Sake Company, and is shown in Table 11.5. In the more traditional, only partially mechanized plants, daily production of sake per employee in the 1969–70 season varied between 1.58 and 2.44 koku. The comparable figure for the Number 2 automated factory in 1966 was 2.74 koku. This factory was closed in 1969, and in the same year the newest factory, automated factory Number 3, was opened. Factory Number 3 was built for a yearly production of up to 50,000 koku of sake. In the April 1969–March 1970 period the Number 3 factory produced 40,000 koku; between April 1970 and March 1971, it produced 45,000 koku. Thus, if we compare the per capita daily productivity of the three traditional plants in N-City and the newest, automated factory Number 3 in 1969–70, we find that the Number 3 factory was 1.6 times more productive than traditional plants Number 2 and 3 (3.97 ÷ 2.44) and 2.5 times more productive than traditional plant Number 1 (3.97 ÷ 1.58). Productivity varies directly with the modernization of technology.

The greater per capita productivity in the automated plant is also based on changes in social organization. We have analyzed these in earlier chapters, and they need only be briefly recapitulated now. The degree of social organizational modernization in Sake Company's new factory is generally greater than that in its partially mechanized, older kura in these respects:

1. The work force is employed in sake brewing all year long.
2. The work force includes both technical experts with scientific knowledge and factory workers who man machines and possess neither expert scientific knowledge nor traditional craft skills.

TABLE 11.5
Productivity of Sake Company's Plants, October 1969–March 1970 Season

Plant	*Number of Employees*	*Days Worked*	*Koku of Sake Produced*	*Per Capita Daily Koku of Sake Produced*
Traditional Social Organization, Partial Mechanization				
In N-City[a]				
No. 1 Plant	38	150	9,000	1.58
No. 2 Plant	15	150	5,500	2.44
No. 3 Plant	15	150	5,500	2.44
In I-City				
No. 4 Plant	35	150	4,200	0.80[b]
No. 5 Plant	15	150	5,500	2.44
No. 6 Plant	15	150	5,500	2.44
No. 7 Plant	15	150	5,500	2.44
Modernized Social Organization, Automated Technology				
In I-City				
No. 2 Factory[c]	28	300	23,000	2.74
No. 3 Factory[d]	36	280	40,000	3.97

Source: Mannari's April 22, 1971 interview with kachō of Technical Section of Sake Company.

[a] The three traditional plants in N-City are the ones studied most intensively.

[b] Productivity is lower in this plant because 20 of the 35 employees do the work of malt development and malt fermentation not only for their own plant but for the other plants as well.

[c] This factory stopped operation in 1969, and its 28 employees were transferred to the new automated factory, Number 3 Factory. The data refer to 1966.

[d] The data cover the period April 1969–March 1970.

3. The foreman has lost some of the recruitment functions performed by a traditional master.
4. The traditional apprenticeship system has declined.
5. The terminology for describing jobs resembles that of a modern factory.
6. The role of master is differentiated into two new roles—that of scientist and that of foreman.
7. Overall supervision of the plant is now in the hands of the scientist.
8. A staff and line system has been established.
9. New staff roles of consumer agent and sales agent have emerged. Individuals in these roles have the authority, on the basis of their market and consumer research, to decide what kind of sake to produce.

10. Newer workers express dissatisfaction with aspects of traditional, paternalistic relationships that have been carried over from the traditional plants to the new plant.

These changes in social organization at the company and plant level, accompanied by technological modernization, have increased productivity. We interpret these findings as partial confirmation for our major proposition that organizational performance is a result of technological and social organizational modernization.[8]

SHIPBUILDING FACTORY

X Plant has the capacity to build 450,000 horsepower of diesel engines a year; its actual production between April 1969 and March 1970 was 27 diesel engines, with a total of 340,000 horsepower, which is 76 percent of its maximum capacity. This is not regarded as a sign of inefficiency, because workers are also assigned to work on products other than diesel engines—industrial machinery, sugar plants, steel casting. Thus, when the output of diesel engines declined from 397,000 horsepower in 1967 to 305,000 in 1969, the output of machinery and other products increased from 22,000 tons to 35,000.[9] A kakarichō explained these shifts of personnel: "Because they produced less horsepower of diesel engines per employee, it doesn't mean they were idle. Instead, they were shifted to other work, such as press machine work for making auto bodies. Auto bodies need big press machines—2,000-ton and 4,000-ton press machines for work for Nissan Jidosha[10]—and we made that product more frequently in 1968. That work brings more profit than many orders for ship diesel engines."

Shipbuilding Factory management is attempting to increase its productive efficiency. As one walks around the Engine Division's production workshops, wall posters enjoining workers to increase productivity (e.g., "*nobaso seisansei*") can be seen, along with posters calling for a happy workshop and fewer defects. The management introduced a system of computerized control for the assembly of diesel engines in April 1970. The complete assembly of a diesel engine is divided into 270 units of work. In 1963 the standard time units allotted for the assembly of the Pattern 62 engine were 7,912 standard hours; by 1968 the actual hours required had been reduced to 6,417, i.e., an 18.9 percent reduction of the 1969 actual from the 1963 standard. Because it actually takes less time to produce the

[8] Since we lack data that would enable us to incorporate the social organizational and technological variables in a multiple regression analysis, we cannot determine the relative influence of the technological variables when the social organizational variables are held constant, and vice versa.

[9] Data refer to the calendar year, not April through March.

[10] A Japanese automobile manufacturer.

engine, the standard time unit has also been adjusted downward:

Year	*Standard Time*	*Base (1963 = 100)*
1963	1.20	100
1968	0.83	70
1969	0.74	62

Other methods by which the company is trying to further reduce standard time include reorganization of the work flow. The goal is to have the parts made in the Machine Section or in subsidiary firms delivered on time to the Assembly Section. The company classifies its subsidiary firms as A, B, or C; A firms have their own trained inspectors and are most likely to meet deadlines and quality control criteria.[11] Improvements in methods of work are also being introduced, for example, doing finishing work on lathes, which, in contrast to hand work, makes for greater precision.

Shipbuilding Factory data on a sample of 10 of the 27 diesel engines assembled in 1969 show differences among han in the General Assembly Section. When the nine han are arranged according to the ratio

$$\frac{\text{actual number of man-hours worked to assemble 10 engines}}{\text{standard time man-hours allotted for assembly of 10 engines}}$$

all of them completed assembly work in less than the allotted standard time. But the most productive han spent only .475 as many man-hours as allotted standard time, while the least productive han spend .821 as many man-hours as allotted. In interviews, the employees' own explanations of these differences at times stressed technology, at other times social factors. An example of the former is that the more productive han are more likely to specialize in assembling the Pattern 84 diesel engine, which is easier than other patterns. An Assembly worker explained: "We do the job faster now, by using new kinds of materials, by extensive use of the block assembly method, prefabrication. For example, the 84-Pattern engine is now assembled in several large blocks, rather than in conventional blocks, and we eliminate some work steps." Other han were slower because their work had less product standardization. As an Assembly Section hanchō noted: "We just finished assembling the Number 74 Base for the first time, so we took longer than the standard time. The second time we assemble the 74 Base we hope to be within the standard time. It takes 4,000 man-hours to do this job." A common remark by those in Shipbuilding Factory who regard technology as more important than social factors is: "Productivity

[11] Shipbuilding Company itself has inspection at every stage of the production process—inspection of raw materials, of pressed materials, of work done in the Machine Section, and of the final assembly stage. Inspection techniques include superfrequent inspection, Roentgen inspection, and magnetic inspection.

can be improved by many factors other than investment for technology, but the effect of all the other factors is small. In order to gain a long-range large profit, it's necessary to have a big equipment investment."

What were the social factors adduced to account for the han differences in productivity? The Assembly Section kakarichō's view was: "The hanchō can influence his han's productivity 30 percent, but 70 percent is determined by whether the parts come on time. But the *quality* of work depends heavily on the hanchō's skill." The production control staff specialist commented specifically on the differences in productive efficiency among the nine han:

> These han differences are kept to a minimum by management, by means of the following practices. First, what is the work environment? How much cooperation is there between tamakake and crane operator and General Assembly workers? The crane transport han and its hanchō are the key men in engine assembly work. Second, there may be differences in attendance and overtime work among han. Third, if the hanchō is young and less experienced, we put a strong boshin in his han.
>
> When differences still exist after these things have been checked, they reflect first the relative experience of hanchō. Second, a forceful boshin. For example, the hanchō in the least efficient han is young and relatively inexperienced. He needs a strong boshin to make his record better than 0.821. Third, workers' motivation reflects the influence of their hanchō and sagyōchō. Workers come here with the motivation to work, but a bad supervisor can discourage it. Fourth, formerly it was the *craft skill* of the hanchō that was respected, now it's how well he can improve work speed, reduce costs, arrange and standardize work methods and the division of labor. We have more work rotation now, in order to reduce workers' dissatisfaction with too specialized work.

Expanding on the role of first- and second-line supervisors, the same staff specialist said: "The hanchō and sagyōchō's leadership ability is reflected in their workers' productivity. This ability [involves] how they persuade the workers they are part of a group, how they listen to the workers' family and personal worries, and so forth." An Assembly hanchō saw the difference between the Company's production control system on paper and the realities of the workshop: "If the production control system worked perfectly the hanchō's job would be easy, but human control on a day-to-day basis by me is still necessary. A Machine Section worker underlined the importance of role models for productivity: "To improve productivity the boshin should become the lead-off man, and work efficiently himself." Another Machine Section hanchō saw the problem as one of fitting the demands of the job to the capacities of workers: "If I ask a 'C-class worker' to do a B-class job, it takes him more time, and he needs more tools to help him. Only an 'A-class worker' can be left alone to do this job."

ELECTRIC FACTORY

Traditionally, workers in Japanese manufacturing firms simply worked hard, but did not have specific production goals. Electric Company now has a more rationalized system in which top management issues specific goals for a given period—a 2.5 percent increase per month, or 30 percent per year, in labor productivity—and workers then try to meet the goal. Production goals are also stated for specific products. For example a 1969 goal was to produce one complete vacuum cleaner motor of a certain type in 6.59 minutes; the daily goal per worker was to produce 72.9 motors. Production records indicate that although these goals were not being met in the first two months of 1969, they were being exceeded in March, April, and May. In May, for example, it took only 5.23 minutes to produce a complete motor, and a worker produced 91.8 motors per day.

The basic data on the dependent variable, labor productivity in Electric Factory, by section, are shown in Table 11.6. The cell entries for a given section are computed as follows. The section's productivity in December is taken as the standard productivity for each month of the following year: as a productivity base line it is standardized to 100 for all sections as of December. For each month from January through the following December, the cell entry is computed as the ratio of:

$$\frac{\text{actual number of units produced per worker that month}}{\text{standard number, i.e., number produced in previous December}}$$

The cell entry, then, indicates whether that month's productivity is higher or lower than the previous December's, shown as greater than or less than 100, respectively. Each cell entry can also be compared with the goal for that month, which is, of course, independent of the ratio of actual to previous December's productivity.

Consider Hardware as an example. Its productivity in January 1968 was 114.0, which indicates that it was 14 percent higher than it had been in December, the previous month. As such, it was also 114.0 − 102.5, or 11.5 percentage points higher than the *goal* for the month of January. In April 1968, on the other hand, Hardware's productivity was 109.0, which although nine percent higher than its previous December level, was 109 − 110 or one percentage point lower than the April productivity goal.

It should be noted that one property of the company's measure of labor productivity is that the productivity race among sections begins with a clean slate each January. No matter how much one section lagged behind another in December, during the following year it was not expected to

TABLE 11.6

Electric Factory's Labor Productivity Per Capita, Standardized to a Base Line Annually, January 1968 through December 1969[a]

Unit	Base Line	Jan.	Feb.	Mar.	Apr.	May	June	July	Aug.	Sept.	Oct.	Nov.	Dec.	Mean	Median
1968 Goal		102.5	105.0	107.5	110.0	112.5	115.0	117.5	120.0	122.5	125	127.5	130		
Factory total	100	101.7	104.7	101.0	103.5	107.7	110.0	109.7	111.0	113.4	113.4	112.6	116.7	108.8	109.0
Hardware	100	114.0	114.3	114.8	109.0	112.3	112.4	110.0	117.4	130.4	112.4	115.4	111.7	114.5	112.0
Motor	100	101.3	105.4	101.7	106.6	114.5	107.8	110.4	116.6	117.2	112.1	124.8	135.8	113.7	112.0
No. 1 Production	100	99.4	100.3	95.6	96.0	104.0	106.4	104.5	104.5	107.6	114.4	106.3	113.0	104.3	104.8
No. 2 Production	100	98.2	100.5	98.3	106.8	101.3	110.0	110.8	109.0	107.4	107.2	113.0	112.3	106.2	107.7
Surface	100	97.3	107.4	105.6	112.4	125.7	124.5	130.3	102.8	108.0	103.0	115.0	120.5	114.2	112.0
1969 Goal		102.5	105.0	107.5	110.0	112.5	115.0	117.5	120.0	122.5	125	127.5	130		
Factory total	100	102.5	101.0	105.0	107.0	110.8	114.0	106.5	108.0	104.0	113.3	113.6	113.2	108.2	108.0
Hardware	100	108.5	107.0	109.7	109.8	119.0	117.3	106.0	115.0	110.0	112.1	122.6	119.8	113.1	111.0
Motor	100	104.0	103.5	103.0	108.5	108.0	113.0	117.0	120.1	116.5	126.2	135.8	130.6	115.5	115.0
No. 1 Production	100	100.0	97.5	101.0	99.5	104.5	109.7	103.8	100.2	100.1	103.4	98.5	101.4	101.6	100.0
No. 2 Production	100	101.5	98.0	110.0	114.6	115.3	118.0	106.5	106.2	95.2	119.1	114.6	110.4	109.1	111.0
Surface	100	106.0	111.5	104.0	102.5	111.0	117.7	103.5	118.1	120.1	123.9	127.8	134.6	115.1	112.0

[a] Cell entries = actual number of units produced/worker/month divided by standard number of units produced/worker/month. Entries of 100+ indicate actual production exceeded standard.

catch up with the more productive sections. Rather it had only to improve upon its own production of the previous December by the 30 percent called for in the new year's goal.

The data in Table 11.6 were summarized by ranking the five sections' productivity according to four measures: (1) the mean and median productivity per worker; (2) the number of months the section's productivity per employee was higher than the production goal (i.e., higher than 102.5 in January, higher than 105.0 in February, and so on); (3) the cumulative amount by which the section was above or below the production goal each month; and (4) the cumulative amount by which the section was above or below the base line, or standard (= 100), i.e., above or below the actual production in the previous December. When we compare each section's rank on the four measures we find some variation in ranks 1–3 among Hardware, Surface, and Motor sections. However, there is a high degree of stability in that regardless of how the productivity data are summarized, these three sections are always higher than Number 1 and 2 Production, and Number 2 Production is always more productive than Number 1 Production. We derived our dependent variable—section productivity during 1968–69[12]—by summing each section's rank on all four measures:

Section	*Sum of Section's Ranks on Four Measures of Labor Productivity*	*Summary Rank*
Surface Treatment	6	1
Motor	7	2
Hardware Components	11	3
Number 2 Production	16	4
Number 1 Production	20	5

Our explanatory task, therefore, will be to account for why Motor, Surface, and Hardware sections have greater productivity increases than Number 1 and 2 Production, even when each section's productivity is compared only with its own previous December base line.

[12] Some readers may raise the methodological problem that our independent variables—social organization—were measured in late September 1969, whereas the dependent variable—labor productivity—was measured over the period from January 1968 through December 1969. Anticipating the objection that our cause comes *after* its effect, rather than prior to it, as is proper, in 1973 we attempted to gather data on labor productivity for the years 1970 and 1971. This proved impossible, since Electric Factory had switched to different, and non-comparable measures of performance after 1969. This problem of the time order of our cause and effect variables, however, is mitigated by three factors. First, using the data in Table 11.6 only for October–December 1969 in no way alters the fact that Motor, Surface, and Hardware sections have higher labor productivity than Number 1 and 2 Production. Second, some of our social organization variables—performance, participation, interfirm mobility—refer to the 12-month period prior to September 1969 or even earlier. Third, for some of the other social organization variables—sex, organizational status—there is obviously a considerable degree of temporal stability.

EXPLANATIONS OF LABOR PRODUCTIVITY

THE MANAGERIAL THEORY OF PRODUCTIVITY

We discussed at length with managers and staff in Electric Factory why Motor, Surface, and Hardware sections had higher productivity than Number 1 and 2 Production sections. Before testing our own explanation, we present a composite version of the managerial theory of productivity, which adduces the following explanatory variables: (1) management planning and decision making; (2) technology; (3) design improvements to eliminate work steps; (4) supply of parts; (5) market demand; and (6) social organization of the factory and other social and human factors.

Management Planning and Decision Making. Not surprisingly, management sees itself as the key dynamic factor in productivity. It is management, the argument runs, that decides what shall be produced and what the production goals are to be. Management views the influence of workers as more or less a given, i.e., as a factor that cannot be improved in the way technological advances can improve productivity. Management sees its role as crucial in the planning and decision making by means of which machine efficiency is improved and technological innovation introduced. This management explanation may be diagrammed as follows:

Management and staff initiative, decisions ——→ technological improvements ——→ increased productivity per worker

We shall have more to say about this aspect of management's theory later.

Technology. It is clear that management assigns much more importance to technology as a determinant of productivity than it does to social organizational and attitudinal factors. As a Hardware kumichō put it when asked what factors determine productivity: "The fifth assembly line produces 10,000 lamp cases, whereas the fourth line produces only 7,000 to 7,500. This is because the fifth line is more mechanized and has a more limited range of products than the fourth line. The variation of 7,000 to 7,500 in the output of the fourth line is due to differences in four factors: supply of raw materials; the conditions of the moulding [pattern for cutting or pressing parts]; the conditions of the machines; and the mood of the workers. The complex programming of the machines is done by the moulding, and *this is the largest influence on productivity.*"

Design Improvements to Eliminate Work Steps. The more work steps that can be eliminated by improvements in the way products are designed, *ceteris paribus*, the more productivity can be increased. In this regard, the situation is basically different as between Number 1 and 2 Production on the one hand, and Surface, Motor, and Hardware, on the other. These differences are apparent in the remarks of a kumichō in Number 1 Production: "Productivity increases can't be attained by *our* efforts [on the assembly line]. Draftsmen, designers do this. The most important factor leading to high productivity is design. There are improvements in work methods that lead to productivity increases, but improvement in design is more important." The design simplifications that are made for work in Number 1 and 2 Production are more likely to come from outside those sections, i.e., from production engineering. This is because, although the jobs are simpler than in Surface, Motor, and Hardware, the personnel in Number 1 and 2 Production have much less seniority, and are less experienced and less skilled. In Motor, Surface, and Hardware, there is more of a craftsman's level of skill—the worker understands his machine well, and, to a greater extent than in Number 1 and 2 Production, can be his own production engineer. Relative to the knowledge of production engineers outside the five line production sections, Surface, Motor, and Hardware workers' knowledge is higher than that of Number 1 and 2 Production workers. Members of the first three sections more often advise designers and make contributions to the elimination of work steps. Although some work steps have been eliminated in all five sections, the greater participation of workers themselves in this simplification process in Motor, Surface, and Hardware has had a larger impact in those sections than in Number 1 and 2 Production. This is turn has enabled these three sections to outdistance the other two in labor productivity.

Supply of Parts. The timing and delivery of raw materials and parts from subsidiary firms and of parts from one section to another within Electric Factory influence productivity. As was shown diagramatically in Figure 3.2, Number 1 and 2 Production sections, the final stage of assembly in Electric Factory, receive parts from Surface, Motor, and Hardware sections. Number 1 and 2 Production are therefore more dependent upon the other three sections for their supply of parts than are these three on Number 1 and 2 Production. A kumichō in Number 1 Production stated: "Delay in meeting the work schedule isn't so much a result of delays in the assembly line as of delays in delivery of parts. I telephone the parts supplier [in this factory]." A kakarichō in Number 1 Production Control had the same story: "We can control our own Department's production [of fans], but we can't control what Hardware or Surface Treatment sections do, yet we get our parts from them. And we can't control productivity of subsidiary firms either, so they can delay us." Thus, another reason for lower productivity in Number 1 and 2 Production is that they have less control over the supply of parts than do Surface, Motor, and Hardware sections.

Market Demand. Management's explanation of the level of a section's productivity over time also includes the factor of market demand. Orders are processed by the Production Control Section (kōmuka) in each department. This section plans two months in advance, on the basis of orders from customers and market information, how many units of which products shall be produced. A kakarichō in Number 1 Production said that from his point of view productivity was influenced 40 percent by the efficiency of the assembly line and 60 percent by demand.

The section with the lowest productivity is Number 1 Production, which specializes in the assembly of various models of electric fans. There are months when demand for all kinds of fans slacks off, and since employees in Number 1 Production cannot easily be shifted to other kinds of work, the section's productivity declines. Number 2 Production has a somewhat greater diversity of product lines—vacuum cleaner motors and cases, and juicers-blenders. This enables Number 2 Production to adjust more easily than Number 1 Production to changes in market demand, thereby maintaining a higher level of labor productivity. This is even more true of the sections with the highest labor productivity increases, Hardware, Motor, and Surface. These sections have the greatest product diversification of all in that they produce cases, motors, and other parts for *all* the product lines in Electric Factory: fans, vacuum cleaners, circulators, blenders, juicers, mixers. When demand and orders shift from one product to another, workers in Hardware, Motor, and Surface can more readily be shifted from one task to another, thereby keeping productivity up. The greater the product diversification of a section, in short, the more its workers can be kept busy on some relevant productive task.

Social Organization of the Factory and other Social and Human Factors. The five determinants of productivity we have reviewed thus far are the ones most often stressed by Electric Factory managers and staff personnel. Analytically, the managerial theory of productivity gives primacy to management planning, technology, and economic-logistic variables. Social organizational and attitudinal variables—prominent in several of the sociological theories of organizational performance (Georgopoulous and Tannenbaum 1957, Etzioni 1964, Dubin et al. 1965, Likert 1967, Price 1968, Neuhauser 1971, Mott 1972, and Heydebrand 1973)—are conspicuous by their absence in Electric Factory's managerial theory. Managers perceive at least four kinds of relationships between the foregoing, analytically non-social elements, and analytically social elements.[13]

[13] It is essential that the distinction between analytical and concrete elements be kept in mind throughout this discussion. Concretely, social and human factors obviously are intertwined with management planning (bounded rationality; non-rational aspects of decision making), technology, etc. Our distinction between social and non-social elements is analytical, not concrete.

First, some managers cite social and human factors as having mainly a negative effect on productivity. In the words of a Hardware kumichō: "The speed of the conveyor belt isn't all important in productivity. Its speed is determined by the feeling of the workers. Workers can press a button and stop their own machine, and can even stop the entire conveyor belt itself. They can let pieces pass them without working on them. . . . If a worker continues to operate the same machine for two hours, he gets tired and careless and brings danger. I let workers leave the conveyor belt, the press machine, and do other work, like arrange boxes for the finished parts. I let them change their work positions sometimes. It is difficult for a person over 40 to do long continuous work. The productivity of workers in their fifties is about 70 percent that of younger wokers."

Second, management initiative in productivity is seen (by managers) as so important that one supervisor recognized human relations (ningen kankei) factors only to point out that they are traceable to the kind of management climate in the factory: "The morale situation is better [in the han of Hardware] these days. There are still some sly, cunning fellows, as a result of bad management. I try to manage them by showing them clear work standards and goals. If management provides clear standards, cunning behavior by workers will be reduced and morale will be higher. It's best to use output per day for all products, and be sure workers know what they are. Without standards, they don't work well."

A third mode of relationship recognized was that although social factors are important in maintaining a given level of productivity, it is technology that brings about *increases* in productivity. This view was expressed by a staff kakarichō in Hardware: "Human motivation has only a maintenance function. In order to *raise* productivity, production engineering staff members have to introduce new methods. Thus, buying new production machines is important for improvement in production. Although we recognize the importance of human motivation in maintaining present human productive efficiency, the continuous advancement of productivity is the responsibility of technical improvement. If we neglect workers, then productivity will be seriously affected. While maintaining workers' motivation, if we also improve technology, then the result will be real improvement."

Finally, a fourth connection stated by some managers was that human relations factors have a positive influence on productivity. The kachō of the Hardware Section said: "My kumichō and hanchō are stressing human relations because if that's good, productivity will be good."

Explaining section differences in labor productivity

The dependent variable is the sum of each of Electric Factory's five production sections' ranks on the four summary measures of increases in labor productivity during 1968–1969, shown in Table 11.6. To explain these section differences, we have two analytical categories of variables—non-social and social organizational. We do not deny that the non-social independent variables identified by management—technology, design improvements, supply of parts, etc.—cause section differences in productivity. Unfortunately, we lack systematic data on these variables, and therefore cannot test their influence. What we can do is answer the question, do social organizational variables have *any* causal significance in explaining these section differences in productivity?

The form of each hypothesis to be tested is: Surface, Motor, and Hardware sections are higher in labor productivity than Number 1 and 2 Production sections *because* of differences in social organization. The data consist of a series of contingency tables in which the five sections are dichotomized into (1) Surface, Motor, and Hardware and (2) Number 1 and 2 Production, and compared with regard to given social organizational variables. In reporting the findings (see Table 11.7) we shall only state (1) whether the two groups of sections have statistically significant differences in the predicted direction in a given social organizational variable, and (2) the strength of the difference between the two groups of sections, as measured by C/C_{max}.

The social organization variables shown in Table 11.7 include several of those tested in relation to performance at the individual level (see Tables D.30 and D.32, p. 404 and p. 406, as well as the following new ones:

1. *Performance*. Labor productivity is positively related to employees' performance index scores; therefore we predict Hardware, Motor, and Surface employees will have higher performance scores than Number 1 and 2 Production employees.

2. *Legitimacy of promotion system*. Labor productivity is positively related to the legitimacy of the promotion system. Conceptually, a promotion system's legitimacy refers to the reasons the members of an organization give for their relatively poor promotion chances. Analytically, these reasons are of two kinds: (a) legitimate reasons for poor promotion chances, i.e., reasons that imply that the fault lies with the person rather than with the company's promotion system and (b) illegitimate reasons, i.e., reasons that imply that the fault lies with the company's promotion system rather than with the individual. Legitimate reasons include inadequate education,[14] being too old, not having the right kind of personality, lack of ability, and a poor record or poor reputation with management.[15] Illegitimate reasons include not having

[14] Inadequate education is the fault of the individual only with regard to the company, not necessarily with regard to the individual's family background.

[15] The list was not presented in this order in the questionnaire.

a chance for training within industry, not being known to the factory managers, and being a woman. In order to measure the legitimacy of the promotion system, we sum the number of times (a) the five legitimate reasons are given by respondents, (b) the three illegitimate reasons are given, and (c) all eight reasons are given. Legitimacy is then measured by the proportion of all reasons given that are legitimate.

3. *Conflict.* A third set of new hypotheses to be tested in Table 11.7 is derived from the general proposition that labor productivity is negatively related to the degree of conflict in an organization. Three questions asked only of personnel in the ranks of first-line foremen and above—whom we shall refer to as managers—form the basis of our three hypotheses here: the labor productivity of a section is negatively related to: (a) the frequency of conflicts among managers concerning authority and responsibility in their jobs; (b) the frequency with which managers report that they get conflicting orders in carrying out their jobs; and (c) the degree to which managers think their subordinates receive communications from superiors with suspicion. Therefore, our prediction is that Hardware, Motor, and Surface sections have higher labor productivity than Number 1 and 2 Production sections because managers in the first three sections report less conflict in each of these three areas than do managers in the other two sections.

4. *Achievement.* The final new hypothesis is based on the questionnaire items, "Which do you think is the most worthwhile section in which to work?" and, "Why is that section most worthwhile?" The latter question was open-ended, and the reasons given by respondents were coded in terms of whether or not they included achievement-related factors. Our hypothesis is that Hardware, Motor, and Surface sections have higher labor productivity than Number 1 and 2 Production sections because employees in the first three sections are more likely to say a given section is most worthwhile to work in because it provides "opportunity for learning new skills, new technology" and other achievement possibilities, as opposed to such non-achievement-related reasons as "I'm accustomed to jobs in that section."

The direction of relationship predicted between each social organization variable and labor productivity in Table 11.7 is the same as that between the social organization variable and performance in Table D.30 and D.32. The difference is that now we are aggregating performance at the level of the section. We follow the earlier practice of listing first the hypotheses in which we make the same predictions as Abegglen and then those in which we make different predictions.

The findings in Table 11.7 support eight of the 11 hypotheses in which we make the same prediction as Abegglen. Employees in the production sections with greater productivity are more likely than those in less

TABLE 11.7
Relation of Labor Productivity to Section Social Organization in Electric Factory

Hypotheses in which Abegglen and Marsh-Mannari Make the Same Prediction	*Confirmed*	C/C_{max}
Hardware, Motor, and Surface sections are higher in labor productivity than Number 1 and 2 Production sections because in Hardware, Motor, and Surface sections:		
1. employees have higher organizational status index scores (v. 138)	Yes	.41**
2. employees have higher index of performance scores (v. 140)	Yes	.40**
3. managers have fewer conflicts among themselves concerning the authority and responsibility of their jobs (v. 127)	Yes	.38**
4. employees perceive their promotion chances as better (v. 109)	Yes	.33**
5. employees are more likely to say a given section "is most worthwhile to work in" because it provides "opportunity for learning new skills, new technology," and other achievement possibilities (v. 38) (Number 1 and 2 Production employees are more likely to say a section is most worthwhile because they are "accustomed to its jobs")	Yes	.30**
6. employees have more knowledge of procedures concerning work problems (v. 39)	Yes	.28**
7. employees are more likely to give primacy to work values over family or pleasure values (v. 31)	Yes	.24**
8. employees have higher job satisfaction index scores (v. 139)	Yes	.22**
9. managers are less likely to think subordinates receive communications from superiors with suspicion (v. 135)	No	N.S.
10. managers say they get conflicting orders less often in carrying out their job (v. 128)	No	N.S.
11. employees ascribe greater legitimacy to the promotion system (v. 110–112; 115–119)	No	N.S.

Findings for which Abegglen, Marsh-Mannari, and Differentiation theory make different predictions	*Supports*	C/C_{max}
1. Hardware, Motor, and Surface employees are less likely than Number 1 and 2 Production employees to express company concerns (v. 121)	Differentiation theory	.30**
2. Hardware, Motor, and Surface employees have higher cohesiveness than Number 1 and 2 Production employees (v. 141)	Abegglen	.24**
3. Hardware, Motor, and Surface employees are more likely than Number 1 and 2 Production employees to	Partial support for both	.23**

TABLE 11.7 *(Cont.)*

Findings for which Abegglen, Marsh-Mannari, and Differentiation theory make different predictions	*Supports*	C/C_{max}
prefer a job-classification wage system and to prefer a seniority wage system (v. 101) (Number 1 and 2 Production employees are more likely to think the two wage systems are the same, or to say they don't know which is better)	Differentiation theory and Abegglen	
4. Hardware, Motor, and Surface employees have had more previous interfirm mobility than Number 1 and 2 Production employees (v. 148)	Differentiation theory	.17**
5. Hardware, Motor, and Surface employees participate more frequently in company recreational activities than Number 1 and 2 Production employees (v. 143)	Abegglen	.15**
6. Hardware, Motor, and Surface employees and Number 1 and 2 Production employees do not differ significantly in lifetime commitment (v. 146)	Marsh-Mannari	N.S.
7. Hardware, Motor, and Surface employees and Number 1 and 2 Production employees do not differ significantly in the proportion who live in company housing (v. 2)	Marsh-Mannari	N.S.
8. Hardware, Motor, and Surface employees and Number 1 and 2 Production employees do not differ significantly in paternalism (v. 144)	Marsh-Mannari	N.S.

** Significant at the .01 level.

productive sections to have higher organizational status, performance orientations, and job satisfaction, to know procedures, perceive their promotion chances as good, give achievement-related reasons for regarding a given section as most worthwhile to work in, and to give primacy to work values. Managers in the more productive sections report fewer conflicts among themselves concerning authority and responsibility in their job than do managers in the less productive sections. The hypotheses not supported concern two other areas of conflict, as seen by managers, and the legitimacy of the promotion system; there are no significant differences between the more and the less productive sections in these respects.

With regard to the independent variables for which Abegglen, differentiation theory, and we make different predictions, there is approximately equal support for each theory. Because employees in the more productive sections are less likely than those in the less productive sections to express company concerns, and have had more previous interfirm mobility, differentiation theory is supported. Because the former employees are more likely than the latter to prefer both a job classification wage system and a seniority wage system, there is partial support for both differentiation theory and for Abegglen. Abegglen is also supported in that employees in the more productive sections have higher cohesiveness and participate more often in company activities than do employees in the less productive sections. Finally, in three other areas it is our hypotheses, not Abegglen's or those of differentiation theory, that are confirmed: the more productive sections do not differ significantly from the less productive sections in lifetime commitment, paternalism, or living in company housing. These aspects of the integration of the employee into the firm would appear to have no causal influence on section labor productivity.

It seems clear that both the technological factors prominent in the managerial theory of productivity and the social factors we have tested are causes of labor productivity. Ideally, the next step would be to regress section labor productivity simultaneously on (1) the non-social independent variables—mechanization, design improvements, supply of parts, market demand; and (2) our social organizational independent variables. This would enable us to specify the independent influences of the two broad sets of independent variables, and to answer the question, to what extent does section labor productivity vary with social organizational variables when the non-social variables are held constant? We cannot do this because we lack systematic data on the non-social variables at the level of sections.

FACTORY DIFFERENCES IN PERFORMANCE

We next test basically the same hypotheses to see if they can account for factory-level differences in our index of performance scores. In Tables D.30 and D.32 we tried to explain performance scores at the level of individual employees within each factory. When these scores are aggregated at the factory level, Shipbuilding Factory employees as a whole are significantly higher in performance than Electric Factory employees as a whole. Although this difference is weak ($C/C_{max} = .12$), we shall nevertheless attempt to explain it. Table 11.8 shows the extent to which this difference is related to social organization. Of the nine hypotheses in which we make the same prediction as Abegglen, five are confirmed. Shipbuilding Factory's higher performance level is related to the fact that its employees have higher status, higher job satisfaction, attribute greater legitimacy to the company's promotion system, think they have better promotion chances, and have more knowledge of the firm's procedures than employees in Electric Factory. On the other hand, the hypotheses that Shipbuilding Factory's higher level of performance is a result of work values, or various aspects of conflict (as perceived by managers), are disconfirmed.

TABLE 11.8

Shipbuilding Factory-Electric Factory Differences in Employee Performance in Relation to Factory Social Organization[a]

Hypotheses in Which Abegglen and Marsh-Mannari Make the Same Prediction

	Confirmed	C/C_{max}
Shipbuilding Factory has higher performance than Electric Factory because in Shipbuilding:		
1. employees have higher organizational status (v. 138/123)	Yes	.54**
2. employees have higher job satisfaction (v. 139/124)	Yes	.43**
3. employees ascribe greater legitimacy to the promotion system (v. 110–112, 115–119/74–76, 81–84)	Yes	.28**
4. employees have better perceived promotion chances (v. 109/73)	Yes	.23**
5. employees have more knowledge of procedures concerning work problems (v. 39/18)	Yes	.14**
6. employees are more likely to give primacy to work values over family or pleasure values (v. 31/8)	No	N.S.
7. managers have fewer conflicts among themselves concerning the authority and responsibility of their jobs (v. 127/113)	No	N.S.
8. managers say they get conflicting orders less often in carrying out their job (v. 128/114)	No	N.S.
9. managers are less likely to think subordinates receive communications from superiors with suspicion (v. 135/121)	No	N.S.

Findings for which Abegglen, Marsh-Mannari, and differentiation theory make different predictions

	Supports	C/C_{max}
1. Shipbuilding employees have had more previous interfirm mobility than Electric Factory employees (v. 148/108) (this finding also holds when age is held constant)	Differentiation theory	.53**
2. Shipbuilding employees are more likely than Electric Factory employees to express personal concerns in relation to the company rather than company concerns (v. 121/86)	Differentiation theory	.42**
3. Shipbuilding employees are more likely than Electric Factory employees to prefer a seniority wage system (v. 101/65)	Abegglen	.27**
4. Shipbuilding employees have higher lifetime commitment than Electric Factory employees (v. 146/132)	Abegglen	.22**
5. Shipbuilding employees participate less frequently in company recreational activities (v. 143/128)	Differentiation theory	.20**
6. Shipbuilding employees have lower cohesiveness than Electric Factory employees (v. 141/127)	Differentiation theory	.18**
7. Shipbuilding and Electric Factory employees do not differ significantly in paternalism (v. 144/129)	Marsh-Mannari	N.S.
8. Shipbuilding and Electric Factory employees do not differ significantly in the proportion who live in company housing (v. 2/102)	Marsh-Mannari	N.S.

** Significant at the .01 level.

[a] Includes male and female employees in both factories. The first variable number refers to Electric Factory, the second to Shipbuilding. For a complete list of variables by number, see Appendix C.

For eight other independent variables, Abegglen, differentiation theory, and we make different predictions; the outcome of these tests appears in the second part of Table 11.8. Differentiation theory is supported four times, Abegglen and we each twice. With regard to interfirm mobility, concerns, participation, and cohesiveness, the more differentiated employees are from the firm, the higher the factory's performance. With regard to wage system preferences and lifetime commitment, Abegglen is supported: higher factory performance is associated with a preference for a seniority wage system and greater support for lifetime commitment norms and values. Finally, with regard to paternalism and residence, our hypotheses that these have no significant independent influence on factory performance are supported.

We offer this explanation with some hesitation in view of the weakness of the difference between Shipbuilding and Electric factories' performance levels. In Table D.30 we see that sex is one of the best predictors of performance at the individual level. Since men in Electric Factory have higher performance than women, and since Electric Factory has so much larger a proportion of female employees than Shipbuilding, it is possible that the factory-level difference in performance is a function of the sex composition of the two factories' personnel. When we compare the index of performance scores of males in the two factories, we see that this is indeed the case: Electric Factory males have significantly higher performance than Shipbuilding males ($C/C_{max} = .29$). Since this difference is stronger than that between all employees in the two factories, an explanation of it in terms of the same independent variables may be more worthwhile.

Abegglen and we make the same predictions for nine independent variables, and four of these are confirmed (see Table 11.9). Employees with higher performance—Electric Factory males—have better perceived promotion chances, higher status, more knowledge of procedures, and give more weight to work values than do Shipbuilding males. Contrary to Abegglen's and our hypotheses, Electric Factory males have less job satisfaction than Shipbuilding males, and are not significantly different from Shipbuilding males in the legitimacy they attribute to their promotion system; nor do managers in the two factories report significantly different levels of conflict in three areas.

Of the eight hypotheses in which differentiation theory, Abegglen, and we make different predictions, Abegglen fares best, with five confirmations (see the second part of Table 11.9). The employees with higher performance—Electric Factory males—are more integrated into the firm than those with lower performance—Shipbuilding males—in the following respects: they have had less previous interfirm mobility, they are more likely to express company concerns, they participate more often in

TABLE 11.9
Differences in Performance among Male Employees in Electric and Shipbuilding Factories[a]

Hypotheses in Which Abegglen and Marsh-Mannari Make the Same Prediction

	Confirmed	C/C_{max}
Electric Factory males have higher performance than Shipbuilding males because Electric Factory males:		
1. have higher job satisfaction (v. 139/124)	No	.34**
2. have better perceived promotion chances (v. 109/73)	Yes	.28**
3. are more likely to give primacy to work values over family and pleasure values (v. 31/8)	Yes	.27**
4. have higher organizational status (v. 138/123)	Yes	.26**
5. have more knowledge of procedures concerning work problems (v. 39/18)	Yes	.24**
6. ascribe greater legitimacy to the promotion system (v. 110–112, 115–119/74–76, 81–84)	No	N.S.
Electric Factory males have higher performance than Shipbuilding males because Electric Factory managers:		
7. are less likely to say there are conflicts among each other over authority and responsibility in their job (v. 127/113)	No	N.S.
8. are less likely to say they get conflicting orders in carrying out their job (v. 128/114)	No	N.S.
9. are less likely to think subordinates receive communications from superiors with suspicion (135/121)	No	N.S.

Findings for which Abegglen, Marsh-Mannari, and differentiation theory make different predictions

	Supports	C/C_{max}
1. Electric Factory males have had less interfirm mobility than Shipbuilding males (v. 148/108) (this difference also holds when age is held constant)	Abegglen	.51**
2. Electric Factory males are more likely than Shipbuilding males to express company concerns (v. 121/86)	Abegglen	.34**
3. Electric Factory males are more likely than Shipbuilding males to prefer a job classification wage system (v. 101/65)	Differentiation theory	.33**
4. Electric Factory males participate more frequently in company recreational activities than Shipbuilding males (v. 143/128)	Abegglen	.30**
5. Electric Factory males are more likely than Shipbuilding males to live in company housing (v. 2/102)	Abegglen	.16**
6. Electric Factory males have higher cohesiveness than Shipbuilding males (v. 141/127)	Abegglen	.15*
7. Electric Factory males have lower lifetime commitment than Shipbuilding males (v. 146/132)	Differentiation theory	.13*
8. Electric and Shipbuilding Factory males do not differ significantly in paternalism (v. 144/129)	Marsh-Mannari	N.S.

* Significant at the .05 level.
** Significant at the .01 level.
[a] The first variable number refers to Electric Factory, the second to Shipbuilding. For a complete list of variables by number, see Appendix C.

company activities, they are more likely to live in company housing, and they have more cohesiveness. The differentiation theory of performance is supported twice: higher performance is associated with greater differentiation from the firm with regard to lifetime commitment, and with a preference for a job classification wage system. Only one of our hypotheses is supported: Electric and Shipbuilding factory males do not differ significantly in paternalism; therefore, paternalism cannot be a cause of performance.

CONCLUSION

In this chapter we have described and attempted to explain several aspects of performance: performance at the individual level in terms of organizational goals, labor productivity at the level of sections in Electric Factory, and performance aggregated at the level of the factory, first for all employees and then for males only.[16] Our explanatory hypotheses have been drawn from a number of theories of performance: Abegglen's, social differentiation theory, human relations theory, Japanese employees' views on how to improve productivity, a Japanese managerial theory of productivity, and our own theory.

The relative importance of the independent variables with regard to all these aspects of performance will be summarized in two ways. First, how many times is an independent variable one of the three best predictors—i.e., has one of the three highest beta weights in the second multiple regression run in Tables D.30 and D.32, or one of the three highest $C/C_{max's}$ in Tables 11.7–11.9?[17] Organizational status is a best predictor of these aspects of performance five times (i.e., in all five tables); knowledge of procedures is a best predictor twice; and the following variables are best predictors once each: sex, job satisfaction, work values, perceived promotion chances, legitimacy of the promotion system, managers' reported conflicts over authority, performance (with regard to labor productivity), and lifetime commitment. Thus, in both factories, whether performance is measured at the individual level, the section level, or the factory level, it is found to vary positively with status. Other independent variables predict better at some levels than at others, or better in one factory than in the other. It is also clear that with the exception of lifetime commitment, all the variables that are best predictors at least once are

[16] Other aspects of performance analyzed in this chapter need not or cannot be included in this summary; the ratio of absences (because it has been incorporated in the index of performance); and productivity in Sake Company (because we lack comparable data on the social organizational independent variables in Sake Company).

[17] Using the zero-order *r*'s in Tables D.30 and D.32 would be somewhat more comparable to the C/C_{max} than are beta weights. However, the same variables emerge as best predictors in these two tables, whether one uses beta weights or *r*'s.

those for which Abegglen and we (and the other theories as well) would make the same prediction concerning performance. Of the variables for which each theory would predict a different relationship to performance—wage preferences and the seven aspects of the social integration of the employee into the firm—only lifetime commitment is ever among the best predictors.

As a second way of summarizing our findings, consider only the hypotheses in which Abegglen's integration theory, our theory, and differentiation theory make different predictions. (1) Abegglen's theory is that performance is positively related to the degree to which employees are socially integrated into the firm, and that in a Japanese setting, employees who favor a seniority-based wage system are higher in performance than those who favor a job classification-based wage system. (2) As we interpret differentiation theory, performance is negatively related to the degree to which employees are integrated into the firm (i.e., the greater the differentiation of the identity and extra-work behavior and orientations of the employee from the firm, the higher the performance), and employees who favor job classification wage systems are higher in performance than those who prefer seniority wage systems. (3) Our own theory is that performance is determined primarily by formal organizational variables, e.g., status and knowledge of procedures, and that when these are held constant, performance will vary independently of the degree of integration of the employee into the firm and of wage system preferences.

If each of the eight controversial independent variables—cohesiveness, paternalism, residence, participation, company concerns, number of previous jobs, lifetime commitment, and wage system preferences—had been run in all five tables concerning performance (D. 30, D. 32, 11.7–11.9) we should have a total of 40 discrete tests of the three competing theories. In fact, we have 34 tests, since not all variables were run in each table. Of these, our hypotheses are empirically confirmed 14 times, Abegglen's 11.5 times, and differentiation theory 8.5 times.[18] Although this outcome does not permit us to reject any of the three theories, it does suggest that the theory that performance varies independently of social integration is somewhat better supported than the theory that it varies positively with social integration; and that the theory that it varies positively with social integration is somewhat better supported than the theory that performance varies negatively with social integration.

Abegglen's theory of performance is therefore not fully disconfirmed, but it is clearly on stronger ground when it makes the same predictions we

[18] Partial support for both Abegglen and differentiation theory (hypothesis 3 in the second part of Table 11.7) is counted as .5 support for each theory.

do than when its predictions differ from ours. Both theories would explain performance on the basis of employees' status, knowledge of procedures, sex, job satisfaction, work values, perceived promotion chances, legitimacy of the promotion system, and absence of organizational conflict, and in this both theories are sustained. On the other hand, when the theories make different predictions, ours is confirmed somewhat more often than Abegglen's.

Finally, performance is determined both by technological and other non-social variables, and by social organizational variables. We are not able to specify the independent effects of social organizational variables when technological variables are held constant (or vice versa).

CHAPTER TWELVE

The Social Organization of Japanese Firms

In this study we have made parallel analyses of three Japanese manufacturing firms' social organization and personnel. This concluding chapter has four objectives. The first is to recapitulate our main findings. Since this chapter is meant to stand alone as a summary of the entire monograph, those who have read the earlier chapters are asked to forgive a certain amount of repetition. The second objective is to make more explicit interfactory comparisons.[1] The third is to compare factories on each aspect of social organization after controlling for the sex composition of personnel. The three firms vary in the percentage of employees who are male: 49 percent in Electric Factory, 96 percent in Shipbuilding, and 100 percent in Sake Company's kura. We saw in earlier chapters that in Electric Factory women have less seniority, and are therefore lower than men on a number of other organizational status variables; because of this, women also differ significantly from men on knowledge of procedures, job satisfaction, work values, promotion chances, participation in company recreational activities, etc. Therefore, it is appropriate to ask whether the results of any comparison of all employees in the two factories will be similar or different when only men in the two factories are compared. If the results are similar, we can conclude with more confidence that the difference is a genuine difference in the social organization of the two factories, independent of their sex composition. If the results differ, we shall conclude that they are more a function of differential sex composition of the factories than of a difference in social organization of the factories.

It should be clear that we are in no way minimizing the importance of the sex composition of firms. The 523 women are as much a part of the concrete social organization of Electric Factory as are its 510 men. We

[1] To summarize the extent to which the factories are similar or vary, we shall use three measures for each aspect of social organization: the χ^2 test of the statistical significance of the difference between each pair of factories, the C/C_{max} measure of the strength of the difference between each pair of factories, and multiple regression analysis to ascertain whether the same independent variables provide the best explanation for given aspects of social organization across factories, or whether there are different patterns of multivariate explanation in each factory.

are banishing the women only in an analytical sense. The rationale is simply that we have already considered sex differences in social organization, and are now concerned to explain differences between firms that are due to factors other than sex composition.

The fourth objective of this chapter is to locate our findings and their implications in terms of larger contexts. This will be done by trying to answer the question, what is the relationship between our findings and more general knowledge of Japanese industry, society, and culture?

CHARACTERISTICS OF THE THREE FIRMS

Electric Company, Shipbuilding Company, and Sake Company, the names given to the three main firms studied, are each among the five leading firms in their respective industries: electrical appliances and machinery, shipbuilding, and rice wine (sake). Each company has plants located in various parts of Japan; the ones studied are in the Osaka–Kobe area, though they vary in location from metropolitan Osaka (Shipbuilding Factory) to a large city in the environs of Osaka (Sake Factory) to a small city with a rural hinterland northwest of Kobe (Electric Factory).

Sake Company, founded in the middle of the sixteenth century, had 525 employees in 1969. It produces sake by means of a centuries-old manufacturing process in its traditional brewing plants and a modern automated technology in its newer plant. The older plants employ dekasegi—rice farmers in the summer who work in the sake plants in the winter—while the newer has year-round personnel. Traditional craft skills are still taught through an apprenticeship system in the older plants, but have become largely obsolete in the automated brewing process. In general, Sake Company is characterized by the coexistence of traditional and modern technology, and of largely traditional but in some ways modern social organization. But this coexistence is not stable; most pressures are operating in the direction of weakening the traditional patterns.

Shipbuilding Company dates from the latter part of the nineteenth century. In 1969 its 18,000 employees were engaged in the construction of a variety of ships (e.g., supertankers and naval craft) and heavy land machinery for industrial uses (e.g., cement mills, sugar refineries, and steel structures). The plant studied has 756 employees and specializes in building diesel and turbine engines for ships and industrial machinery. The technology is not automated, and the work force consists primarily of highly skilled machinists and less skilled assemblers.

The third firm, Electric Company, was established at the end of World War II, but in this short period, as a result of Japan's postwar boom in electrical household appliances, has attained a commanding sales position among all Japanese industrial firms. Electric Company has developed extensive product diversification within the category of household appliances; the plant studied specializes in the production of electric fans, vacuum cleaners, and other small appliances. The company had 13,500 employees in 1969, of whom 1,200 worked in the plant studied. Though moving in the direction of greater automation, in 1969 the factory was still based mainly upon classical assembly line, batch production technology. The majority of workers perform highly fractionated, repetitive tasks, although there are some complex lathes and other machines run only by men.

FORMAL STRUCTURE

Status structure

The status structure in Electric Factory and Shipbuilding Factory is measured in terms of an identical index of organizational status, consisting of six items: pay, rank, job, seniority, education, and the section in which one works. Employees' scores on this index are a continuous variable which ranges from a low status score of 6 to a high status score of 18. Shipbuilding employees have significantly higher organizational status than Electric Factory employees ($C/C_{max} = .54$). However, when only men in the two factories are compared, the difference is reversed: Electric Factory males have significantly higher organizational status than do Shipbuilding males ($C/C_{max} = .26$). Thus, the difference in status structure is predominantly a result of sex composition; if Electric Factory had the same sex composition as Shipbuilding, its personnel would have higher organizational status than Shipbuilding personnel.

Bureaucratization

The tasks performed by personnel in a bureaucratic form of organization follow "general rules, which are more or less stable, more or less exhaustive, and which can be learned" (Weber 1946:198). When asked, "In your shop, when there is an accident such as work trouble, or a fire, do you know how to deal with it?" Shipbuilding employees are significantly, but weakly, more likely to know the procedures ($C/C_{max} = .14$, see Table 3.9, p. 51). This difference, however, is again a function of the sex composition of the two factories. Men in Electric Factory are more likely than those in Shipbuilding always to know the procedure, and this difference is somewhat stronger than that for all personnel in the two factories ($C/C_{max} = .24$). We may conclude that Shipbuilding Factory is more bureaucratized in the sense of widespread knowledge by employees of procedures, but that this is due more to sex composition than to differences in bureaucratization independent of sex.

In Chapter 3 we developed a theory to explain differential knowledge of procedures among employees and tested it in the two factories. We found that a similar explanation holds for both factories. The best predictors[2] of knowledge of procedures among all Electric Factory employees are status, sex, and employee cohesiveness (Table D.1, p. 366) and among Shipbuilding employees, status, age, and cohesiveness (Table D.2, p. 368). Among Electric Factory males the best predictors are age, employee cohesiveness, and status—identical to those in Shipbuilding Factory.

JOB SATISFACTION AND WORK VALUES

Though analytically distinct from the social organization of a firm, employees' job satisfaction and their values concerning work both influence and are influenced by the social organization. For this reason, it is worth comparing factories with regard to their employees' job satisfaction and value orientations.

Job satisfaction

The employees we studied perform specific tasks which vary in level of mechanization and automation, degree of specialization, autonomy, and variety. Jobs also vary in the social structure of the work situation in which they are located—in, for example, the worker's relations with other workers or foremen. Our index of job satisfaction measures employees' subjective responses to six aspects of the job, along a continuum of satisfaction-dissatisfaction (see Table 5.1, p. 100). Shipbuilding employees have significantly higher job satisfaction than Electric Factory employees (C/C_{max} = .43), and this difference largely persists when we compare only male employees. Shipbuilding Factory males are significantly higher in job satisfaction than their Electric Factory counterparts (C/C_{max} = .34). Thus, unlike the comparisons of status structure and knowledge of procedures, the higher job satisfaction of Shipbuilding employees is independent of sex composition.

We can compare Sake personnel with Electric and Shipbuilding personnel only on five specific job satisfaction items in Table 5.1, since not all six items in the job satisfaction index were used in Sake Company, and some job attitude questions that were asked in all three firms were not

[2] By "best predictors" here, as earlier, we mean that set of independent variables that provides the most parsimonious explanation of a given dependent variable; operationally, we refer usually to the variables in the second multiple regression run for a given dependent variable in Appendix D. The best predictors are the variables that account for the largest amount of the variance in the dependent variable, when other independent variables are held constant.

included in the index of job satisfaction. Sake employees are the most satisfied with regard to three items—interesting job, work load, and job satisfaction—and Electric Factory men with regard to the other two—variety and autonomy. For a more concise comparison of the three firms, we summed the number of employees in each who have positive, or favorable, job attitudes on these five questions. The percentage of employees with favorable job attitudes on all five items is 76.9 in Sake Company, 75.5 among Electric Factory men, 70.6 in Shipbuilding Factory, and 62.4 among all Electric Factory employees. When these percentages are compared—Sake versus the other three units[3]—Sake employees have significantly more favorable job attitudes than Electric Factory employees (C/C_{max} = .10) and Shipbuilding employees (.06), but there is no significant difference between Electric Factory males and Sake employees.

In Electric and Shipbuilding factories we went beyond measuring the degree of job satisfaction and attempted to explain why some employees are more satisfied than others, using the same independent variables. The best predictors are again basically similar across factories: status, cohesiveness, perceived promotion chances, and sex in Electric Factory; age, cohesiveness, and perceived promotion chances in Shipbuilding (see Tables D.3 and D.4, pp. 369 and 370). The pattern of explanation for Electric Factory males is also highly similar, with status, cohesiveness, and perceived promotion chances as the best predictors. Job satisfaction is most likely among higher status, older, male employees, who think their promotion chances are relatively good and are cohesive with their fellow employees.

Two recent large-scale surveys of job satisfaction in Japan, while not using the same instruments as ours, suggest that the level of satisfaction in our firms is higher than in a more broadly representative set of firms. One study is a public opinion survey on social consciousness conducted by the prime minister's office in March 1971. It covers men and women 20 years old or over throughout Japan. The study found that "only about 40 percent of the total respondents expressed satisfaction with their jobs, whereas the remaining 60 percent voiced slight discontent or deep dissatisfaction (a little less than 10 percent). Among the causes of such dissatisfaction were wages, working hours, and other working conditions, the lack of the sense that the job was worthwhile and an inadequate assessment of one's ability and achievement" (Japan Report 1972:6–7). A second source of data is a series of attitude surveys administered by Japan's large national unions. These "reported, without exception, that

[3] When Electric and Shipbuilding factory males are compared on these five questions, the results are the reverse of those based on the *index* of job satisfaction, noted above. Because we think the index is the better measure, we shall not comment further on the discrepant findings. However, this suggests that different measures of job satisfaction may yield different results.

the growing loss of positive motivation was spreading among many workers. In the case of Tekkororen [Japanese Federation of Iron and Steel Workers' Unions], only 26.9 percent felt positively motivated toward their work. Twenty percent felt that their work was uninteresting, while the remaining fifty percent seemed to be rather indifferent to their jobs. Other unions indicated that their workers seemed to have higher morale than the workers in Tekkororen's case but reaffirmed the suspicion that the majority of their members felt indifferent to the work" (Okamoto 1971a:9).

A recent analysis of alienation in Japanese industry brings together many of the threads of our own analysis: ". . . The efficiency-oriented management of Nikkeiren (Japan Federation of Employers' Associations) stressed, in its basic policy for the 1970's, coping with alienation, the gradual modification of educational status order, the opening of opportunities for promotion, job enlargement, the introduction of participatory target management, zero-defect study circles and quality control circles, and the use of individual ratings for career development. At large firms, these measures have been tried already for some years. The findings of union attitude surveys tend to report that 'staff control' increases the antagonism of labor and zero-defect and quality control group studies increase the physical and mental burdens while human relations programs tend to result in '*de facto*' cold relations irrespective of the surface outlook" (Ibid.:10).

Work values

We saw in Table 5.7, p. 113, that there are significant, but weak (C/C_{max} = .12) differences between Electric and Shipbuilding factory employees in the value orientations to which they give primacy. Electric Factory employees are somewhat more likely to favor pleasure values ("Work is only a means to get pay to spend on the pleasures of life"), and Shipbuilding employees are slightly more favorable to family values ("A happy family life is more important than a company job"); the same proportion of employees in both factories (27 percent) give primacy to a third value orientation, work ("Work is my whole life, more important than anything else"). When only male employees are compared, the difference in values becomes sharper: Electric Factory males are significantly more likely than Shipbuilding males to give primacy to work values, Shipbuilding males are more likely to give primacy either to family or pleasure values (C/C_{max} = .27).

The same variables provide the best explanation of these value-orientation differences among Electric Factory employees as a whole and Electric Factory males: organizational status and promotion chances, with sex as a third best predictor for all Electric Factory employees (Table D.6, p. 372). Among Shipbuilding employees, organizational status and size of community of origin are the best predictors (Table D.7, p. 374). Thus, work values are most likely to be primary, in all three groups, when the individual has higher status in the firm. In Electric Factory, work values are favored by men who perceive their promotion chances to be good, while in Shipbuilding, work values are somewhat more likely to be given primacy by those who come from smaller communities—towns and villages.

Comparable measures of value orientations over time within Japan are hard to come by. It is often stated that Japanese values traditionally emphasized diligence in work. Okamoto, for example, declares "... in the folklore of agrarian Japan, Japanese adults have been taught that the prosperity of a family is associated with the founders' diligent strivings while its decline can be linked to some successors' indulgence in hedonic pleasure" (Okamoto 1971b:I, 8). Postwar data suggest that the modal value among Electric and Shipbuilding employees—pleasure (see Table 5.7, p. 113)—is typical of Japan's industrial workers at the present time, but there is some question whether work value primacy has been a minority pattern for some time, or has become such only recently. In a study of an automatic steel factory in Japan, Tominaga found that when workers were asked in what respect they felt their life was worth living, "only 27 percent responded to the category of *work*, and most ... replied that they pursued their living goal in matters other than their own work" (Tominaga 1961:114). These data from over a decade ago suggest that the primacy of pleasure over work values for most Japanese workers was not a new phenomenon at the time of our study. A report of the Expert Committee on Labor of the Economic Council, an advisory body to the prime minister, provides longitudinal data on values. The report summarizes surveys by the prime minister's office, the Labor Ministry, the Social Welfare Center of the Economic Planning Agency, and research institutes in universities, conducted in 1971, and compares these with a survey done in 1957. The report finds that "workers, while interested in their jobs, are increasingly showing a greater interest in their home life and recreational pursuits as well, thus giving equal attention to both the public and private aspects of life" (Japan Report 1972:6).

RECRUITMENT[4]

Shipbuilding Factory, located in a major metropolitan area, recruits its personnel from significantly more distant places than does Electric Factory (C/C_{max} = .46). Shipbuilding personnel are much more likely

[4] Space constraints necessitated the omission of a chapter on recruitment and socialization. Here we can only note one or two of the main findings of the multivariate analysis, without the benefit of the larger structural context of recruitment or quotations from interviews.

than Electric Factory personnel to come from urban areas, especially the six largest cities of Japan (.65). Electric Factory employees are more likely to have fathers who are farmers, and less likely than Shipbuilding employees to have fathers who are in business, clerical, and service occupations (.33). These three differences persist when only males in the two factories are compared (.31, .54, and .30 respectively), and are therefore genuine factory differences in recruitment, not simply a function of differences in the sex composition of employees.

Shipbuilding employees have had significantly more previous interfirm mobility than Electric Factory employees ($C/C_{max} = .53$), and Shipbuilding males more than Electric Factory males (.51). These differences persist when age is held constant (see Table 12.1, below). The difference between factories is specified by age: it is greatest among employees (regardless of sex) aged 27 to 32, least among employees (regardless of sex) 33 years old or over. But in the present context, the main point is that there is a basic difference in the kind of employees the two factories recruit: Shipbuilding Factory recruits employees who have had significantly more previous interfirm mobility than Electric Factory employees; this difference is not due to differences in the sex or age composition of the two factories, and appears to be a genuine difference in factory social organization as such.

TABLE 12.1
Previous Interfirm Mobility of Shipbuilding and Electric Factory Employees, by Age and Sex

	Relationship to Previous Mobility (C/C_{max})			
	26 or Younger	*27–32*	*33 or Older*	*All Ages*
All Shipbuilding Factory vs. all Electric Factory	.22**	.75**	.16*	.53**
Shipbuilding Factory males vs. Electric Factory males	.50**	.74**	.18**	.51**

* Significant at the .05 level.
** Significant at the .01 level.

To explain variations among employees in previous interfirm mobility, we tested the hypotheses that mobility varies with five variables that precede it in time. This causal model of social origins and the recruitment process predicts that:

1. Father's occupation exerts no significant influence on interfirm mobility, when other variables (listed below) are controlled
2. The larger one's community of origin, the more interfirm mobility one has had
3. Men have had more interfirm mobility than women

4. The more education one has had, the less one's interfirm mobility
5. Since older employees have been exposed to the labor market longer, they have had more interfirm mobility than younger employees

The results of the regression analysis, shown in Tables D.35 and D.36, p. 409 and p. 410, are that the best predictor of previous interfirm mobility among Electric Factory employees as a whole, and among males (table not shown), is age. Among Shipbuilding personnel, age and education are the best predictors. When these variables are taken into account, the others have virtually no impact on previous mobility. Thus, the older one is—and therefore the longer one has been exposed to the labor market—and (in Shipbuilding) the less education one has had, the more interfirm mobility one has had. Although the inverse relationship between education and mobility confirms the lifetime commitment model, the consistently positive relationship between age and mobility disconfirms it. Since the key element in the model is employees' lifetime commitment norms and values, the model predicts the opposite of what we predicted, namely, length of exposure to the labor market will not significantly increase interfirm mobility. Our data for both factories clearly fit Western labor mobility models, which recognize a positive relationship between age and previous mobility, better than they fit the lifetime commitment model. We shall return to this point.

By "recruitment channel" we refer to the way in which one came in contact with a firm. This variable was coded from low to high values on employers' expectations of likely length of one's commitment to the firm: (1) government employment center or newspaper want ads; (2) knew someone in the firm; and (3) school recommendation. Although the lifetime commitment model minimizes the importance of job ads, and it is true that firms prefer not to recruit through such an impersonal channel, the plain fact is that both Electric and Shipbuilding companies do run ads in the major newspapers. Doing so publicly acknowledges that the firm is no longer relying exclusively on personal introductions and school recommendations and is in fact willing to hire people with previous work experience. Shipbuilding Company advertises an entering pay scale (with overtime) from ¥ 38,100 for 18 year olds to ¥ 56,400–66,700 for 40-year-old applicants. Its ads also offer Sunday interviews, which enable people to search for a new job on their day off, instead of having to quit one firm before looking for a job in another one. These ads clearly signify the firm's willingness to depart from the lifetime commitment pattern in which "recruitment directly from schools into the company is to all intents and purposes the *only* way in which men enter the firm" (Abegglen 1958:29; see also Abegglen 1969:105–107).

Shipbuilding employees were classified according to the year in which they were recruited into the company and the recruitment channel they used.[5] Table 12.2 shows that as we move from earlier (pre-1950) to more recent recruitment cohorts, the proportion relying on someone they knew in the company declined, while the proportion relying on school recommendation increased. The proportion recruited through government employment centers or the mass media shows a curvilinear pattern over time: higher in the earlier period and in the 1960s than in the late 1950s.

TABLE 12.2
Channel of Recruitment into Shipbuilding Company, by Year Recruited (%)

Seniority Cohort (Year Recruited into Shipbuilding Company)	*Channel of Recruitment*				
	Vocational Center or Want Ads	*Knew Someone in Company*	*School Recommendation*	*Total*	N
Pre-1950	29.0	64.5	6.5	100.0	(138)
1950–57	23.3	64.4	12.3	100.0	(146)
1958–61	13.6	59.1	27.3	100.0	(66)
1962–65	33.3	32.5	34.1	99.9	(126)
1966–67	35.1	27.0	37.8	99.9	(37)
1968–69	34.5	25.9	39.7	100.1	(58)
					(571)

$C/C_{max} = .46^{**}$

** Significant at the .01 level.

We attempted to explain the recruitment channel Shipbuilding employees used on the basis of four variables that precede it in time—father's occupation, size of community of origin, education, and number of previous jobs—as well as cohort (the year one was recruited into Shipbuilding Company). Table D.37, p. 410, shows that the most important variables are number of previous jobs and education. When these two variables are held constant, the social origins variables and the year in which one was recruited have no influence on recruitment channel.

Thus, more educated employees are more likely to use school recommendations, but this is not so much because of their education *per se*, as because more educated people are more likely to stay in one firm. If they change firms, then regardless of how far they went in school, the channel of school recommendation is no longer usable; they must instead rely either on being introduced by someone in the firm, or on impersonal recruitment channels.

National data are available on recruitment channels. It is true that larger firms hire more people directly from school than do smaller firms and

[5] Data on recruitment channel for each employee were not collected in Electric Company.

that, of all new graduates hired each year during the period from 1957 to 1964, firms of 500 or more employees have been the choice of an increasing proportion, while firms of under 100 employees have been the choice of a decreasing proportion (Ōhara Shakai Mondai Kenkyūjo 1966:44 and 52). It is also true that firms with 1,000 or more employees are more likely than smaller firms to have recruited a majority of their employees directly from school (Ibid.:59). Having said this, however, it must also be recognized that (1) in the first six months of 1965, only 46 percent of all employees hired in manufacturing firms of all sizes had been recruited directly from school; 37 percent had had previous occupational experience (Rōdō Shō 1966:30–31, Table 20), and (2) even in firms with 500 or more employees, the proportion who had been hired with no previous occupational experience (i.e., mostly new graduates) was only 45 percent in 1956, 41 percent in 1961, and 47 percent in 1963 (Ōhara Shakai Mondai Kenkyūjo 1966:44). In other words, even for the larger firms to which the lifetime commitment model is intended to apply, it is not true that recruitment directly from school is virtually the only mode of entry. Although most graduates may try for the large firms, it is not the case that these firms can fill their labor needs from this source. Thus, while the demand for labor in general decreased in 1971, there were still seven times more job openings for new middle and high school graduates than the supply of these individuals (Japan Institute of Labor 1971a:2).

THE REWARD SYSTEM

Pay

Shipbuilding Factory personnel receive significantly higher pay than Electric Factory personnel (C/C_{max} = .29), but this is largely a function of sex: Electric Factory males actually receive significantly, albeit only slightly, higher pay than do males in Shipbuilding (.17).

In the paternalism–lifetime commitment model of the Japanese firm, pay is based primarily on seniority, not on job classification. This model fit Electric and Shipbuilding Companies as of 1969–70. The best predictors of the amount of pay are seniority and number of dependents in Electric Factory (see Table D.8, p. 375); two other variables—age and rank—are the best predictors of pay in Shipbuilding Factory (Table D.9, p. 377) and among Electric Factory males. The seniority wage system is so highly institutionalized in these two firms that we can predict more of the variance in pay than in any of our other dependent variables.

The error of the lifetime commitment model is its assertion or implication that the seniority wage system is a traditional feature of Japanese work organizations, and that it will continue to be dominant in the future (Okochi and Sumiya 1955). Takezawa points out that: "the seniority basis of the remuneration system only dates back about fifty years as it applies to the blue-collar worker in large industrial concerns. Previously, it was common among staff workers, but job rates predominated in the remuneration system of the blue-collar worker" (1969:184). The seniority wage system, like other aspects of the paternalism model, arose as a result of specific twentieth-century developments, and in this sense is not a traditional Japanese pattern. Chao has shown that the seniority wage system is not necessarily dysfunctional for economic rationality, but also that it is "now in the process of disappearing" as a consequence of an intensified labor shortage which has increased starting wages faster than the wages of more senior workers (1968:16). There is considerable evidence that wage differentials based on education, age, and seniority are narrowing in Japanese firms (Japan Institute of Labor 1964b:2, 5, 1966:8, 10, 1971b:2). Table 12.3 indicates that although

TABLE 12.3
Monthly Wages of New and Retiring Male Employees, by Education and Status, Japanese Manufacturing Firms

Education and Work Status	*Monthly Salary (¥)* *When Hired* (*1*)	*At Age 55*[a] (*2*)	*Ratio* (*2*) ÷ (*1*)
College graduates, white collar			
1958	14,149	80,879	5.7
1960	15,183	85,495	5.6
1965	24,746	116,831	4.7
1967	28,964	136,615	4.7
High school graduates, white collar			
1958	9,544	62,225	6.5
1960	10,145	66,289	6.5
1965	17,589	88,495	5.0
1967	21,189	110,722	5.2
High school graduates, manual, blue collar			
1963[b]	13,796	70,638	5.1
1965	17,605	82,230	4.7
1967	21,260	92,629	4.4
Junior high school graduates, manual, blue collar			
1958	6,827	43,475	6.4
1960	7,263	46,020	6.3
1965	13,892	68,488	4.9
1967	16,977	79,654	4.7

Source: Tōyō Keizai, *Chingin soran* 1969:146–200.

[a] Fifty-five is the normal retirement age in Japanese firms. Many men continue working, but most have to move to another firm in order to do so.

[b] Data for earlier years not given.

seniority wage differentials were still considerable in 1967, they were narrower than in 1958; the same is true for differentials by education.

Electric Company had gone farther than Shipbuilding Company as of 1969–70 in institutionalizing a new wage system, in which job classification will have more importance than in the past, and seniority less. When asked which wage system they prefer, Shipbuilding employees are more likely than Electric Factory employees to prefer a seniority wage system (38.1 percent vs. 21.5 percent), Electric Factory employees are more likely to favour a job classification wage system (50.1 percent vs. 42.4 percent in Shipbuilding); Electric Factory employees are also more likely to think there is no difference between the two wage systems (28.4 percent vs. 19.5 percent in Shipbuilding). This difference is significant (C/C_{max} = .27). When only males in the two factories are compared, the difference becomes slightly stronger (.33). The more modern wage system preferences of Electric Factory personnel, then, reflect a basic difference between the two firms and are not due simply to differential sex composition.

The two factories' employees also differ significantly in the spontaneous reasons they give for preferring one or the other wage system (C/C_{max} = .57, see Table 6.11, p. 146). This difference persists when only males are compared (.53). Whether all employees, or only males, are compared, Electric Factory personnel are more likely to favor a job classification wage system because "it is rational to have wages fit the job;" Shipbuilding personnel, on the other hand, are more likely to favor a seniority wage system because "it provides security to the worker as he gets older." Among employees who favor a job classification wage system, Shipbuilding personnel are more likely than their Electric Factory counterparts to give as their reason that "a merit system encourages motivation, is good for morale."[6]

Employees were asked to rank six factors that influence pay according to their values, "Which factor *should* be most important? Which factor second?" etc. When the weighted points are converted to percentages, the resulting aggregate ranking of the six factors is strikingly similar in the two factories (see Table 6.12, p. 149); this is also true when only males are considered. Regardless of sex (and rank and age) employees think ability should be the most important determinant of pay, then skill or job classification, and then seniority, contribution to company profit, education, and number of dependents, in that order.[7] The similarity in aggregate ranking of these factors is also seen in the extremely weak C/C_{max}

[6] Wage system specialists in the West distinguish between job classification and merit wage systems. In the former, wages are based on the kind of job one does, but employees who perform a given job are not paid differentially on the basis of merit. Our Japanese respondents evidently do not draw this distinction so sharply.

[7] The only difference is that the rank of the two lowest factors in Electric Factory—education and number of dependents—is reversed in Shipbuilding Company.

relationships: .06 for all employees in the two factories, and .05 for males in the two factories.

In both factories, then, employees' values concerning pay are more modern than the actual system by which pay is presently determined, which is still overwhelmingly seniority based. In dynamic terms, we agree with Ono that there are a number of forces in operation that will continue to weaken the seniority wage system. "To what extent can the conditions for 'lifetime employment system' which have been systematized in a buyer's labor market be maintained under full-employment and in the time of extensive technological innovation? The changing worker consciousness will not acknowledge a wage system based upon a seniority . . . system administered according to educational background, with livelihood supplements affixed to the starting wage rate over time" (Ono 1971:4).

PROMOTION

Having considered pay we now turn to a second subsystem of the reward system: promotion. The present distribution of employees in the various ranks is one way of estimating the objective promotion chances in each factory. Table 12.4 shows this for all Electric Factory employees, Electric Factory males, and all Shipbuilding Factory employees. Once again, the difference in sex composition between the two factories is crucial. When all employees in both factories are compared, there are proportionately more staff positions in Shipbuilding than creative jobs in Electric, and more routine and training workers in Electric than rank-and-file workers

TABLE 12.4
Distribution of Employees by Rank and Sex, Electric and Shipbuilding Factories (%)

Rank		*Electric*	*Electric*	
Electric	*Shipbuilding*	*Total*	*Males*	*Shipbuilding*
Kachō	Kachō	2.1	4.4	0.9
Kakarichō	Kakarichō	2.0	4.2	2.6
Hanchō	Sagyōchō	1.8	3.8	1.6
Kumichō	Hanchō	3.3	6.8	6.6
Operative	Boshin	6.8	14.0	6.6
Creative	Staff	11.8	22.4	19.9
Routine, training workers	Workers	72.1	44.4	61.8
		99.9	100.0	100.0
	N	(1,033)	(510)	(573)

Electric total vs. Shipbuilding: $C/C_{max} = .21$**
Electric males vs. Shıpbuilidng: $C/C_{max} = .31$**

** Significant at the .01 level.

in Shipbuilding. There are 11.7 percent supervisory-managerial positions in Shipbuilding, in contrast to 9.2 percent in Electric Factory. But all women in Electric Factory are in the lowest rank, routine and training workers.[8] When only males are compared, objective promotion chances appear to be better in Electric Factory. Almost twice as large a proportion of positions held by males are supervisory-managerial positions in Electric as in Shipbuilding: 19.2 percent vs. 11.7 percent. Among males, it is Shipbuilding that has the significantly larger proportion of lower rank-and-file workers (61.8 percent vs. 44.4 percent). Even creative jobs are slightly more numerous proportionately in Electric than are staff jobs in Shipbuilding, among male employees.

Tables D.11 and D.13, p. 380 and p. 382, show which variables determine present rank in each factory. When the same set of independent variables is also run for Electric Factory males, we find that the same three variables are best predictors of rank for both Electric Factory employees as a whole and for Electric Factory males: rank is a positive function of the number of firms in which one has previously worked, seniority in the present firm, and education. Among Shipbuilding employees rank is a positive function of education, seniority, and performance. The main determinants of rank are, then, basically the same across firms.

How do employees perceive their promotion chances? In Table 7.2, p. 173, we saw that Shipbuilding employees see their chances of promotion as significantly better than do Electric Factory employees (C/C_{max} = .23). But this is a function of sex. Electric Factory males in fact perceive their promotion chances as significantly better than their counterparts in Shipbuilding (.28). These findings parallel those for objective promotion chances, noted above. When all employees are compared, those in Shipbuilding have better objective chances, and perceive their promotion chances as better, than do those in Electric Factory. But when only males are compared, the situation is reversed: those in Electric Factory have better objective chances, and perceive their promotion opportunities as better, than those in Shipbuilding.

Two types of explanation can be considered for these differences in subjective promotion chances. The first is our theory of rewards, which has been tested for each aspect of rewards (pay, job classification, rank, and perceived promotion chances). The second is a list of reasons for poor promotion chances that employees filled out in the questionnaire. With regard to the first, rank is the only variable that is consistently a best predictor for Electric Factory employees as a whole, for Electric Factory males, and for Shipbuilding employees. Promotion chances are seen as better when one is already in a higher rank, and therefore has been

[8] The 13 female respondents in Shipbuilding are all B-Group personnel and clerical; they are excluded from Table 12.4.

promoted at least once in the past (see Tables D.14 and D.15, p. 383 and p. 384). Other best predictors of perceived promotion chances—education, participation in company activities, performance, etc.—work for one or two of these groups, but not all three. Thus, the pattern of explanation of perceived promotion chances is rather different across factories.

The second explanation of perceived promotion chances is based on this questionnaire item, "(If in the previous question you said your chances of becoming a managerial-supervisory person are 'not too good' or 'no chance at all') What, then, is the reason you cannot become a manager or supervisor? Check as many of the following items as apply to you." Table 12.5 shows that the most frequently checked reasons in both factories are lack of education, many capable senior workers, not having the right kind of personality, and lack of ability.[9] Relatively unimportant

TABLE 12.5
Reasons Cited by Employees for Poor Promotion Chances[a]

Number of Employees Who Give Each Reason for Poor Promotion Chances[b]		*Electric Factory Total*		*Electric Factory Males*		*Shipbuilding Factory*	
		N	%	N	%	N	%
110/74.	Lack of education	395	19.0	196	24.1	205	23.5
119.	Because I am a woman	363	17.5	—	0.0	—	0.0
117/83.	Lack of ability, merit (*jitsuryoku*)	332	16.0	121	14.9	77	8.8
114/80.	Many capable senior workers	285	13.7	136	16.7	125	14.3
116/82.	I don't have the right kind of personality	271	13.0	108	13.3	129	14.8
111/75.	No chance to get "training within industry"[c]	162	7.8	78	9.6	68	7.8
113/77.	No openings for promotion exist	112	5.4	75	9.2	52	5.9
118/84.	I have poor record, reputation with management	65	3.1	52	6.4	50	5.7
112/76.	I am not known to any of the factory managers	58	2.8	16	2.0	32	3.7
115/81.	I am too old	34	1.6	30	3.7	136	15.6
	Total number of reasons[b]	2077	99.9	812	99.9	874	100.1

Electric Factory vs. Shipbuilding C/C_{max} = .49**
Electric Factory males vs. Shipbuilding males (not shown here): C/C_{max} = .32**

** Significant at the .01 level.
[a] When two variable numbers are given, the first refers to Electric Factory, the second to Shipbuilding. For a complete list of variables by number, see Appendix C.
[b] Respondents could check more than one reason.
[c] A short package course on how to handle people, how to teach them, and how to improve work.

[9] Another way to compute this is to divide the number of employees who check a given reason by the number of employees who think they have poor promotion chances. The

barriers in both factories include lack of opportunity to get training within industry, a poor record or reputation with management, and not being known by the factory managers.

Despite these similarities, there are significant differences between all employees in the two factories ($C/C_{max} = .49$), and also between male employees (.32). Shipbuilding employees are more likely than either all Electric Factory employees, or Electric Factory males alone, to attribute their poor chances to their age. This difference is reflected in the fact that Shipbuilding employees, and males, are older than Electric Factory employees, and males. When all employees in both factories are compared, there are two other significant differences: Electric Factory employees are more likely than Shipbuilding employees to say their poor promotion chances are due to their lack of ability or to being a woman. When only males are compared, the latter reason by definition is irrelevant but even males in Electric Factory are more likely than their Shipbuilding counterparts to ascribe their poor promotion chances to their lack of ability.

Since the variables in our theory of rewards are not matched except in a few instances (age, sex, education) by the list of reasons employees could check for their perceived poor promotion chances, the two explanations cannot closely correspond. Among Electric Factory personnel, however, there is somewhat more overlap between our best predictors of perceived promotion chances and employees' own reasons than among Shipbuilding personnel. Sex and education are both best predictors of and major reasons for perceived promotion chances among Electric Factory employees, while among Shipbuilding employees there is virtually no correspondence between the two explanations. It is perhaps best to conclude that these two sets of factors provide complementary, rather than mutually exclusive, explanations of subjective promotion chances.

To summarize our discussion of pay and promotion, it is clear that there has been a close functional interdependence between seniority wages, seniority-based promotions, and the permanent employment pattern. Some proponents of the lifetime commitment model (1) insist this is still dominant, and (2) tend to dismiss changes as more apparent than real. Our data support (1) but indicate that, with regard to (2), the changes Japanese firms are institutionalizing in their reward systems are more than merely "seniority categories under another name" (Abegglen 1969:113). Our conclusion is much closer to that of Cole (1971a: Chapters 3, 4) and Bennett (1967) than to Abegglen. Referring to a medium-sized optical firm in Tokyo, Bennett notes its plan to change from paternalism to

denominator is 845 for all Electric Factory employees, 341 for Electric Factory males, and 417 for Shipbuilding. Thus, among all Electric Factory employees, 395 of the 845, or 46.7 percent of those who could check education did so. However, when reasons are ordered by this computation, the ranking of reasons is unchanged.

"a higher standard wage and a more competitive system of promotion according to skill" and goes on to conclude that "the picture of Japanese industrial social organization presented in such books as James Abegglen's *The Japanese Factory* is seriously out of date, or at least not representative of many sectors of the economy" (Bennett 1967:426).

THE SOCIAL INTEGRATION OF THE EMPLOYEE INTO THE COMPANY

The central element in the paternalism–lifetime commitment model of the Japanese firm is that there is a high degree of social integration of the employee into the company, what Abegglen calls "the extent and nature of the involvement of the firm in the life of the worker" (1958:9 and Chapter 6). Dore, similarly, speaks of the Japanese firm as a "community," such that "an Hitachi man's consciousness of being an Hitachi man is a more salient part of his sense of identity than belongingness to the firm is for an English Electric worker. This is also a difference in degrees of individualism" (1973:221).

If the differentiation of the identity and interests of the employee from those of the firm were measured on a continuum from minimum to maximum, to the extent that the Japanese employee manifested minimum differentiation, the model would be confirmed. Specifically, the model would be confirmed to the extent that one observes high social integration in the following areas: high cohesiveness with fellow employees, high preference for paternalistic rather than functionally specific relationships with superiors and the company in general, a high degree of participation in company recreational activities, spontaneous expression of company (rather than personal, private) concerns on the part of the employee, a low degree of intraorganizational conflict, a low degree of previous interfirm mobility, and a high level of support for lifetime commitment norms and values. What have we found?

Employee cohesiveness

Electric Factory employees as a whole, and Electric Factory males have significantly higher index of cohesiveness scores than their Shipbuilding counterparts (C/C_{max} = .18 and .15, respectively; see Table 8.1, pp. 186–87). Although this difference is a factory difference, not simply an artifact of differential sex composition, it is relatively weak. We could only compare these two firms with Sake Company on two cohesiveness items: warmth of social relationships in the factory, and perceived teamwork (Table 8.1). On both items, Sake personnel are significantly more cohesive than all Electric Factory employees (.26 and .14) and Shipbuilding Factory employees (.31 and .21). On the first item Sake employees are significantly more cohesive than Electric Factory males (.31), but on the second item there is no significant difference.

When we regressed cohesiveness index scores on the same set of independent variables in Electric and Shipbuilding factories, a relatively common subset of variables emerged as the best predictors (see Table D.16 and D.17, p. 386 and p. 387). Among all Electric Factory employees, Electric Factory males, and all Shipbuilding employees, cohesiveness varies directly with job satisfaction and lifetime commitment, and inversely with organizational status variables (education and the informational level of one's section in Electric Factory and rank in Shipbuilding). Finally, among Shipbuilding employees and males in Electric Factory, cohesiveness also varies directly with the frequency of participation in company recreational activities. In general, therefore, the factors that conduce to high cohesiveness in both factories, regardless of sex, are lower status in the company combined with higher job satisfaction, lifetime commitment, and participation in company activities.

Although we could not test the same explanation of cohesiveness in Sake Company, its higher cohesiveness (on the basis of our partial indicators) would seem to be a result of the smaller number of workers in each kura, the greater similarity in their social backgrounds, and their more intensive and frequent social interaction, both in work and leisure hours.

COMPANY PATERNALISM

Paternalism was defined in Chapter 1 as present when "the managerial element assumes responsibilities for workers over and beyond the basic contractual provisions for wages and routine working conditions. The responsibilities include . . . the practice of carrying workers on the payroll in periods of business decline; housing; religious facilities . . . and many others (Bennett 1968:475).

Paternalism is widely recognized by Japanese social scientists as an analytically traditional pattern of social interaction between management and workers. Okochi and Sumiya (1955) find the roots of paternalistic labor relations in Japan's traditional rural areas, which have supplied the industrial labor force. Paternalism in the form of a familistic ideology (fictive kinship) has been described in a variety of settings: among iron workers (Odaka 1952) and merchants (Nakano 1964); in the mutual aid associations among traditional miners (Matsushima 1951); and in labor relations and "familistic management" in business firms during the Meiji-Taishō periods (Sumiya 1959, Hazama 1964); see also Bennett and Ishino (1963).

Paternalism as an historically traditional pattern of actual social relationships in Japan is somewhat more open to question. Abegglen's claim that paternalism in modern industry represents a continuity with earlier social patterns has been criticized by Noda (1963). In the Tokugawa period there was a *hō-kō* system, in which the master was paternally protective of his apprentices, who reciprocated with a sense of obligation. But this pattern was displaced in early industrial firms during the later nineteenth and early twentieth centuries. Noda argues that a more accurate picture of labor relations in that era is provided by novels such as Wakizo Hosoi's *Jōkō Ai-shi* (Pathetic Episodes in the Lives of Female Mill Hands). Noda also points out that the revival of paternalism in the post–World War I period applies more to how managers treated their white-collar employees (*shoku-in*) than their blue-collar workers (*kō-in*). A paternalistic relationship could be extended to the former group because it was much less numerous than the latter.

We initially sought to measure paternalism in terms of (1) the proportion of employees who live in company housing, and (2) paternalistic preferences, i.e., the degree to which employees prefer diffuse, paternalistic relationships to functionally specific relationships with (a) superiors and (b) management in general, (see Table 8.5, p. 194). We found that in both Electric and Shipbuilding factories, (1) varies independently of (2), and therefore our index of paternalism contains only items 2a and 2b. Let us analyze these first, and then company housing.

Paternalistic Preference. Preferences for a paternalistic relationship with one's superior (item 2a) are so widespread among all employees, regardless of sex, that there are no significant differences in this respect between Electric Factory, Sake Company, and Shipbuilding Factory. Nor are there any significant differences in paternalism index scores between Electric Factory and Shipbuilding Factory, again regardless of sex. Because of this relatively low variance in paternalistic preferences, no one of our explanatory variables, nor all of them together, predicted more than a small amount of the variance. The best predictors of paternalism are only partly similar in the two factories (see Tables D.18 and D.19, p. 388 and p. 389). Among all Electric Factory employees, Electric Factory males, and Shipbuilding employees, paternalism varies inversely with education. Among all Electric Factory employees paternalism also varies directly with rank, participation, and lifetime commitment; among Electric Factory males it varies inversely with seniority and directly with rank; and among Shipbuilding employees it varies directly with participation and pay. To put this another way, we can say that although not all these variables are best predictors in all factory/sex comparisons, *when* they are best predictors they have a uniform direction of relationship to paternalism. Paternalism increases with rank and pay, participation, and lifetime commitment, but decreases with education and seniority. The most paternalistic employees, in general, are those who have less education, less seniority, higher rank and pay, who participate in company recreational activities, and subscribe to lifetime commitment norms and values. We stress, however, that all these relationships are relatively weak.

The question on preferences for the type of supervisor under whom one would want to work was also asked in Japanese national sample surveys in 1953, 1958, 1963, and 1968 (see Table 12.6). The results are striking in

TABLE 12.6
Paternalistic Preferences, as Shown in Japanese National Surveys in 1953, 1958, 1963, and 1968 (%)

Question (#5.6) Supposing you are working in a firm. There are two types of department chiefs. (Card shown) Which of these two would you prefer to work under?

A. A man who always sticks to the work rules and never demands any unreasonable work, but on the other hand, never does anything for you personally in matters not connected with the work.

B. A man who sometimes demands extra work in spite of rules against it, but on the other hand, looks after you personally in matters not connected with work."

	1953	1958	1963	1968
Type A [Non-paternalistic]	12	14	13	12
Type B [Paternalistic]	85	77	82	84
Other & don't know	3	9	5	4
	100	100	100	100
N	(2,254)	(1,449)	(2,698)	(3,033)

Source: Adapted from Suzuki 1970: Part IV, p. 7, Table 2.

NOTE: "Each of the four surveys . . . consisted of face-to-face interviews with Japanese nationals aged 20 or over, who were selected by a stratified, three-stage probability sampling method."

two respects. First, they show that the high paternalism in the three firms we studied reflects a national cultural pattern. One may infer that Japanese employees bring paternalistic preferences with them from the wider society into the organizations of which they are members. Moreover, the "overwhelming preference runs consistently through all age groups, urban-and-rural breakdowns, and levels of education. There is a slight difference according to occupation, between those in professional and management categories preferring the independent, non-traditional type and those employed in small and medium-size enterprises; but the difference is no more than that between an 8-in-10 proportion and a 9-in-10 proportion for the traditional" (Suzuki 1966:29). The second noteworthy aspect of these findings is the high degree of stability of paternalistic preferences from 1953 to 1968. We shall return to this point below.

Housing. Sake Company's kura workers are much more likely than either Electric or Shipbuilding employees to live in company housing (Table

8.5, p. 194, first panel, C/C_{max} = .76 for Sake vs. Electric employees as a whole, .83 for Sake vs. Shipbuilding, and .74 for Sake vs. Electric Factory males). Considering all employees in Electric and Shipbuilding factories, only a small and not significantly different proportion live in company housing. However, males in Electric Factory are significantly, though only weakly, more likely than males in Shipbuilding to live in company housing (.16).

There is a rather similar pattern of explanation of housing in Electric and Shipbuilding factories (see Tables D.20 and D.21, p. 390 and p. 391). When other variables are held constant, living in company housing is most likely when the employee has migrated from a distant community (and therefore may not be able to afford expensive, private local housing within commuting distance of the factory) and is more highly educated (such employees are more likely to be recruited from more distant locations). This is true in both factories and regardless of sex. Among Shipbuilding employees two additional best predictors of residence emerged in the multiple regression analysis: more seniority makes one more likely to live in private housing, while higher rank makes one slightly more likely to live in company housing.

To live in company housing is to experience a form of company paternalism. The paternalism model implies that this experience will have positive consequences for employees' performance, job satisfaction, participation in company activities, cohesiveness, paternalism, and lifetime commitment. Our findings rather uniformly disconfirm these expectations of the model, in both Electric and Shipbuilding factories. Employees who live in company housing either do not differ from those who live in private residences in these respects, or if they do, the differences are found to be a function of other variables; when the other variables are controlled in multiple regression analysis, the differences apparently due to residence disappear.

To put our findings on housing and other forms of paternalistic company benefits in a broader perspective, we can consider theories of modernization and Japanese data on the national level. Convergence theories of modernization usually assert that extensive company housing is associated with the early stages of industrialization in a society (Dunlop 1958: 357–58). In later stages there is a process of differentiation of place of work from place of residence; modern life styles are such that workers prefer higher pay with which to rent or buy a private residence, rather than company-subsidized housing and lower pay (Smelser 1968:130–32, 138–39). Does Japan, with its rapid industrialization and modernization, fit this proposition?

A 1961 survey by the Japan Employers' Federation asked companies how much they spent on welfare services, defined as grants of money on occasions of marriage, illness, death; on workers' housing and subsidized housing facilities; and on grants or loans in support of various cultural activities. The average amount for firms in all the industries covered was ¥ 4,554 per worker per month. As a proportion of average wages per month (¥ 39,041), welfare services comprised 13 percent (Matsushima 1968:227). Ministry of Labor data indicate that in Japan in 1964 about 13 percent of the 11.3 million employees in the manufacturing sector benefited from some form of company housing (Japan Ministry of Labor 1965: 258–59, quoted in Cole 1967:212). Table 12.7 provides data on housing and other company benefits for Japanese firms, by size of firm, at three points in time—1958, 1963, and 1968. Among firms of all sizes (columns 1–3) the general trend has been toward an increasing incidence of welfare benefits, i.e., more rather than less paternalism, over time. Although in the largest firms (over 5,000 employees) there has been a decline over time in the incidence of company housing, in most other respects, and for other sizes of firms, the general trend has been toward greater paternalism.

Thus, the evidence indicates that in our firms as well as in other Japanese firms and among the Japanese people in general, there has been a persistence of preferences for paternalistic relationships, company housing, and other benefits. In our view, the area of paternalism is the major instance in our study of "partial modernization" (Rueschemeyer 1969), i.e., the stable, long-term coexistence of traditional and modern social patterns. Although many other aspects of Japanese society and factory social organization are becoming more modernized, paternalism appears to be a major stronghold of traditionalism that has not given way to more functionally specific role relationships.

Among the reasons for this persistence is the fact that the housing shortage in metropolitan areas exerts pressure on firms to provide company housing. On the other hand, the nature of this paternalism must be very precisely understood, lest it be exaggerated. Although a large proportion of firms *offer* housing, we have seen that only a very small proportion of employees live in company housing. Abegglen's data on 25 large Japanese firms reveal a decline between 1956 and 1966 in the proportion of employees living in company housing (1969:111). Moreover, our data on the *effect* of company housing on other aspects of factory social organization and employee behavior fit Cole's position better than Abegglen's: "Undoubtedly, [Japanese] workers get more from their job than money but this is not grounds for assuming *a priori* that notions of hierarchy, paternalism, non-competition, and group loyalties explain worker behavior" (Cole 1967:17–18).

Participation in company recreational activities

A third dimension of the employee's integration into the firm is the frequency with which he belongs to company athletic and cultural clubs and participates in recreational activities sponsored by the company, the

TABLE 12.7

Japanese Firms Offering Various Welfare Benefits, 1958, 1963, and 1968, by Size of Firm[a] (%)

Size of Firm	*All Sizes*			*1,000–2,999*			*3,000–4,999*			*5,000+*		
Year	*1958*	*1963*	*1968*	*1958*	*1963*	*1968*	*1958*	*1963*	*1968*	*1958*	*1963*	*1968*
Number of Firms	*1,053*	*1,061*	*875*	*247*	*300*	*248*	*71*	*85*	*77*	*107*	*127*	*127*
Type of Welfare Benefit												
1. Company house	83	76	83	92	85	92	94	94	91	99	98	89
2. Dormitory for single employees	70	83	91	89	92	100	94	99	97	94	97	91
3. Money lent for purchasing a house	43	51	74	56	66	84	68	87	92	83	94	87
4. Money presented for congratulations and condolences	94		97	93		96	94		92	95		100
a. marriage		91	97		90	97		84	96		86	96
b. childbirth		82	90		79	91		75	81		76	78
c. scholarships for children		14	18		11	15		22	25		21	21
d. accidents		87	93		89	91		88	97		87	87
e. sickness		85	92		84	90		84	83		83	90
f. retirement		43	48		42	46		49	49		41	57
g. death		93	94		92	100		89	97		92	91
5. Hostels for recreation and clubs	49	51	61	66	63	74	80	75	86	96	93	97

Source: Nihon Keieisha Dantai Renmei 1971:19–21.

[a] Size of firm is measured by number of employees. Size classes are: under 100, 100–299, 300–499, 500–999, 1,000–2,999, 3,000–4,999 and 5,000+. All size classes are included in columns 1–3. Specific breakdowns for size classes smaller than 1,000 are not given in this table. For complete data see original sources. Empty cells indicate no data.

more cohesive than Electric Factory males (.31), but on the second item there is no significant difference.

When we regressed cohesiveness index scores on the same set of independent variables in Electric and Shipbuilding factories, a relatively common subset of variables emerged as the best predictors (see Table D.16 and D.17, p. 386 and p. 387). Among all Electric Factory employees, Electric Factory males, and all Shipbuilding employees, cohesiveness varies directly with job satisfaction and lifetime commitment, and inversely with organizational status variables (education and the informational level of one's section in Electric Factory and rank in Shipbuilding). Finally, among Shipbuilding employees and males in Electric Factory, cohesiveness also varies directly with the frequency of participation in company recreational activities. In general, therefore, the factors that conduce to high cohesiveness in both factories, regardless of sex, are lower status in the company combined with higher job satisfaction, lifetime commitment, and participation in company activities.

Although we could not test the same explanation of cohesiveness in Sake Company, its higher cohesiveness (on the basis of our partial indicators) would seem to be a result of the smaller number of workers in each kura, the greater similarity in their social backgrounds, and their more intensive and frequent social interaction, both in work and leisure hours.

Company paternalism

Paternalism was defined in Chapter 1 as present when "the managerial element assumes responsibilities for workers over and beyond the basic contractual provisions for wages and routine working conditions. The responsibilities include . . . the practice of carrying workers on the payroll in periods of business decline; housing; religious facilities . . . and many others (Bennett 1968:475).

Paternalism is widely recognized by Japanese social scientists as an analytically traditional pattern of social interaction between management and workers. Okochi and Sumiya (1955) find the roots of paternalistic labor relations in Japan's traditional rural areas, which have supplied the industrial labor force. Paternalism in the form of a familistic ideology (fictive kinship) has been described in a variety of settings: among iron workers (Odaka 1952) and merchants (Nakano 1964); in the mutual aid associations among traditional miners (Matsushima 1951); and in labor relations and "familistic management" in business firms during the Meiji-Taishō periods (Sumiya 1959, Hazama 1964); see also Bennett and Ishino (1963).

Paternalism as an historically traditional pattern of actual social relationships in Japan is somewhat more open to question. Abegglen's claim that paternalism in modern industry represents a continuity with earlier of all others, at any given point in the life cycle. Thus, school-age Japanese exhibit high identification with their school (or with school-related collectivities to which they belong); later, the individual transfers his primary allegiance from school to his company. Although loyalty to one's company is not unknown in American society, there are probably stronger counteracting forces. Clark Kerr speaks of the American belief that one should have multiple loyalties simultaneously and that only in this way can the individual be free.

In our field work we encountered such manifestations of company identification as employees' using the term "our company" (*uchi no kaisha*) and speaking invidiously of some employees' "ego-centrism." On a more systematic level, our main measure of company identification is an open-ended questionnaire item, "Recently, what concerns [worries] you the most?" Responses were coded as follows, from a low to a high degree of spontaneous identification with company instrumental goals and problems: (1) personal, private concerns unrelated to the company, (2) personal concerns in relation to the company, and (3) company concerns. Electric Factory employees are significantly more likely than Shipbuilding employees ($C/C_{max} = .42$) or Sake employees (.44) to express company concerns. These differences are independent of the sex composition of the factories: Electric Factory males are significantly more likely than either Shipbuilding males (.34) or Sake males (.44) to express company concerns (see Table 9.5, p. 212). There is no significant difference between Shipbuilding and Sake on this item. It is worth noting that Sake Company, with the most traditional social organization of the three firms, has the lowest level of employee identification with company instrumental goals and problems.

In the multiple regression of the concerns variable on a number of independent variables, we were unable to explain more than a small amount of the variance (see Tables D.24 and D.25, p. 397 and p. 398). Expressing company concerns varies positively with rank, among Electric Factory employees as a whole, among Electric Factory males, and among Shipbuilding employees; it varies positively with job satisfaction among Electric Factory employees as a whole and among Electric Factory males. Otherwise, the best predictors of concerns are different for each collectivity: among all Electric Factory employees, expressing company concerns is more likely if one is younger and has less lifetime commitment; among Electric Factory males, it is positively related to performance and negatively related to age; and in Shipbuilding Factory it is more likely when the employee has higher lifetime commitment and has had a larger number of previous jobs. Not too much confidence can be placed in these relationships, however, because the betas are all weak.

Once again, it is our view that the lifetime commitment model has exaggerated the degree of employee identification with his company's instrumental goals and problems. Cole found that urban, high school educated workers were more independent in their orientation to the company than those with rural backgrounds and less education. "The cynicism and skepticism of the urban high school educated and mobile Tokyo diecast workers, their willingness to exercise civil rights, their willingness to oppose management . . . their competitiveness, and their use of *giri* in an instrumental fashion for their own self-interest are all characteristics that can hardly be described as traditionalistic. It is precisely these characteristics which are increasingly common" (Cole 1967:509).

ORGANIZATIONAL CONFLICT

Two of our three firms—Electric and Shipbuilding companies—have unions, and in tracing the origin of the former company's union we uncovered several examples of conflict between management and the proponents of the new union. In more recent years Electric Company and its union appear to have evolved a number of accommodations, and the level of overt conflict is low. We tested conflict and consensus models by comparing managerial-supervisory personnel with rank and file in Electric and Shipbuilding factories (see Table 9.7, p. 223). In both factories there were several latent grounds of conflict: for example, managers and rank and file differ significantly, and moderately strongly, with regard to perceived promotion chances, knowledge of procedures, identification with company instrumental goals and problems, wage system preferences, and performance. In this sense there is support for a conflict model. Yet on other aspects of social organization, behavior, and attitudes, the differences between those who have authority and those who do not are weak or non-significant, and there is thus also support for a consensus model. Presumably, had we used more discriminating measures of conflict, and asked about overt conflict rather than only latent bases of conflict, there would have been even more support for a consensus model.

Though our measures of organizational conflict leave something to be desired, we can conclude that, for what they are worth, they indicate a relatively similar degree of conflict in Electric Factory and Shipbuilding Factory. One indication of this is that the mean C/C_{max} over 16 common areas of latent conflict is virtually identical in the two factories (Table 9.7, p. 224: $\bar{X}$ C/C_{max} is .28 for Electric Factory and .29 for Shipbuilding). Three other measures of conflict, based on questionnaire items asked only of those in supervisory-managerial ranks, are presented in Table 12.8. All three factory differences are non-significant, since the N's are small.

TABLE 12.8
Organizational Conflict in Electric and Shipbuilding Factories[a] (%)

Variable			Electric Factory	Shipbuilding Factory
127/113.	In your job are there conflicts among supervisors (*uwayaku*) and colleagues (*dōryō*) over authority and responsibility?			
	Often, sometimes		59.1	52.4
	Rarely, not at all		40.9	47.6
			100.0	100.0
		N	(88)	(63)
	N.S. $C/C_{max} = .09$			
128/114.	How often do you get conflicting orders and directions in carrying out your job?			
	Often, sometimes		61.4	49.2
	Rarely, never		38.6	50.8
			100.0	100.0
		N	(88)	(63)
	N.S. $C/C_{max} = .17$			
135/121.	In your section (ka) or department (bu) what do you think about the manner with which communications from superiors are received by (your) subordinates?			
	1. Subordinates view them with great suspicion		0.0	1.7
	2. They may or may not be viewed with suspicion		9.2	20.3
	3. They are usually accepted, but at times are viewed with suspicion		39.1	30.5
	4. They are generally accepted, but if not, then subordinates openly and candidly question the communications		51.7	47.5
			100.0	100.0
		N	(87)	(59)
	N.S. $C/C_{max} = .27$			

[a] The first variable number refers to Electric Factory, the second to Shipbuilding. For a complete list of variables by number, see Appendix C. Responses of managers and supervisors only.

But the strength of the relationships (C/C_{max}) is also low, and we therefore conclude that Electric and Shipbuilding factories do not experience different degrees of conflict. In both factories, somewhat over half of the managers say there are conflicts over authority and responsibility among supervisors and colleagues; half or somewhat more than half of them get conflicting orders relatively often; and about half report that their subordinates, generally accept their communications. Although on this last item a somewhat larger percentage of Shipbuilding managers report some suspicion by subordinates, the overall pattern is clearly a similar level of conflict in Electric and Shipbuilding factories.

Lifetime commitment

The last, and from the point of view of the model, most important aspect of the social integration of the employee into the firm is lifetime commitment. Defined as the practice of entering a firm directly after completing school and staying in that firm until retirement, lifetime commitment also presupposes a reciprocal obligation on the part of the employer not to fire his employees.

Our analysis began by distinguishing two aspects of lifetime commitment. The first is role behavior, as indicated by interfirm mobility; the second is the degree to which one subscribes to lifetime commitment norms and values. We have already seen that Shipbuilding employees have had significantly more interfirm mobility than Electric Factory employees (.53), and Shipbuilding males more than Electric Factory males (.51). Although interfirm mobility varies positively with age, and Shipbuilding personnel are older than Electric Factory personnel, Table 12.1 shows that when age is held constant, Shipbuilding employees (and males) still have had more interfirm mobility than Electric Factory employees (and males).

We measured lifetime commitment norms and values by questionnaire items on perception of other employees' intentions to stay in their present firm until retirement, one's own intentions, attitude toward employees who voluntarily change firms, and beliefs concerning whether males should stay in one firm until retirement (see Table 10.3, p. 235). These items formed an acceptable index of lifetime commitment, on the basis of which Shipbuilding Factory males were found to have significantly, though only weakly, higher lifetime commitment than Electric Factory males[10] (C/C_{max} = .13). The paternalism–lifetime commitment model predicts an inverse relationship between previous interfirm mobility and lifetime commitment, i.e., employees who have never changed firms should affirm lifetime commitment norms and values more than those who have changed firms in the past. Our data consistently disconfirm this prediction: there is in fact no correlation between previous mobility and present lifetime commitment norms and values (r = .06 for Electric Factory males and −.01 for Shipbuilding). Although our interpretation of this finding can only be *ex post factum*, and requires further research, it suggests that there are forces in operation that raise the lifetime commitment of those who have worked elsewhere previously, and lower it among those who have always worked for the same firm. These counterpressures may at least partly explain the observed lack of relationship.

[10] The analysis of lifetime commitment is restricted to males, in order properly to test the lifetime commitment model. Therefore in this section it is unnecessary to compare all employees in our factories.

In general, this finding is less difficult to understand when one rejects the formulation that lifetime commitment is an absolute value of unconditional loyalty to one's firm. Our findings are much more consonant with the view that lifetime commitment is more of a contingent, conditional norm: under certain conditions I shall stay in this firm and give it my loyalty; but under other conditions, I shall leave it.

The multivariate explanation of lifetime commitment is only partially similar for the two factories (see Tables D.28 and D.29, p. 402 and p. 403). Males profess lifetime commitment norms and values to the extent that they have high job satisfaction and are cohesive with fellow employees. Among Electric Factory males a third best predictor is perceived job autonomy: employees with less autonomy have somewhat higher lifetime commitment than those with more autonomy. Among Shipbuilding employees, lifetime commitment is also positively related to age and to perceiving few other firms as offering as good conditions as one's present firm.

Another inference from the lifetime commitment model can be tested. If the employee who voluntarily changes firms thereby violates norms and values that enjoin moral loyalty to *one* firm, we might expect him to encounter negative sanctions in the new firm. The new employer would hire him grudgingly and would share with his new fellow employees a suspicion of his loyalty to *any* firm. If these mechanisms operate, we would expect him to encounter various forms of discrimination in the new firm, for example (other things being equal): lower pay, lower rank, poorer promotion chances, lower cohesiveness with fellow employees, and lower participation in company recreational activities, than employees who have always worked for the same firm. We are *not* interpreting the model as predicting so much discrimination against those who enter a new firm after leaving another firm that, for example, a manager would be hired at the wages of a worker; that would be an absurd interpretation. Rather, "other things being equal" means the manager would be hired at lower pay than other managers of his rank who had always worked for the same firm. The same logic applies to our inferences with regard to the other dependent variables.

That we are not setting up a straw man is indicated by the fact that Japanese scholars have noted that chūto saiyōsha ("midterm entrants" into a firm, who have worked elsewhere previously) are discriminated against. When a firm needs to reduce its number of employees, temporary and outside workers are the first to be let go; if more personnel cuts are required, chūto saiyōsha are let go (Tsuda 1973:405; see also Okamoto in Okochi, Karsh, and Levine 1973:192). Moreover, "the midterm employees had little expectation of promotion to higher positions, however long they might serve the company" (Tsuda 1973:414). To test this, we ran a series of multiple regression analyses (see Table D.38, pp. 411–12). Each of the dependent variables—pay, rank, promotion chances, cohesiveness, and participation—was in turn regressed on (1) number of previous jobs, and (2) other theoretically important independent variables that had already been shown to be best predictors of the dependent variable. The results are clearly negative for the lifetime commitment model. In seven of the 10 multiple regressions run in Electric and Shipbuilding factories the number of previous jobs has the lowest zero-order correlation, and in nine of these 10 regressions the number of previous jobs has the lowest beta weight of the set of independent variables. Virtually all the explained variance in each dependent variable is accounted for by variables other than one's previous interfirm mobility. For example, pay is more a function of seniority or age, number of dependents or rank, than of the number of firms for which one has worked. Sex, rank, and education explain more of the variance in perceived promotion chances than does one's previous mobility. Education, sex, and cohesiveness are better predictors of participation in company activities than is number of previous jobs; and so on. In the one instance (in Electric Factory) in which previous mobility is a better predictor than other independent variables—i.e., in relation to rank—its direction of relationship with rank is positive, rather than negative. It is employees who have worked for more previous firms who have higher rank, not those who have always worked for the same company, as the lifetime commitment model would have it. Nor is this simply because older employees are more likely to have worked for other firms. The relationship between previous mobility and present rank persists when age is held constant (as Table D.11 shows).

If the presumed negative consequences of changing firms are in fact not observed, the inference to be drawn is that the alleged moral disapproval of interfirm mobility and the negative sanctions applied to the mover are not in fact operative. Once again, a theoretical proposition derived from the lifetime commitment model is disconfirmed by the evidence.

Major analyses of trends in interfirm mobility in Japan as a whole confirm our conclusions as against those presented in terms of the lifetime commitment model. Mobility rates for a given point in time underestimate the extent to which a cohort of the work force changes jobs during its entire work life. If mobility analyses fail to take into account the length of time an individual has been in the labor force, they will reflect only differences in length of labor market exposure, rather than differences in attachment to one firm. Tominaga (1962, Table 13) presents data based on a cross-section sample of 1,227 males above 20 years old, in all occupational strata, living in Tokyo in 1960. The percentage who have always worked for the same firm, analysed by the period in which they entered the labor force, is as follows: 1956–60 :73.0; 1951–55 :52.5; 1946–50 : 34.6; 1941–45 :19.1; 1936–40 :20.0; 1931–35 :13.7; 1926–30 :11.6; 1921–25 :16.0; and 1896–1920 (grouped because of small N) :12.9. Thus, even if we consider only those who have entered the labor force since the war (1946–60 period), 49 percent have experienced interfirm mobility. Moreover, among those who have moved, multiple mobility is not negligible, although the number of moves is probably lower in the Tokyo sample than in comparable American samples. Of the total Tokyo sample, 30.6 percent had always worked for one firm, 22.6 percent had made one interfirm move, 18.8 percent two moves, 13.4 percent three moves, 6.3 percent four moves, and 8.2 percent five or more moves. Although interfirm mobility rates in Tokyo are probably higher than for Japan as a whole, one must conclude that, in purely behavioral terms, lifetime commitment is far from universal in the Japanese economy. Thus, Tsuda concludes that "it is a safe general assumption that the lifetime employment system is beginning to crumble as midterm employees become more numerous" (1973:436).

Cole analyzed the distribution of newly hired male employees by size of present and previous establishment, in manufacturing industries in 1965 and 1969. Interest centers on firms of 500 or more employees, for it is these that the model argues are the mainstay of the lifetime commitment pattern. Citing data from the Office of the Prime Minister, *Employment Status Survey*, Cole finds: "Those leaving firms of 500 or more rose from 42,600 employees in 1965 to 61,400 in 1969. This was an increase of 44 percent; from 1965 to 1968, the number of male employees in manufacturing firms of 500 or more employees increased by only 10.5 percent. Comparable data do not exist for 1969, but there is no indication that the increase was especially remarkable for that year" (1972:621). Cole also shows that ". . . those leaving firms of 500 or more rose from 15 percent (42,600) of all those reported changing jobs in 1965 to 20 percent (61,400) in 1969" (Ibid.:621). Moreover, the percentage of employees who leave firms of 500 employees and enter other firms of that size class also increased, from 33 percent in 1965 to 47 percent in 1969.

These national interfirm mobility rates for males in the larger manufacturing firms are extremely significant for the lifetime commitment controversy. Heretofore, trend data had suggested a quite constant (low) rate of mobility (Marsh and Mannari 1971). Abegglen's comparison of average exit rates from 25 large firms in 1956 and 1966 led him to the similar conclusion that exit rates had not significantly increased for males (Abegglen 1969:114). For generalizations about large manufacturing firms Cole's data are obviously superior to Abegglen's. The significance of these national findings Cole has brought to the attention of Western social scientists cannot be stressed too strongly. Heretofore, critics of

Abegglen, including ourselves (Marsh and Mannari 1971), were able to present data that indicated a relative absence of moral loyalty to the firm and a lower level of normative and value support for lifetime commitment than that asserted by Abegglen. However, these findings could always be countered with the statement, "It may be that Japanese employees stay in a firm for reasons other than moral and normative commitment to permanent employment, but the fact remains that they do stay in one firm, and there is no trend toward increasing interfirm mobility." In other words, the model could be defended with regard to overt role behavior (non-mobility), even if it could no longer be supported with regard to norms and values. Cole's data destroy even this version of the model. There is in fact a trend toward higher levels of interfirm mobility. One may infer that Japanese employees' role behavior and norms and values are becoming consistent, i.e., converging in the direction of a lower level of lifetime commitment. Whereas earlier findings implied a split between behavior and values, the new findings appear to bring behavior and norms and values into a closer, and more modern, integration.

Government policy in Japan is closer to the outlook Cole, we, and others have stated than to the lifetime commitment model. The 1971 White Paper on Labor of the Ministry of Labor identifies as one of the nation's most urgent problems the facilitation of labor mobility: ". . . A positive manpower policy is needed . . . considering *increased labor mobility particularly among middle and older aged workers*, there is a need to help workers adjust themselves smoothly to new working environments when they change their jobs" (Japan Institute of Labor 1972:7, italics added). Similarly: "During the five-year economic boom from 1965 to the autumn of 1970, the rate of job change climbed mainly in the big enterprises. Moreover, the increase in the job change rate among employees of big businesses was seen not only among young workers, but also among middle-aged people centering on those in their late 'thirties. Furthermore, job change registered an upward curve among highly educated employees" (Japan, Office of the Prime Minister 1971; Japan Report 1972:7). In short, there is now evidence of increasing interfirm mobility among the very group usually identified as the core of the lifetime commitment pattern—older, more educated male employees in large firms.

We have argued that Japanese employees' commitment to their firm is contingent and conditional, rather than absolute. A number of objective macrostructural factors influence variations in mobility rates. Among these are the condition of the labor market and the general national economy. In the mid-1950s Japan's unemployment rate was higher than it is now. The economy was just beginning to emerge from its postwar period of sluggish growth. By the late 1960s capital and labor markets were quite different. A falling birth rate and higher educational aspirations reduced the number of new workers entering the manufacturing sector, although demand for labor increased. There is now a labor shortage for young and skilled workers. In Japan, as in the United States, employees are more likely to leave their firm voluntarily in periods of prosperity and favorable labor market conditions. In the United States, during the prosperous 1920s, about three-fourths of all separations in manufacturing firms were voluntary, in contrast to one-fourth or less in the depressed 1930s (Parnes 1968:484). Similarly, of Japanese employees who quit, the proportion who do so for "personal" (*jiko*) reasons increased from 68 percent in 1956 to 86 percent in 1967 (Rōdō Shō 1969:19, Table 49). Contrary to the lifetime commitment thesis, the Japanese response is not fundamentally different from the American. Moreover, from this evidence it is hard to accept Abegglen's conclusion that "the worker . . . is bound, *despite potential economic advantage*, to remain in the company's employ" (1958:17, italics added).

We have examined some national data that bear upon the overt role behavioral aspects of lifetime commitment. The further question is, how well do our findings agree with other studies and national level data on the normative, value, and attitudinal aspects of commitment? To repeat, the heart of the lifetime commitment thesis is not simply that Japanese exhibit the role behavior of staying in one firm longer than Americans. It is that they stay because they hold certain beliefs and have internalized certain norms and values.

In 1965 the Japan Institute of Labor sponsored an attitude survey of 35,000 organized workers in Tokyo and Yokohama. Table 12.9 shows that the "I would move depending on the conditions" response is more common than the lifetime commitment response, "I have no intention of moving," in the total sample, among both sexes, and for most age groups. As in the factories we studied, the decision between commitment and mobility is contingent and conditional, rather than based on absolute moral loyalty to the present firm.

A survey called "Consciousness and Action of Salaried Men," which was published in 1970 in the *Asahi Shimbun*, received considerable public attention. Based on a questionnaire completed by 550 Japanese men in their twenties, employed in 50 private firms, the results showed that employees' "feeling of being one with their companies has decreased. . . . [They] place a high value on their lives as individuals. . . . Asked if they would change over to a competing company if they were offered better conditions, 38 percent answered in the affirmative. *This is a high percentage considering that the firms for which the young workers are working are all listed in the first section of the Tokyo Stock Market*" (*Asahi Evening News*, 1970a, italics added). An editorial comment on the findings was that "the survey has clarified [these young salaried employees'] view that passion

TABLE 12.9
Responses of Japanese Workers to the Question, "The other day Mr. A moved to another company because of 'better conditions.' What do you think about moving to another company?" (%)

	I Have No Intention of Moving	*I Would Move Depending on the Conditions*	*I Want to Move by All Means*	*Undecided*	*No Answer*	*Total*
Total	40.8	49.0	1.5	7.4	1.2	99.9
Male	40.4	50.1	1.7	6.5	1.3	100.0
Female	42.6	45.5	1.2	10.0	0.8	100.1
Age (male)						
-19	37.5	40.6	—	21.9	—	100.0
20–21	29.7	55.0	3.3	12.1	—	100.1
22–23	31.2	56.2	3.1	9.4	—	99.9
24–25	39.6	48.5	1.0	9.9	1.0	100.0
26–27	40.7	55.8	1.2	2.3	—	100.0
28–29	33.3	56.1	1.5	6.1	3.0	100.0
30–31	34.3	56.7	3.0	3.0	3.0	100.0
32–33	29.0	61.3	—	6.5	3.2	100.0
34–35	56.5	37.0	2.2	4.3	—	100.0
36–39	40.7	50.6	1.2	6.2	1.2	99.9
40–49	59.1	35.2	1.1	1.1	3.4	99.9
50–	68.6	31.4	—	—	—	100.0
Age (female)						
-19	40.0	31.4	—	28.6	—	100.0
20–21	35.2	51.9	1.9	11.1	—	100.1
22–23	31.8	54.5	2.3	11.4	—	100.0
24–27	26.3	63.2	—	7.9	2.6	100.0

Source: Japan Institute of Labor, *Research and Study Report Series*, No. 74, 1966, quoted in Okamoto 1970:9.

for work is based on a person's individuality, knowledge and technique and that *it is not prompted by such a sentiment as loyalty to the company or a sense of belonging*" (*Asahi Evening News* 1970b, italics added; for a Japanese version of the report on the survey see *Asahi Shimbun* 1970; 21–23).

In conclusion, there are clearly factors that continue to operate so as to maintain and strengthen commitment to one firm, and the Japanese interfirm mobility rate is not likely in the immediate future to become as high as in the United States. Nevertheless, considering the net balance of functional and dysfunctional consequences (Marsh and Mannari 1972), we expect a further weakening of the lifetime commitment pattern.

To paraphrase Marx, the lifetime commitment system may contain the seeds of its own destruction. The more successful a company is in eliciting lifetime commitment, the higher the average seniority of its personnel will be, and the greater the company's direct and indirect wage and welfare costs will be. Therefore, the company will be compelled to modify its seniority basis of payment in the direction of a job or skill classification system. As it does this it may weaken the degree of lifetime commitment. Japanese industry applied liberalization (*jiyūka*) and rationalization (*gōrika*) first to technology, then to marketing. It is likely that it will now increasingly extend these policies and practices to personnel administration (Imai 1968).

UNIFORMITY AND VARIATION IN THE SOCIAL ORGANIZATION OF JAPANESE FIRMS

The paternalism–lifetime commitment model asserts that despite variations in product, technology, or age, Japanese firms have a uniform social organization. Abegglen (1958:13) says his model describes "the general rule in the large factories of Japan," and Dore declares "the features which make up what we shall call the 'Japanese system' . . . are generally shared by all Japanese *large* corporations" (1973:301). We shall refer to this as the uniformity of social organization hypothesis, with reference to males in firms with over 1,000 employees. Our data for males in Electric and Shipbuilding factories are therefore relevant. Sake Company has only 525 employees. However, the model's emphasis on the historical roots and the traditionalism of the social organization depicted encouraged us to select at least one firm that would approximate as closely as possible the technology and social organization characteristic of firms in Japan's past. We soon discovered that any such firms that still exist are small scale. The inclusion of Sake Company represents our best efforts in this direction. We shall not however, argue that variations in *its* social organization are as relevant to the model as variations in the other, larger firms studied.

Table D.39, pp. 413–14, presents 39 comparisons of males in the three firms—22 for Electric and Shipbuilding and 17 for Electric and Sake or Shipbuilding and Sake. Contrary to the uniformity hypothesis, we find strong or moderate differences that are statistically significant between firms on the following aspects of social organization. Firms vary in the proportion of their personnel who live in company housing. Firms vary in the degree to which their employees prefer a job classification wage system or a seniority wage system, and in the reasons employees give for these preferences. Firms vary in how much previous interfirm mobility their employees have had. Firms vary in the degree of employee identification with the company, as indicated by spontaneous expression of company (rather than personal, private) concerns. Firms have varying levels of employee job satisfaction. Firms vary in the reasons employees give for poor perceived promotion chances, in their objective promotion chances, in their employee cohesiveness, and in how much their employees participate in company recreational activities.

Table 12.10 summarizes these 39 factory comparisons, and reveals that two show a very strong difference between factories ($C/C_{max} > .60$), 14 show moderately strong differences (.30–.59), and 23 show only weak differences (<.30). These last 23 do support the uniformity of social organization hypothesis, but 16/39 or 41 percent of the comparisons indicate moderately strong or very strong *differences* between firms. As already noted, the Electric-Shipbuilding comparisons are more pertinent as tests of the model than comparisons that involve Sake Company. Of the 22 Electric-Shipbuilding comparisons in Table 12.10, 8 (36 percent) indicate moderately strong differences.

Needless to say, these findings are only suggestive with regard to the uniformity of social organization hypothesis. We cannot disconfirm it on the basis of variations among three firms; an adequate test requires data from a larger and more representative sample of large Japanese firms, and ideally, comparable data from firms in other societies, to see whether the degree of interfirm variation differs across societies. An important step in this direction has been made by Azumi and McMillan's comparison of 29 Japanese manufacturing organizations, each with over 1,000 employees, which found great variation in such social organizational characteristics as specialization, centralization, formalization, and stratification. They conclude: "Obviously no unitary image of the Japanese organization is justified" (Azumi and McMillan 1973:15).[11]

Another assessment of the uniformity of social organization hypothesis uses as the criterion of uniformity the extent to which the independent variables that make for the greatest change in each aspect of social organization, behavior, and attitudes are identical in Electric and Shipbuilding factories. For each of 14 dependent variables in this study, Figure D.2, pp. 417–22, reexamines the multiple regression analysis and presents a causal model.[12] For each dependent variable we are interested in those independent variables that produce statistically significant amounts of change in the dependent variable, when other hypothesized independent variables are held constant.

Since we are now primarily interested in whether the same causal processes operate under varying conditions with regard to a given dependent

[11] Azumi and McMillan's paper reports on 50 Japanese manufacturing firms, of which 21 have fewer than 1,000 employees. Separate tabulations for the 29 firms with more than 1,000 employees were kindly provided by the authors. Although the range of variation in social organizational characteristics is slightly less for the larger firms, it is still considerable.

[12] Figure D.2 also includes a causal diagram for a fifteenth dependent variable, performance, but this is not conceptualized as a social organizational variable and therefore its causal variables are not included in the cumulated totals for the other 14 dependent variables.

TABLE 12.10

Variation in Social Organization between Electric Factory, Shipbuilding Factory, and Sake Company[a]

Domain of Social Organization	Strength of Relationship between Factory and Aspect of Social Organization: $C/C_{max} < .30$		C/C_{max} .30–.59		$C/C_{max} > .60$		Total
	EM vs Ship.	*EM vs Sake or Ship. vs. Sake*	*EM vs. Ship.*	*EM vs. Sake or Ship. vs. Sake*	*EM vs. Ship.*	*EM vs Sake or Ship. vs. Sake*	
Status	1			2			3
Bureaucratization (knowledge of procedures)	1						1
Job satisfaction	1	2	1				4
Values	1						1
Rewards	3		4				7
Cohesiveness	1	2		2			5
Paternalism	1	2					3
Residence	1					2	3
Participation			1				1
Company identification (concerns)		1	1	1			3
Conflict	3						3
Previous interfirm mobility		1	1	1			3
Lifetime commitment	1	1					2
Total	14	9	8	6	0	2	39

[a] EM = Electric Factory males.

variable, ideally it is necessary to vary these conditions one at a time. The closest we can come to this is to compare only male employees in Electric and Shipbuilding factories, using the same set of independent variables for a given dependent variable. Doing this will enable us to see if the same causal processes operate under the set of conditions that vary across these two factories—technology, skill structure, etc.—but with at least the sex composition of personnel held constant.

To understand the procedure, consider the example of the dependent variable, job satisfaction, in Figure D. 2. The significant causes (as measured by partial regression coefficients) of job satisfaction among Electric Factory males are status, perceived promotion chances, and cohesiveness; among Shipbuilding personnel they are age, perceived promotion chances, and cohesiveness. Thus, there are two common causes of job satisfaction—promotion chances and cohesiveness—and two unique causes—status and age. The procedure is repeated for each dependent variable in turn. Then the number of times a variable is a common cause is cumulated, as is the number of times it is a unique cause.

Cumulating for all 14 dependent variables, we find 26 times when the causal variables are common to the two factories, and another 34 times when they are unique to one or the other factory. This means that for every independent variable that is a significant cause (when other variables are held constant) in both factories, there are another 1.3 variables (34 ÷ 26) that are significant causes in one factory but not in the other.

Because the explanation of these 14 aspects of social organization—the particular combination of best independent variables—is different more often than it is the same in the two factories, even when only male employees are compared, we again conclude that our data provide at least suggestive negative evidence on the uniformity of social organization hypothesis. Not only do our factories vary in social organization, but the causes of the various aspects of social organization also vary across factories.

The ideal-typical formulation of the paternalism–lifetime commitment model has served a useful purpose in focusing attention on certain uniformities across large Japanese firms, especially those uniformities that may stem from Japanese cultural traditions and societal patterns. Nor would the proponents of this model argue that all large Japanese firms are alike in *all* aspects of their social organization. However, insofar as work in terms of this model is interested in the sources of cross-national organizational variation, it tends to focus on only one source: national cultural setting. Abegglen, Dore, and others have been asserting that to the extent organizations vary, they vary because they are Japanese or American or British.

This is in sharp contrast to a second group of comparativists, who have initiated a new trend in organizational research in the West. Burns and Stalker (1961), Blauner (1964), Woodward (1965), Fullan (1970), Perrow (1970), Blau and Schoenherr (1971), and Heydebrand (1973) have identified such variables as size, technology, environment, and goals as the main bases of variation among organizations.

The challenge of the future is to integrate these two lines of comparative organizational research. One point of articulation is that the national cultural setting variable emphasized by Abegglen (1958, 1969), Dore (1973), Whitehill and Takezawa (1968), and others can be subsumed under the concept of the "environment" of an organization. The studies of the second group of comparativists tend to deal with organizations within one society, and consequently "environment" refers to such variables as the demographic structure of the community, state, or region in which an organization is located, the degree of stability or change in market demand for the products or services of the organization, and the competitiveness of R and D departments in other firms in the same industry. When genuinely cross-national, cross-organizational comparisons are made in the same study, various elements of national culture can be specified and included as part of the organization's environment.

Putting the two kinds of comparative studies together, then, will enable us to seek answers to a central question in organizational theory, to what extent do organizations and their members vary as a result of differences in their cultural setting and to what extent as a result of structural factors—size, technology, goals, and other aspects of their environment besides culture—that cut across differences in national culture?

CONCLUSION

Earlier research on Japanese firms has centered on the question of how the typical Japanese firm differs from typical Western firms. It is now time to take seriously a different question, how much variation is there among firms within Japan, and how much of this fits the paternalism–lifetime commitment model? If one wants to make a cross-national aggregate comparison of *the* Japanese factory and *the* American (British, etc.) factory, the paternalism–lifetime commitment model is useful. Dore (1973), Whitehill and Takezawa (1968), and others have shown that the social organization of Japanese firms is in fact closer than that of comparable Western firms to the paternalism–lifetime commitment model.

But proponents of this model commit an error not uncommon in comparative analysis. In striving to report what is most distinctive or unique about the Japanese factory they fail to keep what is most *important* in perspective. The lifetime commitment pattern may be distinctive to Japan—relative to the United States or Britain—but it may not be the most important characteristic of Japanese firms. Several of the important characteristics have been elaborated in this study.

First, the cultural differences between Japanese and Western firms are evidently not so all-powerful that they overrride the influence of structural differences at the level of the firm (factory, etc.), when these are compared *within* one society. There is more empirical variation in the social organization of Japanese firms than the paternalism–lifetime commitment model recognizes. Second, a number of important predictions from the model concerning the relationship of social organization variables within given firms are disconfirmed empirically. Third, Abegglen asserts a positive causal relationship between the paternalistic type of social organization in Japanese firms and their performance. Our findings lead to a different conclusion. The social organizational variables that are more distinctly Japanese–paternalism, company housing, participation in company activities, company identification, and lifetime commitment—have *less* causal impact on performance than do the more universal social organizational variables that are also part of the model—the employee's status in the company, sex, job satisfaction, and knowledge of organizational procedures. In other words, we argue, contrary to Abegglen and some others, that the more distinctively Japanese a social organizational variable is, the less it has to do with performance; performance in Japanese firms appears to have the same causal sources as in Western firms.

Finally, what of the convergence theory of modernization? Over time have the differences in paternalism–lifetime commitment between Japanese and Western firms been increasing, decreasing, or remaining essentially constant? On this point experts differ. Nearest to the anti-convergence extreme is Abegglen. Cole sees convergence as continuing up to a point, but argues that beyond the point viable functional alternatives will develop. Societies will not have to have the same structures to handle similar functional problems associated with high industrialization; different structures can, within limits, effectively meet the same functional requirements. Dore and we agree that a number of changes have occurred in the paternalism–lifetime commitment system in Japan since the early 1960s, and that it is difficult as yet to evaluate their long-range significance. We are less convinced than Dore, however, that there are institutional obstacles to the continuation of these trends. Interestingly, in the last analysis Dore does not deny convergence; quite the contrary. He sees British and Japanese factories as converging toward a pattern he calls "welfare corporatism," although "there is more evidence of the British system becoming more like the Japanese than vice versa" (Dore 1973:302).

The study of Japanese industrial firms is much more relevant to convergence theory than many studies that purportedly refute the theory on the basis of data from societies at lower levels of modernization. It is our view that because the paternalism model exaggerates the uniformity, uniqueness, and traditionalism of the social organization of Japanese firms, it cannot be regarded as evidence for anti-convergence theories of modernization. Convergence theory, in our view, provides at least as good an understanding of Japanese industrial organizations as any presently available alternative theory.

APPENDIX A

Research Methods

FIELD WORK

Our field work lasted a total of three months (July 7–October 10, 1969) in Electric Factory, three months (November 15, 1969–February 15, 1970) in Shipbuilding Factory, and two months in Sake Company (by Mannari from December 1966 to January 1967 and by Marsh and Mannari in several visits spread through November 1969 and February and March 1970). Our overall calendar is shown in Table A.1. In each factory, the early weeks were spent mainly in studying company personnel and production data; in observing, first the plant as a whole and then, in more detail, specific sections and settings; and in conducting individual interviews with specific employees. We gradually increased the amount of informal interaction we had with employees during leisure periods. We deliberately overlapped these phases of data collection. For example, on on a given day we might spend the first hour in our research office (provided by the company in one of the factory buildings) examining company records, then conduct two interviews before lunch. After lunch we might spend an hour or two observing work settings, and then end the work day back in our office, consulting company records again, editing the morning's interviews, and discussing the day's research.

The general structure of an interview was as follows. We would obtain management's permission to interview one or more employees in specific departments, sections, jobs, etc. Those employees willing to be interviewed—and we had almost no refusal—would be given approximately one and a half hours off during work hours, at a prearranged time. The employee would come to our office, and after introductions would be told the purpose of our study. We explained our presence by making it clear we were not efficiency experts or in any other way company investigators who would ever report any personal information that would be traced to the individual respondent. We described our affiliation as professors in Kwansei Gakuin University and Brown University, interested in comparing the overall operation of Japanese and American industrial organizations. The interview situation was private in that only Mannari, Marsh, and the interviewee were present.

Although Marsh had studied Japanese one year immediately prior to the field work, his competence in Japanese limited him to asking simple

TABLE A.1.
Research Calendar

1966–67	Mannari and his research associates study Sake Company
April 1968	Marsh-Mannari correspondence begins
May-June 1968	Mannari visits companies listed in Table 2.1 to get permission for field work
September–October 1968	Marsh and Mannari write project proposal and submit it to foundations
1968–69 academic year	Mannari visits more firms, collects more published statistical data on Japanese firms; Marsh and Mannari develop research plan further
June 25, 1969	Marsh arrives in Japan
July 1, 1969	Visit D Denki Company
July 3, 1969	Visit Tetsuko Company
July 12, 1969	Visit Chain Manufacturing Company
July 7–October 10, 1969	Living in Electric Factory company apartment, doing field work in factory. Evening visits made to several small, local factories in same city
October 7, 1969	Visit Kinzoku Company
November 10, 1969	Visit Boseki Company
November 12, 1969	Visit Sake Company
November 15, 1969–February 15, 1970	Field work in Diesel Engine Division of Shipbuilding Company's X Plant
February 16 and 24, 1970; March 31, 1970	More visits for observation and interviewing in Sake Company
April 9, 1970	Visit Petroleum Company

questions, so that Mannari's bilingual fluency was essential. Both investigators put questions to the interviewee, Mannari directly, and Marsh indirectly, by first stating the question in English for Mannari to restate in Japanese. Mannari would translate the more complicated answers from Japanese back to English, and both investigators took notes during the interview. We did not use tape recorders. Nor did we follow a fixed pattern of questioning. We suited the questions to the status, role, and other characteristics of whomever we were interviewing, to explore in greater depth areas he or she could tell us most about. We probed liberally. We tried out alternative wordings of questions we eventually planned to incorporate in the mass questionnaire, so that interviews in part had a pretest function for questionnaire items.

Field work in Electric Company was greatly aided by the fact that Mannari, Marsh, and one Japanese graduate student research assistant lived in an employee apartment complex a block away from the factory. We ate lunch every noon with the workers in the factory cafeteria and on several occasions visited with individual or small groups of managers and workers at drinking parties, in our apartments or local restaurants, at after work baseball and sporting contests, and on weekend trips to the seashore. In addition, Marsh gained some rapport with a group of managers who requested that he teach them English conversation two afternoons a week after work. We had complete freedom to move about the entire factory complex.

We did not develop as much rapport with the employees in Shipbuilding and Sake Companies because there was less opportunity for off-the-job interaction. We commuted to the plants from our own homes in Nishinomiya City, rather than living in company housing. Shipbuilding employees either worked overtime (as well as six days a week) or commuted to their own homes—dispersed widely over the greater Osaka–Nara area. The greater distance of work from residence in Shipbuilding Factory meant we did not see its employees as much after work as we had Electric Factory employees.

Our period of field work at Sake Company in the winter of 1969–70 was briefer than that in Electric and Shipbuilding companies. The men lived in dormitories, the 20 or so sake workers in a given kura sharing a single room. As older men they had more sedentary leisure habits, and as farmers they retired early at night. For all these reasons we did not interact with them as much outside work as we did with the men in Electric Factory.

THE QUESTIONNAIRE

Knowledge obtained from our first three sources of data—company records, observations, and interviews—helped in the construction of the questionnaire. For example, data on the division of labor came from company manuals which list job titles; these were checked by observing what the employees were actually doing on their job; through open-ended questions and probes we explored the objective and subjective aspects of the job with those employees we interviewed. Finally, on the basis of all these preliminary sources of data on the division of labor, we drafted the questionnaire, which gave us our most systematic and comprehensive data on employees' attitudes toward their jobs and on other subjective aspects of the division of labor.

A questionnaire instrument was developed first in Electric Factory. Items were taken from previous studies or were formulated *de novo*. In its final form, the Electric Factory questionnaire had 44 questions to be answered by all personnel plus eight questions specifically for managerial personnel. All but three of the 44 questions were used again in the Shipbuilding questionnaire, but the total number of questions in the latter was 68, plus the eight managerial questions. The additions were a result of new questions we had developed between October 1969 and February

1970, to tap the same areas of attitudes and behavior, and also to tap some areas more distinctive to Shipbuilding. Another change is that, in order to keep the questionnaire short, we asked Electric Company employees to write their identification number on the questionnaire. We could then use our company personnel data for the standard background information on age, sex, seniority, rank, etc., instead of having each respondent tell us. We then integrated the two sources of information on each individual in the coding sheets, punched cards, and tape. In Shipbuilding and Sake Companies, we decided we would allow respondents total anonymity in the questionnaire: no identification number or name was requested. This meant that we had to include all the "face sheet" questions *in* the questionnaire.

Given that Electric Factory respondents had less anonymity, they might have been more likely to exhibit a pro-management bias. We are unable to detect any evidence of such bias. For example, Electric Factory respondents are *less* satisfied with their jobs than Shipbuilding Factory employees, perceive their promotion chances as *poorer*, and express *less* support for lifetime commitment norms and values.

Questionnaire construction in Sake Company posed a new set of problems. We knew the respondents, as farmers and older men, were less educated. We accordingly had to jettison many of the questions asked in Electric and Shipbuilding companies, and to modify and simplify the wording of some of the questions we retained. The Sake Company questionnaire contained only 20 attitudinal questions and five background questions; none of the managerial questions was included. Fortunately, we had supplementary data from the questionnaire Mannari and his associates had used in Sake Company in early 1967.

In all three companies the questionnaires were distributed to employees at work by their supervisors, who urged their cooperation. Employees were allowed to take the questionnaire home to fill out, and were given from a few days to a little over a week to return the completed questionnaire.

All managers, staff and clerical personnel, and production workers received a questionnaire. The response rates are high—97 percent in Sake Company, 86 percent in Electric Factory, and 79 percent in Shipbuilding Factory, for an average rate of 84 percent. Since they are not 100 percent, however, the question arises, how representative are those who answered of those who did not? In short, can we generalize from the questionnaire data to all personnel in the three factories?

Electric factory

At the end of September 1969, Electric Factory employed 1,201 people, all of whom received our questionnaire on October 3. Returns were received from 1,120, of which 1,033 were usable.[1] We found respondents to be not significantly different from non-respondents on standard background characteristics, which we could obtain by knowing the I.D. number of those who responded and taking the characteristics of all other employees from the personnel records (see Table A.2). This similarity was also found with respect to: section (ka), residence (company housing vs. private housing), whether hired through headquarters (*honsha*) or in the local factory area, age, line vs. staff section, absenteeism, and number and grade of suggestions (teian). These results are most reassuring, and suggest that the 1,033 employees can be taken as representative of all employees in Electric Factory.

Shipbuilding factory

Our questionnaire was distributed to all Diesel Engine Division personnel on February 10, 1970 and collected approximately a week later. Of the 756 employees in the division, 596 returned usable questionnaires. Table A.3 shows that the respondents are not significantly different from all personnel with regard to sex, rank, age, seniority, and education. Statistically significant differences exist, however, with regard to section and job. Engine Assembly workers and General Affairs employees are somewhat underrepresented, and Number 2 Machine Kakari and Diesel Engine Staff Sections are somewhat overrepresented. With regard to job, storage and procurement workers and frame-making, crane, and transportation workers are slightly underrepresented among our respondents, while machine finishing and finishing assembly workers are somewhat overrepresented. With these rather slight exceptions, then, the 596 Shipbuilding respondents can be regarded as representative of all employees in the Diesel Engine Division.

Sake company

Since 97 percent of the employees in the three kura filled out usable questionnaires, we can take them as representative of all employees in those kura without making χ^2 tests.

In conclusion, our questionnaire data can be taken as relatively representative of all personnel in the plants or division studied in each of the three companies.

[1] We could not use questionnaires that did not give the employee's identification number, since we needed that to collate the questionnaire attitude data with background data on each worker, taken from company records. Of the 1,120 returns, 87 could not be used for the following reasons: no I.D. number given (57 cases), wrong I.D. (7), temporary worker (6), not listed in May 1969 personnel records, the source of our background data (16), and no answer to questions (1).

TABLE A.2.
Comparison of All Electric Factory Employees and Those Who Returned Usable Questionnaires (%)

Characteristic	*All Electric Factory* (N = 1,212)[a]	*Respondents* (N = 1,033)
Sex		
Male	50.6	49.4
Female	49.4	50.6
	100.0	100.0
Education		
College or university	5.3	4.8
High school	29.5	30.0
Middle school	65.3	65.2
	100.1	100.0
Rank		
Kachō, kachō-dairi	2.2	2.0
Kakarichō	1.9	2.0
Hanchō	1.6	1.8
Kumichō	3.3	3.3
Rank-and-file workers	91.0	90.8
	100.0	99.9
Amount of Total Monthly Pay		
¥ 80,000 and over	3.0	2.6
60,000–79,999	9.3	10.0
40,000–59,999	25.9	25.4
20,000–39,999	61.8	62.1
	100.0	100.1
Job Classification		
Kachō	2.3	2.1
Creative (C)	13.4	13.1
Operative (O)	12.5	12.7
Routine (R)	37.2	35.8
Training (T)	34.6	36.3
	100.0	100.0
Seniority		
Less than 2 years	24.3	25.3
2–3 years	24.1	24.7
4–7 years	11.0	10.7
8–11 years	10.7	9.7
12–15 years	16.5	16.0
16–23 years	13.4	13.6
	100.0	100.0

[a] As of May 1969, the date of our personnel records. By the time our questionnaire was distributed, the total had declined from 1,212 to 1,201.

TABLE A.3
Comparison of All Shipbuilding Company Diesel Engine Division Personnel and Those Who Returned Usable Questionnaires (%)

Characteristic	*All Diesel Engine Division Personnel*	*Respondents*
Sex		
Male	98.0	98.0
Female	2.0	2.0
	100.0	100.0
N	(756)	(590)
Education		
College or university	4.0	7.0
High school	32.0	33.0
Middle school	64.0	60.0
	100.0	100.0
N	(756)	(584)
Rank		
Kachō	0.7	1.1
Kakarichō	1.8	2.6
Sagyōchō	1.5	1.5
Hanchō	5.4	6.6
B-Staff, A-Staff	15.6	19.8
Rank-and-file workers	75.0	68.4
	100.0	100.0
N	(756)	(573)
Section		
General Affairs	6.0	1.7
Machining No. 1 Kakari	41.4	40.2
Machining No. 2 Kakari	9.7	15.1
Machining: Tool Engineering	6.3	6.7
Turbine Research	1.5	1.6
Diesel Engine: Piping	5.6	4.6
Diesel Engine: Assembly	21.8	14.9
Diesel Engine; Staff	3.3	8.7
Design	4.4	6.5
	100.0	100.0
N	(756)	(582)
Age		
20 or under	9.0	6.0
21–26	19.0	20.0
27–32	21.0	20.0
33–38	13.0	11.0
39–44	14.0	17.0
45–50	14.0	13.0
51+	10.0	13.0
	100.0[a]	100.0
N	(821)	(585)

TABLE A.3 (*Cont.*)

Characteristic		All Diesel Engine Division Personnel	Respondents
Seniority			
Under two years		13.0	10.0
2–3 years		7.0	7.0
4–7 years		21.0	23.0
8–11 years		12.0	11.0
12–19 years		22.0	25.0
20–27 years		23.0	22.0
28+ years		2 0	2.0
		100.0	100.0
	N	(728)[a]	(582)
Job			
B-Clerical and B-Engineering		14.9	16.3
Drilling, piping		4.6	5.4
Measuring and machining		35.7	37.8
Machine finishing and assembly		18.7	22.9
Frame making, crane, and transportation		8.5	4.9
Storage, procurement		10.7	4.0
Other (including 44 hanchō)		6.9	8.7
		100.0	100.0
	N	(756)	(572)

[a] The total number of employees in the division at the time our questionnaire data were collected was 756. Distributions for this N base were available for all characteristics except age and seniority. Hence, the percentage distributions for all personnel on age and seniority are only estimates of the actual distribution as of February 1970. The estimates were made in July 1971, when the Diesel Engine Division had 869 employees. In the case of age, we subtracted the number of personnel recruited in April of 1970 and April of 1971 from 869, leaving a total of 821, and deducted one year from everyone's age. The difference between 821 and 756 is due to transfers within X-Plant or within the company, and other factors. In the case of seniority, we subtracted the number of employees with less than two years' seniority as of July 1971, as the closest estimate of how many were not in the division as of February 1970, and deducted one year from everybody's seniority. Of the 869, 121 had less than two years' seniority, giving the 728 base N on which the percentages are calculated.

Publication costs of a book as long as this one kept us from printing the complete questionnaires. The exact wording of the English version of the questions we have used in the analysis has usually appeared at the appropriate place in the text. Readers interested in obtaining a complete copy of the English or Japanese version of the questionnaire need only write to:

Professor Robert M. Marsh
Department of Sociology
Brown University
Providence, R. I. 02912
USA

APPENDIX B

Construction of Indexes

The conceptual domains of interest in this study were outlined in Chapter 1. When a conceptual domain was measured by more than one variable (questionnaire items or information from company records), the immediate question is that of the statistical association among these variables; are they in fact measuring the same thing? To the extent that they are, it is useful to reduce the larger number of variables to a single index, on which each respondent can be assigned a summary index score over all the items combined in the index. The technique of scale construction used was an item analysis program, ITEMA, adapted to an IBM 1130 computer in the Brown University sociology department computer laboratory.

ITEMA is usable only for ordinal scale variables. Ideally, all items included in a given index should have the same number and range of coded categories. In our data this is the case for both the Electric and Shipbuilding factory indexes of organizational status, employee cohesiveness, and paternalism. Most of the exceptions have been handled by the following procedure: if most items in an index have, say, three categories, but one item has only two categories, these two are scored 1 and 3. If the standard number of categories per item is four, but one item has only two, these are scored 1 and 4. In this way, all items receive the same weight in the total index.[1] Any respondent who does not answer one or more of the items in a given index is scored 0 (N.A.) and dropped from that index. Each index, therefore, contains only respondents who answered every item in the index.

The ITEMA program first computes the correlation between each item in the set of items and the sum of the respondent's score on all other items in that set. This is trial 1. In trial 2 (and as many subsequent trials for that index as are appropriate or possible), the item with the lowest item-total correlation in the previous trial is eliminated; the remaining items in the set are run again in the same manner, until the investigator

[1] Two exceptions remain: (1) the "relation between job and ability" item in the job satisfaction index has only three categories, while the other items have four each; (2) the "amount of award money for suggestions" item has four categories in Shipbuilding, as do all the other items in the performance index in both factories, but has only three categories in Electric Factory.

achieves the best outcome with regard to balancing (1) the number of items retained in the index, (2) their item-total correlations, and (3) the resulting reliability of the index. The estimate of the latter from the items' intercorrelations is expressed as coefficient alpha.

The categories for each item are ordered so that 1 equals the low value on the variable. Consequently, the lower a respondent's total score on an index, the less he or she has of the characteristic measured by the index. Low scores can be interpreted as follows on the seven indexes we constructed (with high scores having the opposite meaning).

Index	*Meaning of Low Score*
Organizational status	Low status in the company
Job satisfaction	Low job satisfaction
Employee cohesiveness	Low degree of cohesiveness with fellow employees
Paternalism	A preference for functionally specific relationships with superiors, and with company management in general (rather than for diffuse, paternalistic relationships)
Participation	A low frequency of participation by the employee in company-sponsored recreational activities
Lifetime commitment	A low degree of norm- and value-commitment to staying in one firm (as opposed to changing firms)
Performance	A low level of individual performance with regard to company instrumental goals, and the means by which those goals are attained.

There are two major limitations to our indexes. First, although the questionnaires contain relatively large numbers of items—44 questions in Electric Factory, 68 in Shipbuilding, and an additional eight questions for managers in each factory—these pertain to a number of different conceptual domains. Hence, none of our indexes has more than six items, in contrast to the ideal number of 30 or more items per index (Nunnally 1967). Second, as a result of the small number of items per index, some of our indexes fall well below the desired level of reliability, alpha = .80 (see Table B.1).

The items used in the construction of each index are presented in Tables B.2 and B.5–B.10. For each item the item-total correlation shows the correlation between the item and the total score for the given index when the item score is not contributing to the total.

We also constructed a subindex, the index of status enhancement, for Electric Factory males, consisting of variables 5, 8, and 9. This index is compared with the index of organizational status in Table B.3. The differences in the distribution of respondents on the two indexes are shown

TABLE B.1

Indexes Constructed for Electric and Shipbuilding Factory Personnel

		Electric Factory			Shipbuilding Factory			N	
Index	*Number of Items*	*Mean Index Score*	*Standard Deviation*	*Reliability of Index (alpha)*	*Mean Index Score*	*Standard Deviation*	*Reliability of Index (alpha)*	*Electric*	*Shipbuilding*
Organizational status	6	9.222	2.944	.78	10.742	1.963	.47	1032	539
Job satisfaction	6	13.442	3.103	.82	15.485	3.755	.76	977	583
Employee cohesiveness	6	12.491	1.772	.30	12.201	2.196	.47	899	512
Paternalism	2	3.439	0.690	.41	3.485	0.678	.38	964	533
Participation in company recreational activities	3	4.647	1.241	.40	4.412	1.606	.64	985	516
Lifetime commitment	4[a]								
All employees		8.204	1.486	.38	8.702	1.579	.59	880	500
Males only		8.465	1.452	.51	8.726	1.570	.59	417	486
Performance	5	12.462	2.473	.57	12.868	2.801	.57	1024	549

[a] In addition to the identical four-item index of lifetime commitment for each factory we also have a seven-item index of lifetime commitment for Shipbuilding Factory. For all Shipbuilding employees, the mean on the seven-item index is 14.876, the standard deviation 2.705, reliability (alpha) = .69, N = 479. For Shipbuilding males, the mean is 14.920, standard deviation 2.693, reliability (alpha) = .69, N = 465.

TABLE B.2
Index of Organizational Status

	Electric (v. 138)	*Score*		*Shipbuilding (v. 123)*	*Score*
78.	Rank		91.	Rank	
	Rank-and-file employees	1		Workers, A-Staff, B-Staff employees	1
	Kumichō	2		Hanchō, boshin	2
	Hanchō, kakarichō, kachō	3		Sagyochō, kakarichō, kachō	3
	item-total correlation[a] r = .474			item-total r = .361	
8.	Monthly pay		110.	Monthly pay	
	¥ 20,000–39,999	1		¥ 20,000–39,999	1
	¥ 40,000–59,999	2		¥ 40,000–59,999	2
	¥ 60,000+	3		¥ 60,000+	3
	item-total r = .786			item-total r = .456	
9.	Job classification		89.	Job	
	Trainees	1		Finishing, framing	1
	Routine	2		Drilling, piping; measuring, assembly, storage, and other	2
	Operative, creative, kachō	3		B-clerical, B-engineering	3
	item-total r = .716			item-total r = .283	
84.	Seniority		94.	Seniority	
	0–3 years	1		0–3 years	1
	4–11 years	2		4–11 years	2
	12–23 years	3		12+ years	3
	item-total r = .607			item-total r = .202	
13.	Education		95.	Education	
	Middle school	1		Middle school	1
	High school	2		High school	2
	University	3		University	3
	item-total r = .206			item-total r = −.006	
24.	Informational level of section		88.	Section	
	Low	1		Assembly, turbine	1
	Medium	2		Machining	2
	High	3		Design, staff, General Affairs	3
	item-total r = .407			item-total r = .173	

NOTE: Respondent's total organizational status index score is his or her total score over these six items, scored as above. With six items and three categories per item, scores range from 6 to 18, with the latter score indicating the highest organizational status.

[a] For each item this correlation shows the correlation between the item and the total score for the given index when the item score is not contributing to the total.

in Table B.4. The main difference between the two indexes ($r = .90$) is that there is a larger proportion of respondents with medium high scores in the status index than in the status enhancement index. We are indebted to Hesook S. Kang, a graduate student in sociology at Brown University, for suggesting and carrying out this analysis in a seminar in the spring of 1972.

TABLE B.3
Comparison of Status Index and Status Enhancement Index, Males in Electric Factory

		Item-Total Correlations	
Component Items		*Status Index*	*Status Enhancement Index*
8. Pay		.665	.696
9. Job classification		.651	.532
5. Rank		.425	.524
12. Seniority		.328	
13. Education		.259	
24. Informational level of section		.222	
Alpha reliability coefficient		.68	.75
	N	(509)	(510)

TABLE B.4
Distribution of Electric Factory Male Respondents on Status Indexes

Status Index Score	*Percent*	*Status Enhancement Index Score*	*Percent*
Low (6–8)	14.9	(3–4)	21.3
Medium low (9–11)	39.0	(5–6)	53.9
Medium high (12–14)	34.7	(7–8)	13.1
High (15–18)	11.4	(9)	11.7
	100.0		100.0
N	(509)		(510)

TABLE B.5
Index of Job Satisfaction

Electric (v. 139)	*Score*	*Shipbuilding (v. 124)*	*Score*
25. Feeling about job		2. Feeling about job	
Dull all the time	1	Dull all the time	1
Dull most of the time	2	Dull most of the time	2
Interesting most of the time	3	Interesting most of the time	3
Interesting all the time	4	Interesting all the time	4
item-total r = .682		item-total r = .625	
26. Job in relation to interest and desire		3. Job in relation to interest and desire	
Job not at all identical to interest	1	Job not at all identical to interest	1
Not very identical	2	Not very identical	2
Fairly identical	3	Fairly identical	3
Exactly identical	4	Exactly identical	4
item-total r = .734		item-total r = .662	
27. Does job have enough variety?		4. Does job have enough variety?	
Not enough variety	1	Not enough variety	1
Enough variety	4	Enough variety	4
item-total r = .565		item-total r = .507	
29. Relation between job and ability		6. Relation between job and ability	
Job requires less than my ability	1	Job requires less than my ability	1
Job is just right for my ability	2	Job is just right for my ability	2
Job requires more ability than I have	3	Job requires more ability than I have	3
item-total r = .493		item-total r = .384	
34. How often do you want to be moved to another job?		13. How often do you want to be moved to another job?	
Often	1	Often	1
Sometimes	2	Sometimes	2
Rarely	3	Rarely	3
Not at all	4	Not at all	4
item-total r = .464		item-total r = .469	
36. Job satisfaction		15. Job satisfaction	
Very dissatisfied	1	Very dissatisfied	1
Somewhat dissatisfied	2	Somewhat dissatisfied	2
Satisfied	3	Satisfied	3
Very satisfied	4	Very satisfied	4
item-total r = .676		item-total r = .622	

NOTE: A respondent's total job satisfaction index score over the six items falls between 6 and 23.

TABLE B.6
Index of Employee Cohesiveness

Electric (v. 141)	Score	Shipbuilding (v. 127)	Score
64. How would you feel if moved to another job in the factory . . . away from the people who now work near you?		42. How would you feel if moved to another job in the factory . . . away from the people who now work near you?	
Not troubled, not very troubled	1	Not troubled, not very troubled	1
Fairly troubled	2	Fairly troubled	2
Very troubled	3	Very troubled	3
item-total r = .107		item-total r = .175	
66. Are your best friends in the company or outside?		43. Are your best friends in the company or outside?	
Most outside the company	1	Most outside the company	1
Half in and half outside	2	Half in and half outside	2
Most inside the company	3	Most inside the company	3
item-total r = .161		item-total r = .313	
67. How many of the people who work near you are friends?		44. How many of the people who work near you are friends?	
None	1	None	1
1–2 persons	2	1–2 persons	2
3+ persons	3	3+ persons	3
item-total r = .180		item-total r = .252	
68. (If you have 1+ close friends inside the company) What opportunity do you have to see them outside the company?		45. (If you have 1+ close friends inside the company) What opportunity do you have to see them outside the company?	
No association outside work	1	No association outside work	1
When I happen to see them, we associate	2	When I happen to see them, we associate	2
We make a special date (to do things outside work)	3	We make a special date (to do things outside work)	3
item-total r = .047		item-total r = .139	
69. Atmosphere among people who work in this factory		46. Atmosphere among people who work in this factory	
Cold	1	Cold	1
Half and half	2	Half and half	2
Warm	3	Warm	3
item-total r = .181		item-total r = .330	
70. What do you think about the teamwork in your shop (kumi or kakari or ka)?		47. What do you think about the teamwork in your shop?	
Not good at all, not too good teamwork	1	Not good at all, not too good teamwork	1
Fairly good teamwork	2	Fairly good teamwork	2
Very good teamwork	3	Very good teamwork	3
item-total r = .123		item-total r = .249	

NOTE: A respondent's total employee cohesiveness index score over the six items falls between 6 and 18.

TABLE B.7
Index of Paternalism

	Electric (v. 144)	*Score*		*Shipbuilding (v. 129)*	*Score*
63.	Suppose in this company there are two types of superiors. Which would you prefer to work under?		40.	Suppose in this company there are two types of superiors. Which would you prefer to work under?	
	A man who always sticks to the work rules and never demands any unreasonable work, but on the other hand never does anything for you personally in matters not connected with work.	1		A man who always sticks to the work rules and never demands any unreasonable work, but on the other hand never does anything for you personally in matters not connected with work.	1
	A man who sometimes demands extra work in spite of rules against it, but on the other hand looks after you personally in matters not connected with work.	2		A man who sometimes demands extra work in spite of rules against it, but on the other hand looks after you personally in matters not connected with work.	2
	item-total $r = .264$			item-total $r = .237$	
97.	There are two companies. From your experience which would you choose?		61.	There are two companies. From your experience which would you choose?	
	Management thinks its relationship to the worker is simply work; therefore management doesn't find it necessary to take care of worker's personal matters.	1		Management thinks its relationship to the worker is simply work; therefore management doesn't find it necessary to take care of worker's personal matters.	1
	Management thinks they are like parents to workers; therefore they regard it as better to take care of the personal affairs of workers.	2		Management thinks they are like parents to workers; therefore they regard it as better to take care of the personal affairs of workers.	2
	item-total $r = .264$			item-total $r = .237$	

NOTE: A respondent's total paternalism index score over the two items falls between 2 and 4.

TABLE B.8
Index of Participation in Company Activities

Electric (v. 143)	Score	*Shipbuilding (v. 128)*	Score
98. Do you belong to cultural or athletic clubs sponsored by the company?		62. Do you belong to an athletic or cultural club sponsored by the Cooperative Association?	
No	1	No	1
Yes	3	Yes	3
item-total r = .102		item-total r = .407	
99. Within the last year how many times did you attend activities sponsored by your section, or kakari, or by cultural or athletic clubs?		63. Within the last year how many times did you attend a recreational meeting, or athletic meeting sponsored by the factory or section, kakari, shop, han, or dormitory, or by cultural or athletic clubs?	
0–3 times	1	0–3 times	1
4–6 times	2	4–6 times	2
7+ times	3	7+ times	3
item-total r = .313		item-total r = .540	
100. Within the last year how often did you attend all kinds of company-, section-sponsored, athletic, and cultural clubs' activities?		64. How often did you take part in recreational activities sponsored by your shop, or athletic and cultural clubs, during the last year?	
Never, sometimes	1	Never, sometimes	1
Most times	2	Most times	2
Always	3	Always	3
item-total r = .342		item-total r = .430	

NOTE: The respondent's total participation index score over these three items falls between 3 and 9.

TABLE B.9
Index of Lifetime Commitment

Electric (v. 146)	Score	Shipbuilding (v. 132)	Score
73. Do you think male employees intend to work here until retirement?		138. Do you think male employees intend to work here until retirement?	
None or some intend	1	None or some intend	1
Majority intend	2	Majority intend	2
All intend to stay	3	All intend to stay	3
item-total r =		item-total r^a = .329	
males only .287		item-total r^b = .383	
all employees .192			
93. What do you think about an employee who voluntarily changes firms?		57. What do you think about an employee who voluntarily changes firms?	
I'd do the same if I had the chance	1	I'd do the same if I had the chance	1
I can understand his behavior	2	I can understand his behavior	2
His behavior is not "Japanese," disloyal, or opportunistic	3	His behavior is not "Japanese," disloyal, or opportunistic	3
item-total r =		item-total r^a = .454	
males only .344		item-total r^b = .502	
all employees .273			
94. Should male employees work for the same company until retirement?		58. Should male employees work for the same company until retirement?	
Should not	1	Should not	1
It depends	2	It depends	2
Should	3	Should	3
item-total r =		item-total r^a = .430	
males only .339		item-total r^b = .515	
all employees .283			
95. Do you intend to continue to work in Electric Company until retirement?		60. Do you intend to continue to work in Shipbuilding Company until retirement?	
Intend to quit	1	Intend to quit	1
Intend to continue	3	Intend to continue	3
item-total r =		item-total r^a = .312	
males only .296		item-total r^b = .326	
all employees .191			
		48. In company life there are two ways of thinking. . . . which do you think is better?	
		Moving from one company to another	1
		Staying in one company[c]	3
		item-total r^b = .337	

TABLE B.9 (*Cont.*)

Electric (v. 146)	*Score*	*Shipbuilding (v. 132)*	*Score*
		137. To what extent do you feel you are "one body" with the company?	
		Not at all, not too much	1
		Fairly much	2
		Very much	3
		item-total r^b = .418	
		54. Have you ever thought of quitting Shipbuilding Company?	
		Seriously looked for another job	1
		Thought of quitting, but didn't look for another job	2
		Not thought of quitting	3
		item-total r^b = .423	

NOTE: A respondent's total lifetime commitment index score over the four items falls between 4 and 12. Three additional items, used only in Shipbuilding Factory, were developed after the Electric Factory questionnaire data had been collected. They produce a second, seven-item index for Shipbuilding. Because of the difference in the sex composition of personnel in the two factories (97 percent male in Shipbuilding, 49 percent in Electric), lifetime commitment index scores are given separately for all employees and for males in Electric Factory. This is also done because the lifetime commitment model is held, by Abegglen and others, to apply only to men.

[a] Item-total r in the four-item lifetime commitment index, comparable to the Electric Factory lifetime commitment index.

[b] Item-total r in the seven-item lifetime commitment index.

[c] For full wording, see variable 48, Table 10.3.

TABLE B.10
Index of Performance

Electric (v. 140)	Score	Shipbuilding (v. 125)	Score
17. Ratio of absences		103. Ratio of absences	
Absent 1+ more days than eligible vacation	1	Absent 1+ more days than eligible vacation	1
Days absent = eligible vacation, or worked 1–3 days of vacation	2	Days absent = eligible vacation, or worked 1–3 days of vacation	2
Worked 4–7 days of vacation	3	Worked 4–7 days of vacation	3
Worked 8+ days of vacation	4	Worked 8+ days of vacation	4
item-total r = .296		item-total r = .193	
18. Number of suggestions contributed in previous year		133. Number of suggestions contributed in previous year	
0–2	1	0–2	1
3–4	2	3–4	2
5–11	3	5–11	3
12+	4	12+	4
item-total r = .351		item-total r = .456	
19. Amount of award money for suggestions		134. Amount of award money for suggestions	
¥ 0	1	¥ 0	1
¥ 200–499	2	¥ 499 or less	2
¥ 500–3000	3	¥ 500–999	3
		¥ 1,000+	4
item-total r = .238		item-total r = .428	
28. Anxious to fulfill each day's production goal?		5. Anxious to fulfill each day's production goal?	
Not at all anxious	1	Not at all anxious	1
Not very anxious	2	Not very anxious	2
Fairly anxious	3	Fairly anxious	3
Very anxious	4	Very anxious	4
item-total r = .359		item-total r = .270	
33. How often do you think about making suggestions?		·10. How often do you think about making suggestions?	
Never	1	Never	1
Rarely	2	Rarely	2
Sometimes	3	Sometimes	3
Always	4	Always	4
item-total r = .405		item-total r = .417	

NOTE: The total performance index score over the five items is between 5 and 19 in Electric Factory and between 5 and 20 in Shipbuilding Factory. (The difference is due to the fact that variable 19 in Electric Factory has a maximum score of 3, in contrast to the standard maximum of 4 for the comparable item in Shipbuilding.)

APPENDIX C

Correlation Matrices

All variables used in the multiple regression analyses (Appendix D) appear in the following three zero-order correlation matrices (Tables C.2–C.4): one for all employees in Electric Factory, one for males only, and one for all employees Shipbuilding Factory. Variables that became components of an index are omitted, since the index appears as a variable in the matrix. The correlations are Pearson product-moment *r*'s.

TABLE C.1.
Means and Standard Deviations for all Major Variables

	Electric Total		*Electric Males*		*Shipbuilding*	
Variable	$\bar{X}$	*s*	$\bar{X}$	*s*	$\bar{X}$	*s*
Sex	1.493	0.500			1.977	0.146
Residence	1.920	0.270	1.843	0.364	1.865	0.341
Rank	1.151	0.496	1.305	0.672	2.069	1.676
Number of dependents	1.432	0.724	1.868	0.824	2.813	1.520
Pay	1.505	0.708	1.988	0.717	3.099	1.512
Job classification	1.915	0.797	2.405	0.748	5.288	2.082
Age	2.881	1.356	3.776	1.199	4.052	1.842
Seniority	1.796	0.868	2.386	0.776	4.089	1.656
Education	1.396	0.580	1.460	0.664	1.460	0.615
Ratio of absences	3.076	0.938	3.435	0.849	2.927	1.084
Years in previous jobs	1.258	0.597	1.303	0.669	3.047	1.606
Information level of section	1.458	0.757	1.744	0.852	1.957	0.612
Variety	1.422	0.494	1.671	0.470	2.568	1.499
Perceived work load	2.240	0.563	2.259	0.589	2.145	0.566
Values	1.753	0.853	2.074	0.907	1.820	0.829
Move jobs	2.293	0.790	2.378	0.794	2.436	0.946
Job autonomy	2.860	0.740	3.043	0.677	3.013	0.747
Knowledge of procedures	2.175	0.641	2.495	0.575	2.285	0.593
Perceived advantages	2.168	0.748	1.990	0.823	1.286	0.525
Wage preferences	1.811	0.660	1.741	0.768	2.030	0.761
Promotion chances	1.589	0.821	2.076	0.876	1.872	0.958
Concerns	3.456	0.890	3.413	0.912	2.291	0.793
Distance	1.536	0.735	1.731	0.835	2.106	0.889
Size	1.801	0.840	1.969	0.971	2.908	1.115
Father's occupation	1.883	0.988	1.970	1.043	3.005	1.893
Status	9.222	2.944	11.292	2.767	10.742	1.963
Job satisfaction	13.442	3.103	14.946	2.840	15.485	3.755
Performance	12.462	2.473	13.814	2.200	12.868	2.801
Cohesiveness	12.491	1.772	12.509	1.816	12.201	2.196
Participation	4.647	1.241	4.995	1.274	4.412	1.606
Paternalism	3.439	0.690	3.475	0.683	3.485	0.678
Lifetime commitment	8.204	1.486	8.465	1.452	8.702	1.579
Number of previous jobs	1.391	1.301	1.752	1.741	2.077	1.052

TABLE C 2

Correlation Matrix for all Employees in Electric Factory

		2	3	7	8	10	11	13	16	17	21	24	30	31	32	34	35	39
2	Residence	1.00	-28	-04	-14	-17	-14	-44	14	-30	.03	-.41	03	-17	-.03	-06	-11	-07
3	Sex		1.00	59	67	64	65	11	11	38	08	37	03	37	-05	11	24	50
7.	Number of dependents			1.00	82	60	74	-01	18	37	26	18	10	31	-.14	14	.25	47
8	Pay				1.00	.71	85	13	16	43	28	30	07	37	-.14	13	29	54
10	Job classification					1.00	80	04	31	37	22	33	06	32	-11	.13	29	54
11	Age						1.00	14	30	36	33	31	07	34	-13	14	29	54
13	Education							1.00	-26	06	-06	58	-.05	19	06	02	16	04
16	Total number days absent								1.00	-39	13	-06	-00	-00	-.01	02	01	13
17	Ratio of absences									1.00	.03	11	.03	18	-07	.06	14	31
21	*Years in previous jobs*										1.00	.03	.08	.08	-.03	.04	.08	10
24	Section											1.00	-.01	24	01	14	16	14
30	Perceived work load												1.00	-.01	06	-05	02	-00
31	Values													1.00	-08	12	16	31
32	Alienation														1.00	-10	-02	-11
34	Move jobs															1.00	-01	08
35	Job autonomy																1.00	27
39.	Knowledge of procedures																	1.00

		71	72	78	84	101	109	121	122	123	124	138	139	140	141	143	144	146	148
2	Residence	01	20	-15	-02	17	-31	-02	-43	-23	-12	-30	-16	-11	13	-29	04	-04	-03
3	Sex	14	-23	28	65	-.11	57	-05	26	20	09	69	49	54	01	.27	.05	17	21
7.	Number of dependents	10	-04	46	75	-01	36	.01	10	13	05	.70	.42	50	08	07	05	.05	38
8	Pay	.10	-08	56	82	-06	48	03	.22	19	11	87	48	56	09	16	.05	12	46
10	Job classification	11	-10	42	81	-12	.44	-.02	.21	19	.08	81	48	56	09	.16	02	14	34
11	Age	12	-08	.42	88	00	39	-01	18	18	09	86	.46	54	05	15	01	14	40
13	Education	02	-16	20	-15	-26	27	07	39	20	21	39	13	08	-24	36	-04	10	04
16	Total number days absent	.05	01	-09	39	15	-03	-05	-13	-01	-03	13	01	-06	05	-17	-07	-06	03
17	Ratio of absences	02	-.05	24	38	-10	25	01	10	09	04	39	25	.62	09	12	06	06	15
21	*Years in previous jobs*	-00	01	17	18	10	-03	-06	07	05	-.00	19	.11	11	06	-08	-.00	.07	51
24	Section	.02	-17	19	15	-26	37	02	38	21	25	61	38	16	-17	31	-01	02	06
30	Perceived work load	01	-05	09	07	03	-03	02	.03	00	-05	05	05	.09	00	-01	01	-.03	11
31	Values	10	-03	28	30	-17	36	05	16	12	05	40	36	31	09	21	07	19	.19
32	Alienation	04	-08	-06	-16	-00	-12	-11	-04	-06	-02	-11	-18	-09	-16	-01	-07	-05	-06
34	Move jobs	-03	13	10	14	-10	14	05	09	04	03	16	65	05	15	02	02	19	11
35	Job autonomy	04	-03	27	23	-09	26	-04	11	06	10	34	16	23	01	16	03	-07	19
39.	Knowledge of procedures	11	-.07	32	55	-10	40	-05	14	14	.03	53	30	47	16	20	02	12	24
71	Union participation	1.00	01	03	11	-08	11	11	-02	06	04	12	08	12	05	18	07	.10	05
72	Perceived advantages of firms		1.00	02	-04	-.07	-.14	-02	-18	-09	-06	-13	05	-07	18	-06	05	.15	05
78	Rank			1.00	.38	-19	41	.11	19	.15	.07	.59	.34	27	07	16	11	12	.56
84	Seniority				1.00	.06	36	-03	.05	.12	02	77	44	53	13	04	02	05	10
101	Wage preferences					1.00	-31	-12	-15	-08	-.09	-20	-25	-14	-02	-26	-08	-22	-05
109	Promotion chances						1.00	03	27	18	16	57	45	40	08	32	07	11	26
121	Concerns							1.00	-03	03	-06	04	10	02	-01	02	11	.00	05
122.	Distance of community of origin								1.00	46	24	33	22	17	-07	.24	-06	11	11
123	Size of community of origin									1.00	31	25	15	12	-05	.17	02	01	09
124	Father's occupation										1.00	18	11	04	00	17	-00	-02	05
138	Organizational status											1.00	56	54	01	28	04	15	40
139	Job satisfaction												1.00	39	20	19	12	27	26
140	*Performance*													1.00	12	.24	09	14	23
141	Employee cohesiveness														1.00	03	07	22	13
143	Participation															1.00	07	09	04
144	Paternalism																1.00	07	06
146	Lifetime commitment																	1.00	11
148	Number of previous jobs																		1.00

TABLE C.3
Correlation Matrix for Males Only in Electric Factory

	2	7	11	13	39	72	101	109	121	122	123	124	138	139	140	141	143	144	146	148
2. Residence	1.00	.17	.10	−.49	.12	.20	.16	−.22	−.04	−.47	−.21	−.11	−.13	−.04	.12	.19	−.32	.06	−.04	.06
7. Number of dependents		1.00	.72	−.12	.36	.16	.08	.04	.16	−.09	.01	.01	.53	.27	.39	.14	−.17	.04	.03	.34
11. Age			1.00	.04	.45	.13	.09	.10	.06	−.01	.03	.02	.76	.28	.47	.16	−.05	−.01	.11	.37
13. Education				1.00	−.01	−.23	−.30	.42	.07	.53	.30	.30	.48	.15	.03	−.25	.30	−.11	−.14	−.01
39. Knowledge of procedures					1.00	.08	−.08	.28	.09	.01	.09	−.02	.41	.17	.41	.28	.10	−.03	.10	.20
72. Perceived advantages of firms						1.00	−.12	.05	−.01	−.24	−.04	−.07	.02	.23	.11	.31	−.03	.11	.16	.13
101. Wage preferences							1.00	−.38	−.14	−.21	−.12	−.11	−.23	−.33	−.13	−.07	−.28	−.13	−.21	.03
109. Promotion chances								1.00	.15	.20	.15	.16	.39	.35	.19	.12	.28	.03	.12	.26
121. Concerns									1.00	−.05	−.03	−.17	.17	.21	.24	.03	.09	.12	−.02	.06
122. Distance of community of origin										1.00	.52	.28	.22	.08	−.03	−.13	.26	−.08	.07	.03
123. Size of community of origin											1.00	.38	.14	.03	−.02	−.01	.21	.02	−.02	.03
124. Father's occupation												1.00	.19	.06	−.05	−.08	.22	−.05	−.01	.01
138. Organizational status													1.00	.44	.41	.05	.15	−.03	.14	.40
139. Job satisfaction														1.00	.23	.29	.14	.13	.27	.28
140. Performance															1.00	.18	.10	.05	.13	.19
141. Employee cohesiveness																1.00	.10	.13	.23	.20
143. Participation																	1.00	.09	.02	−.00
144. Paternalism																		1.00	.07	.08
146. Lifetime commitment																			1.00	.08
148. Number of previous jobs																				1.00

TABLE C 4

Correlation Matrix for all Employees in Shipbuilding Factory

	7	8	13	14	18	49	65	73	86	88	89	91	93	94	95	98	99	101	102	103	104	105	106	108	109	110	123	124	125	127	128	129	131	132
7 Perceived work load	1 00	03	- 10	-.03	04	- 05	05	12	09	01	- 05	16	- 01	02	02	01	- 03	01	.01	- 04	.02	- 03	05	03	04	02	06	06	- 01	- 10	.00	-.01	- 08	-.08
8 Values		1 00	11	.08	11	04	11	10	15	01	09	24	10	05	10	02	03	09	- 04	14	07	- 07	- 03	- 04	01	15	18	.23	19	03	.07	.06	19	.14
13 Move jobs			1 00	07	06	16	- 07	- 07	- 03	01	- 08	- 05	36	19	- 13	- 18	06	.11	09	11	.07	- 07	00	.11	- 19	19	05	.65	10	24	- 04	05	42	39
14 Job autonomy				1 00	13	05	06	15	09	06	- 02	17	21	26	- 02	- 03	02	13	09	03	- 06	05	11	- 01	- 04	.25	22	.14	19	06	-.01	05	13	13
18 Knowledge of procedures					1 00	05	09	20	23	- 04	- 03	30	28	31	- 07	- 02	11	28	08	17	- 12	11	05	06	- 05	.32	28	15	.33	15	.09	13	.21	19
49 Perceived advantages of firms						1 00	03	04	03	03	- 05	00	04	04	- 12	- 11	- 01	05	07	- 01	- 07	01	03	.04	01	05	01	.22	05	12	06	.04	31	.30
65 Wage preferences							1 00	37	13	- 01	14	30	- 25	- 15	23	18	- 07	- 11	- 13	01	- 04	05	.04	- 18	20	- 04	10	.06	11	04	18	02	04	.01
73 Promotion chances								1 00	19	- 03	.14	51	- 09	09	25	26	07	08	- 11	19	- 04	05	06	-.18	21	16	.35	.16	28	06	18	03	15	13
86 Concerns									1 00	- 02	- 03	28	09	10	02	10	- 10	15	05	08	- 07	11	19	10	- 10	13	13	11	14	06	05	01	21	19
88 Section										1 00	23	13	03	07	16	10	- 18	08	05	15	- 09	.04	03	- 11	08	11	46	06	01	04	02	- 04	06	05
89 Job											1 00	27	- 18	- 19	30	12	- 24	- 13	- 11	03	- 02	04	06	- 16	18	- 01	26	04	- 06	- 04	08	00	-.05	- 06
91 Rank												1 00	19	29	20	14	- 10	25	- 06	29	- 05	07	09	- 05	10	49	69	24	.36	-.04	08	08	.18	15
93 Age													1 00	79	- 34	- 33	18	53	21	.30	- 03	02	08	32	- 47	71	47	31	.29	06	- 16	13	36	29
94 Seniority														1 00	- 29	- 12	20	56	21	33	- 15	12	06	08	19	68	59	20	.37	09	- 12	08	.27	24
95 Education															1 00	39	- 13	- 20	- 25	04	02	01	15	- 31	42	- 08	31	- 06	-.04	- 06	.16	-.12	- 10	-.10
98 Recruitment channel																1 00	- 06	- 17	- 19	04	- 05	05	05	- 46	52	- 11	.13	- 11	02	- 04	.18	- 09	- 13	- 13
99 Sex																	1 00	14	- 05	03	07	- 07	- 07	07	-.11	16	- 05	12	08	04	.01	08	09	.09
101 Number of dependents																		1 00	12	24	- 10	06	05	19	- 27	52	.41	16	.32	03	- 14	09	17	15
102 Residence																			1 00	06	- 17	14	06	15	-.19	.11	.01	05	12	- 06	- 07	01	11	09
103 Ratio of absences																				1 00	- 03	02	06	- 03	01	37	43	17	55	06	03	08	30	.25
104 Distance of community of origin																					1 00	- 69	- 15	07	07	- 09	- 12	03	- 03	00	06	05	- 01	- 02
105 Size of community of origin																						1 00	25	- 08	06	05	12	03	03	- 00	01	- 08	03	05
106 Father's occupation																							1 00	- 04	06	11	15	06	05	- 08	11	02	02	01
108 Number of previous jobs																								1 00	- 72	11	-.11	.06	06	- 06	- 16	06	02	06
109. Years in previous jobs																									1 00	- 22	06	-.10	- 06	04	17	-.15	- 13	-.13
110 Pay																										1.00	72	.31	35	01	- 11	.12	29	.26
123 Organizational status																											1 00	24	35	02	01	05	21	.21
124 Job satisfaction																												1 00	22	28	04	05	50	42
125. Performance																													1.00	10	14	13	33	30
127 Employee cohesiveness																														1 00	19	08	.33	28
128 Participation																															1 00	10	.02	03
129 Paternalism																																1.00	.09	09
131 Lifetime commitment—7																																	1 00	.88
132 Lifetime commitment—4																																		1 00

TABLE C.5
Variables not Included in Tables C.2–C.4

	Variable Number		
	Electric	*Shipbuilding*	*Sake*
Number of suggestions	18	11 or 133	
Awards for suggestions	19	12 or 134	
Interesting job	25	2	2
Job and desire	26	3	
Variety	27	4	4
Production goal	28	5	
Job and ability	29	6	3
Think of suggestions	33	10	
Job satisfaction	36	15	6
Best section in which to work	37	16	7
Why a section is best in which to work	38	17	
Move jobs	40	13	
Influence over section chief	55	33	
Influence over subsection chief	56	34	
Influence over second-line foreman	57	35	
Influence over first-line foreman	58	36	
Two kinds of superiors (paternalism)	63	40	14
If moved to another job in the factory	65	42	
Best friends inside or outside factory	66	43	
Number of close friends in factory	67	44	
Opportunity to see friends outside	68	45	
Relationships in factory	69	46	16
Teamwork	70	47	17
Males intend to work here until retirement	73	53 or 138	
Employee who voluntarily changes companies	93	57	
Should males work for same company	94	58	
Intend to continue or quit	95	60	
Why do you intend to continue?	96		
Ever thought of quitting?		54	
Two ways of thinking about company life		48	18
Feel "one body" with the company		50 or 137	
Two companies (paternalism)	97	61	
Belong to club	98	62	
Frequency of participation	99	63	
Rate of participation	100	64	
Reasons for wage system preferences	102	66	
Six factors that determine pay	103–108	67–72	
Reasons for poor perceived promotion chances	110–119	74–84	
Conflicts among supervisors	127	113	
Frequency of getting conflicting orders	128	114	
Subordinates receive communication	135	121	
First reason for working in Shipbuilding Company		51	
Second reason for working in Shipbuilding Company		52	
First reason for thinking of quitting		55	
Second reason for thinking of quitting		56	
Perceived work load			5
Job autonomy			9
Concerns			23
Job			24
Age			26
Seniority			30
Residence			32

APPENDIX D

Multiple Regression Analyses

For each dependent variable we have hypothesized two or more independent variables as its causes. We first examine the zero-order correlation between a dependent variable and one independent variable at a time. These correlations, however, do not necessarily provide a precise overall view of how well each variable accounts for the dependent variable, because they do not take into account how the relationships between the independent variables themselves affect predictions of the dependent variable. We therefore turn to multiple regression analysis in order to find out how much each variable is contributing to the variance in a dependent variable when other variables are held constant. Multiple regression analysis will eliminate the double counting of the variance that two variables may share jointly with the dependent variable.

Multiple regression analysis also enables us to state the joint effect of all the independent variables simultaneously acting on the dependent variable, and to state the proportion of the explained variance in the dependent variable that is accounted for by all the independent variables taken together. Stepwise multiple regression analysis is a special case of this method, in which the independent variables are selected in the order of their importance. The criterion of importance is based on the reduction of sum of squares, and the independent variable most important in this reduction in a given step is entered in the regression. After the first step, the program then ascertains which one of the remaining independent variables is most important in explaining the remaining unexplained variance, and so on until all independent variables have been entered.

Until the last section of this Appendix we use standardized partial regression coefficients (beta weights). These are appropriate when one wants to compare different beta weights within a given population or to generalize to a specific population, i.e., Electric Factory or Shipbuilding. In the last section, we shift to unstandardized partial regression coefficients, since our purpose there is to compare directly these two populations in order to see if the underlying causal processes are basically similar for given dependent variables. For more on this point, see footnote 16 of this Appendix.

It is essential that the reader understand that Appendix D is not intended to be read by itself, without reference to the relevant portions of Chapters

1–12 that deal with a given dependent variable. The multivariate regression tables in Appendix D must be read in the context of the more discursive background discussion, the development of the theory, and the statement of the hypotheses to be tested, which appear in the body of the book and which for the most part are not repeated in Appendix D.

KNOWLEDGE OF PROCEDURES

In Electric Factory (Table D.1), we see that knowledge of procedures is most strongly correlated with age (r = .54), status (.54),[1] and sex (.50);

TABLE D.1
Multiple Regression of Knowledge of Procedures in Electric Factory

Independent Variable	Standardized Partial Regression Coefficient: First Run	Standardized Partial Regression Coefficient: Second Run	Zero-Order Correlation[a]
138. Organizational status	.23**	.38**	.54**
11. Age	.21**		.54**
3. Sex	.20**	.22**	.50**
141. Employee cohesiveness	.17**	.15**	.17**
139. Job satisfaction	−.09**		.31**
31. Values	.07**		.31**
Multiple R	.60	.58	
R^2	.363	.341	
N	835	887	

** Significant at the .01 level. There are at least two reasons for testing for the statistical significance of relationships between variables. The first is to estimate fixed population parameters from a sample of that population. Since our data are based on all employees in selected factories, rather than a probability sample from a universe of firms or employees, the first reason does not apply. The second reason to test for significance is to estimate structural parameters, i.e., to state the probability that a given set of causal mechanisms could have produced the observed data. It is for this reason that we shall report significance levels for all zero-order correlations, based on the F test, and for all partial regression coefficients, based on the t-distribution. Because most of our hypotheses state a direction of relationship (e.g., "X varies positively with Y") rather than the sheer existence of a relationship, the one-tailed test is used.

[a] For zero-order correlations between independent variables, see Appendix C.

[1] The zero-order correlations reported in the tables of Appendix D and in the text do not necessarily agree with those in Tables C.2–C.4. This is because the zero-order correlations in Appendix C are based on all employees for whom there are data on the two variables being correlated, whereas the zero-order correlations in the multivariate analyses are based on only the employees for whom there are data on *all* the variables in a given multiple

it is less strongly related to job satisfaction (.31) and work values (.31) and least related to cohesiveness (.17). All relationships are in the predicted direction.

Knowledge of procedures was regressed on these six independent variables simultaneously (see Table D.1). Note first that when the other variables are held constant, the partial coefficient for job satisfaction becomes negative. This disconfirms our hypothesis that knowledge of procedures is positively related to job satisfaction (even when other independent variables are held constant). However, while the beta weight is significant, job satisfaction is at most only a weak predictor of knowledge of procedures (−.09), after other variables have been allowed to operate. The other hypotheses continue to hold, though again in varying degrees. The multiple correlation coefficient (R) is .60 but this is not much of an improvement over the zero-order correlation between knowledge of procedures and status alone (.54). Age and status are to a large extent measuring the same thing ($r = .87$)[2] and consequently have almost the same zero-order correlation with the dependent variable, as well as very similar beta weights.[3] Therefore, the only grounds for eliminating one of them in the second run of the multiple regression analysis are theoretical. It makes more theoretical sense to say employees know procedures because they have higher status in the company than because they are older. Consequently, age was dropped. The second run in Table D.1 is based on a stepwise multiple regression analysis. It shows that knowledge of procedures can be predicted virtually as well from three independent variables—status, sex, and cohesiveness—(multiple $R = .58$) as from all six original independent variables ($R = .60$). When only status, sex, and cohesiveness are deployed as independent variables, the relative influence of status on knowledge of procedures increases (from a beta weight of .23 in the first run to .38 in this run). We conclude that knowledge of procedures results somewhat more from having (high) status in the factory than from being male or having high cohesiveness

regression run. For example, the correlations in Table D.1 are based on N's of 835 (first run) and 887 (second run), while those for the same variables in Table C.2 are based on a larger portion of the 1,033 employees in Electric Factory.

[2] This is the problem of multicollinearity, i.e., the situation in which the independent variables are themselves highly correlated. The reliability of estimates of regression coefficients declines as the correlation between independent variables increases. The size of the correlation between independent variables that one considers to be serious multicollinearity is somewhat arbitrary; in all the multiple regression analyses in this study we shall use $r = .80$ as the cutoff point. Any correlations between independent variables of .80 or higher will be noted, and generally all but one of the set of variables so correlated will be dropped from the given regression analysis.

[3] Beta weights "indicate *how much change* in the dependent variable is produced by a standardized change in one of the independent variables when the others are controlled" (Blalock 1972:453).

with fellow employees. Thirty-four percent of the explained variance in knowledge of procedures—$(.58)^2$—is due to a combination of status, sex, and employee cohesiveness.

In Shipbuilding Factory, knowledge of procedures has a zero-order correlation of .32 with status and .30 with age, but is at best only weakly correlated with the other three independent variables—job satisfaction, employee cohesiveness, and values. (see Table D.2). When knowledge of procedures is regressed on these five variables, status, age,[4] and cohesiveness are somewhat better predictors than are (work) values and job satisfaction, but no variable has a beta weight greater than .22 when the other four variables are held constant. In the second multiple regression run, when knowledge of procedures is regressed on only the three better predictors, the multiple R is as high (.38) as when all five variables were used (.39). But for that matter, status alone predicts knowledge of procedures almost as well ($r = .32$) as when age and cohesiveness are added as independent variables. Thus, insofar as Shipbuilding employees profess to know factory procedures, this is because they have somewhat higher status in the organization, are somewhat older, and have slightly higher cohesiveness with their fellow employees. When these variables are controlled, knowing an employee's values (work, family, pleasure) and degree

TABLE D.2
Multiple Regression of Knowledge of Procedures in Shipbuilding Factory

Independent Variable	Standardized Partial Regression Coefficient		Zero-Order Correlation[a]
	First Run	*Second Run*	
123. Organizational status	.22**	.23**	.32**
93. Age	.16**	.17**	.30**
127. Employee cohesiveness	.12**	.14**	.14**
8. Values	.04		.11*
124. Job satisfaction	.03		.18**
Multiple R	.39	.38	
R^2	.149	.142	
N	453	466	

* Significant at the .05 level.
** Significant at the .01 level.
[a] For zero-order correlations between independent variables, see Appendix C.

[4] Whereas age and status are highly positively correlated in Electric Factory, their r is only .47 in Shipbuilding Factory. Therefore, they do not present the same problem of multicollinearity, and both are more often used in multiple regression analyses for Shipbuilding employees.

of job satisfaction adds virtually nothing to the explanation of the variance in knowledge of procedures.

JOB SATISFACTION

The zero-order correlations for Electric Factory (Table D.3) reveal that job satisfaction is most strongly correlated with status (r = .56), sex (.50), perceived promotion chances (.47), and age (.46). It is only weakly correlated with number of previous jobs (.25), participation (.20), and employee cohesiveness (.19), and it varies independently of work load (.03).

TABLE D.3
Multiple Regression of Job Satisfaction in Electric Factory

Independent Variable	Standardized Partial Regression Coefficient: First Run	Standardized Partial Regression Coefficient: Second Run	Zero-Order Correlation[a]
138. Organizational status	.49**	.35**	.56**
3. Sex	.19**	.14**	.50**
11. Age	−.17**		.46**
141. Employee cohesiveness	.17**	.18**	.19**
109. Perceived promotion chances	.12**	.16**	.47**
148. Number of previous jobs	.04		.25**
143. Participation	−.02		.20**
30. Perceived work load	.01		.03
Multiple R	.62	.62	
R^2	.388	.381	
N	791	825	

** Significant at the .01 level.
[a] For zero-order correlations between independent variables, see Appendix C.

The independent variables are themselves intercorrelated to some extent, and this affects the interpretation of the beta weights. When the other variables are held constant, in the first run, status emerges as by far the best predictor, with a beta weight of .49. When status is controlled, the effect of the previously strong variables of sex, promotion chances, and age is greatly reduced. The variables that had weaker zero-order correlations with job satisfaction also have low beta weights. The multiple regression of job satisfaction on all eight independent variables is .62. By eliminating the weak predictors as well as age, which has a .88 correlation with status and is therefore measuring essentially the same thing, we find that there is as high a multiple *R* (.62) using only the four best independent variables—status, promotion chances, sex, and cohesiveness—as when all eight original independent variables were used (see the second run in Table D.3).

In Shipbuilding (Table D.4) the zero-order *r*'s indicate that job satisfaction varies directly with age ($r = .33$), organizational status (.25), employee cohesiveness (.27), and perceived promotion chances (.19). It varies independently, however, of perceived work load, participation in company recreational activities, and number of previous jobs. Thus, employees with higher job satisfaction tend to be older, of higher status in the company, and more cohesive with fellow employees; they are also somewhat more likely to see their promotion prospects as good.

In the first multiple regression run, age, cohesiveness, and promotion chances continue to be the best predictors, when other variables are held constant. When these three variables are held constant, however, the influence of status on job satisfaction reduces to zero.[5] This indicates that employees who are satisfied with their jobs are satisfied not so much because of their status in the factory as because they are older, enjoy cohesive relationships with fellow workers, and see their future promotion

TABLE D.4
Multiple Regression of Job Satisfaction in Shipbuilding Factory

	Independent Variable	*Standardized Partial Regression Coefficient*		*Zero-Order Correlation*[a]
		First Run	*Second Run*	
93.	Age	.34**	.34**	.33**
127.	Employee cohesiveness	.24**	.23**	.27**
73.	Perceived promotion chances	.17**	.20**	.19**
7.	Perceived work load	.07		.05
108.	Number of previous jobs	−.06		−.00
128.	Participation	−.02		.03
123.	Organizational status	.00		.25**
	Multiple R	.46	.46	
	R^2	.215	.212	
	N	359	478	

** Significant at the .01 level.
[a] For zero-order correlations between independent variables, see Appendix C.

[5] There is relatively little multicollinearity to cloud this issue. The correlation between status and each of the best predictors is: .51 with age, .36 with promotion chances, and .01 with cohesiveness. As noted before, unlike the situation in Electric Factory, age and status are *not* measuring essentially the same thing in Shipbuilding Factory. We can therefore say with some assurance that job satisfaction depends more on age than on status in Shipbuilding Company.

chances as good. In both factories participation in company activities, previous interfirm mobility, and perceived work load explain virtually none of the variance in satisfaction. The original seven independent variables have a multiple *R* with job satisfaction of .46, but the three strongest variables—age, cohesiveness, and perceived promotion chances—alone have as high a multiple *R*. It should be noted, of course, that no single variable has more than a moderately strong predictive power with regard to job satisfaction, and that the combination of all our hypothesized independent variables explains only $(.46)^2$ or 22 percent of the variance in job satisfaction.

Job satisfaction and technological implications theory

Table D.5 shows that job satisfaction's highest zero-order correlations are with seniority ($r = .46$), sex (.43), and job classification (.43); it is distinctly less strongly correlated with section level of mechanization (.21). The beta weights are of the same order of relative importance: seniority, sex, and job classification are more important, in that order, and section mechanization level is unrelated to job satisfaction when the other three variables are held constant. In short, job satisfaction among production section workers is more a function of sex, seniority, and job classification than of section level of mechanization. One key variable in technological implications theory—level of mechanization—thus fails to explain satisfaction; the second variable in this theory—job classification—is so highly correlated with seniority ($r = .84$) that it is eliminated in the second multiple regression run. We find that sex and seniority, and even seniority alone, have virtually as high as correlation with job

TABLE D.5
Multiple Regression of Job Satisfaction in
Five Production Sections in Electric Factory

	Independent Variable	*Standardized Partial Regression Coefficient*		*Zero-Order Correlation*[a]
		First Run	*Second Run*	
84.	Seniority	.19**	.32**	.46**
3.	Sex	.18**	.20**	.43**
10.	Job classification	.14**		.43**
23.	Section (mechanization level)	.05		.21**
	Multiple *R*	.49	.48	
	R^2	.241	.233	
	N	682	682	

** Significant at the .01 level.
[a] For zero-order correlations between independent variables, see Appendix C.

satisfaction as the multiple R of all four independent variables with job satisfaction. The same is true in two other multiple regressions not shown here. When only sex and section mechanization are run against job satisfaction, the beta weights are .41 for sex and .05 for section, with a multiple R of .43. When only seniority and section are run against job satisfaction, the beta weights are .44 for seniority and .09 for section, with a multiple R of .47. Thus, among line production workers in Electric Factory seniority and sex have more influence on job satisfaction than the mechanization level of the section in which they work.

VALUE ORIENTATIONS

Value orientations, coded as (1) pleasure, (2) family, and (3) work, are regressed on the seven independent variables (six in Shipbuilding) in Tables D.6 (Electric Factory) and D.7 (Shipbuilding Factory). In the former factory, zero-order correlations reveal that primacy of work values is most strongly related to organizational status ($r = .40$), sex (.37), promotion chances (.37), age (.33), and number of dependents (.30); it is weakly related to residence (−.17) and size of community of origin (.11). This means that favoring work values over family and pleasure values is more likely among higher status, older, male employees, who see their promotion chances as good, and among those with dependents. Employees who live in company housing are somewhat more likely than those living in private residences to espouse work values. All these

TABLE D.6
Multiple Regression of Work Values in Electric Factory

	Independent Variable	*Standardized Partial Regression Coefficient*		*Zero-Order Correlation*[a]
		First Run	*Second Run*	
138.	Organizational status	.19**	.19**	.40**
109.	Perceived promotion chances	.17**	.18**	.37**
3.	Sex	.12**	.14**	.37**
7.	Number of dependents	.07		.31**
11.	Age	−.03		.33**
2.	Residence[b]	−.03		−.17**
123.	Size of community of origin	.01		.11**
	Multiple R	.44	.44	
	R^2	.196	.189	
	N	935	954	

** Significant at the .01 level.
[a] For zero-order correlations between independent variables, see Appendix C.
[b] Dichotomous dummy variable: whether employee lives in company housing or private residence.

findings support our hypotheses, though in varying degrees. One hypothesis is not confirmed: we found that employees from villages (*mura*) and towns (*chō*) are *not* more likely than those from cities to favor work values.

Let us now consider the influence of each independent variable upon value orientations when the other independent variables are held constant (see Table D.6). Of the seven variables included in the first run, status, promotion chances, and sex emerge as the best predictors. Their beta weights are .19, .17, and .12, respectively. Number of dependents and age have moderately strong zero-order *r*'s with work values, but their relationship is reduced virtually to zero when status and the other predictors are held constant. The influence of residence and size of community of birth is slight, both in the zero-order *r*'s and the beta weights. The multiple regression coefficient between values and these seven independent variables considered together is .44. In the second run, we regressed values on only the three strongest predictors. Although their relative strength is essentially the same as in the first run, we find that the multiple *R* of values with status, promotion chances, and sex alone is as high (.44) as it is with all seven independent variables in combination.

It should also be noted that in a stepwise multiple regression analysis of the influence of status, promotion chances, and sex on values, after the first two of these variables had been introduced, the multiple *R* was .42; adding sex increased the multiple *R* only to .44. In other words, women have pleasure, rather than work values, not primarily because they are women, but because they are in low status positions in the company. Male employees in the same low status positions have a similar propensity to favor pleasure over work values.

In conclusion, the most important determinants of favoring work values over family and pleasure values in Electric Factory are higher status in the company, perception of one's promotion chances as good, and being male. However, these variables leave over 80 percent of the variance in value orientations unexplained. Other variables, on which we lack data, presumably account for these differences in value preferences.

Table D.7 shows that value orientations in Shipbuilding Factory have only weak zero-order correlations with status (.16), perceived promotion chances (.11), age (.09), number of dependents (.09), and size of community of origin (−.09); they vary independently of residence. Employees of higher organizational status, older, with more dependents, from smaller communities of origin, and with better perceived promotion chances are *slightly* more likely than their opposite numbers to give primacy to work values over family or pleasure values. When value orientations are regressed on all six independent variables simultaneously, status and size of community of origin are found to be the only predictors that explain any significant portion of the variance; and even these best predictors have beta weights of only .12 and −.10, respectively. When other variables

TABLE D.7
Multiple Regression of Work Values in Shipbuilding Factory

	Independent Variable	*Standardized Partial Regression Coefficient*		*Zero-Order Correlation*[a]
		First Run	*Second Run*	
123.	Organizational status	.12*	.19**	.16**
105.	Size of community of origin	−.10*	−.10**	−.09*
73.	Perceived promotion chances	.07		.11*
102.	Residence[b]	−.05		−.06
93.	Age	.04		.09*
101.	Number of dependents	.02		.09*
	Multiple R	.21	.20	
	R^2	.044	.041	
	N	495	522	

* Significant at the .05 level.
** Significant at the .01 level.
[a] For zero-order correlations between independent variables, see Appendix C.
[b] Dichotomous dummy variable: whether employee lives in company housing or private residence.

are controlled, perceived promotion chances, age, and number of dependents lose even the minimal influence they had on values in the zero-order correlations. As hypothesized, primacy of work values varies positively with status, and is more likely if one came from a small community. But the hypotheses that older employees, those with more dependents, those with better promotion chances, and those who live in company housing are more likely to espouse work values are disconfirmed.

The multiple R between value orientations and all six independent variables is .21, but the two strongest variables—status and size of community of origin—yield a multiple R of .20. One can predict primacy of work values over family and pleasure values on the basis of having higher status in the company and coming from a town or a village as well as one can on the basis of all six original independent variables. However, only four percent of the predicted variance in values can be explained by these two variables acting together.

PAY

The zero-order correlations in Table D.8 reveal that pay is strongly positively correlated with several of the independent variables: age ($r = .85$), seniority (.82), number of dependents (.80), job classification (.72), and sex (.67), and moderately correlated with rank (.56), performance (.56), and attendance (.41). On the other hand, pay is only weakly correlated with number of years in previous jobs (.19), distance (.21) and

TABLE D.8
Multiple Regression of Monthly Pay in Electric Factory

Independent Variable	Standardized Partial Regression Coefficient		Zero-Order Correlation[a]
	First Run	Second Run	
84. Seniority	.35**	.48**	.82**
11. Age	.25**		.85**
7. Number of dependents	.23**	.46**	.80**
78. Rank	.17**		.56**
13. Education	.11**		.14**
3. Sex	.10**		.67**
10. Job classification	−.09**		.72**
122. Distance of community of origin from Electric Factory	.04**		.21**
17. Attendance (ratio of absences)	.04*		.41**
124. Father's occupation	.04*		.14**
21. Number of years in previous jobs	.03*		.19**
140. Performance	.03		.56**
146. Lifetime commitment	.02		.18**
141. Employee cohesiveness	.02		.05
143. Participation	−.02		.16**
144. Paternalism	.01		.05
123. Size of community of origin	.00		.21**
Multiple R	.92	.88	
R^2	.852	.766	
N	688	1,033	

* Significant at the .05 level.
** Significant at the .01 level.
[a] For zero-order correlations between independent variables, see Appendix C.

size of community of origin (.21), lifetime commitment (.18), participation (.16), father's occupation (.14), and education (.14). Finally, pay varies independently of employee cohesiveness (.05) and paternalism (.05).

Keeping in mind the multicollinearity among several of the independent variables, we turn next to the multiple regression of pay on all 17 independent variables simultaneously. When other variables are controlled, the best predicators of pay are seniority (beta weight .35), age (.25), number of dependents (.23), and rank (.17). When these four variables are taken into account, the influence of sex, job classification, attendance, and performance on pay is greatly reduced. This indicates that a person's seniority, number of dependents, age, and rank, more than sex, job classification, attendance, or performance, determine pay. All the variables that had weak zero-order correlations with pay continue to have low beta weights. The multiple R for pay and all 17 independent variables considered together is .92.

To obtain a more parsimonious explanation of pay, we excluded the independent variables that (1) have little effect on pay, as shown by zero-order *r*'s or beta weights, and (2) are redundant with other independent variables. Since seniority, the best predictor when other variables are controlled, was being retained, we dropped the following variables because of their high correlations with seniority: age ($r = .87$) and job classification ($r = .82$). In other words, age and job classification are measuring essentially the same thing as seniority and therefore would not contribute much to the explanation of pay after seniority had been taken into account. In a stepwise multiple regression analysis not reported in Table D.8 we regressed pay on the remaining five best independent variables: seniority, number of dependents, rank, education, and sex. As already seen, seniority alone has a .82 correlation with pay; it emerged again as the best predictor. In the second step, adding number of dependents increases the multiple *R* to .88; adding the third step, education, increases the multiple *R* to .90, adding rank increases *R* to .91, and adding sex brings the *R* up to .92. One can explain as much of the variance in pay on the basis of these five variables as when all 17 independent variables were used.

It is clear that once seniority and number of dependents have been controlled, the other main independent variables exert only a slight further effect on pay. Thus, in the second run in Table D.8, we use only these two variables. Their beta weights are almost equal (.48 for seniority and .46 for number of dependents); together they account for $(.88)^2$ or 76.6 percent of the variance in pay. Although the importance of seniority is not unexpected, that of number of dependents is, in view of our earlier remarks about its minimal influence.[6]

The problem of multicollinearity is particularly serious with regard to the two variables we regard as theoretically the most important—seniority and job classification, which are the main variables of the old and the new Electric Company wage policy, respectively. In a multiple regression, the beta weight of a predictor variable depends not only on its relation to the dependent variable but also on its correlation with other predictors. Since seniority has the higher zero-order correlation with pay, it is the first variable to enter the regression equation. Though job classification is also highly correlated with pay, its high correlation with seniority

[6] It should be added, however, that when number of dependents is omitted, and pay is regressed on seniority and *rank*, there is as high a multiple *R* (.87) as when pay is regressed on seniority and number of dependents (.88). Although number of dependents has both a larger *r* and a larger beta weight than rank, in relation to pay, one can predict as much of the variance in pay on the basis of seniority and rank as on the basis of seniority and number of dependents. There is a zero-order *r* of .46 between rank and number of dependents, and to this extent personnel in higher ranks have more dependents than those in lower ranks.

prevents it from explaining much of the variance in pay once seniority has been taken into account; consequently, job classification has only a low beta weight in the regression equation. In this sense, the relative beta weights do not necessarily reflect the relative influence of seniority and job classification. When pay is regressed on only seniority and job classification, the beta weight for seniority, with job classification held constant, is .71, the beta weight for job classification, with seniority held constant, is .14. The multiple R of pay on seniority and job classification is .82. Nevertheless, we cannot infer that seniority has more than a somewhat larger influence on pay than does job classification because these two predictor variables are themselves so highly correlated.

The results of the test of our theory of rewards in Shipbuilding Factory appear in Table D.9. The zero-order relationships indicate that pay is

TABLE D.9
Multiple Regression of Monthly Pay in Shipbuilding Factory

Independent Variable	*Standardized Partial Regression Coefficient* First Run	*Standardized Partial Regression Coefficient* Second Run	*Zero-Order Correlation*[a]
93. Age	.41**	.64**	.70**
91. Rank	.25**	.36**	.47**
94. Seniority	.25**		.69**
101. Number of dependents	.10*		.48**
95. Education	.08*		−.13*
105. Size of community of origin	−.08		.10
106. Father's occupation	.08*		.15*
103. Attendance (ratio of absences)	.07		.36**
125. Performance	−.07		.29**
127. Employee cohesiveness	−.06		−.03
98. Recruitment channel	.05		−.17**
131. Lifetime commitment	.05		.31**
104. Distance of community of origin from Shipbuilding Factory	−.04		−.08
128. Participation	−.04		−.11
17. Number of years in previous jobs	.03		.24**
89. Job	.01		−.04
129. Paternalism	−.01		.11
Multiple R	.80	.80	
R^2	.644	.643	
N	278	524	

* Significant at the .05 level.
** Significant at the .01 level.
[a] For zero-order correlations between independent variables, see Appendix C.

most strongly correlated with age (r = .70), seniority (.69), number of dependents (.48), and rank (.47). It is moderately correlated with attendance (.36), performance (.29), lifetime commitment (.31), and number of years of work experience prior to coming to Shipbuilding Company (.24). All these relationships are positive. Pay is only weakly (negatively) correlated (−.17) with recruitment channel—i.e., those recruited through a vocational guidance center or newspaper job ad earn somewhat higher pay than those recruited on the basis of knowing someone in the company or school recommendations. Pay is weakly positively correlated with father's occupation (.15) and weakly negatively correlated with education (−.13). Pay varies independently of other social origins variables (size and distance of community of origin from the factory), job, employee cohesiveness, paternalism, and participation.

To explore more precisely the interaction of pay with other variables, we ran a multiple regression of pay on all 17 independent variables. Age emerges as the single best predictor (beta = .41). The other variables that account for more than a marginal amount of the variance in pay are rank and seniority, each with a beta of .25. The variables that had weak zero-order r's with pay continue to explain little or none of the variance in pay and can essentially be discounted. This is true of education, size of community of origin and its distance from Osaka, recruitment channel, job, father's occupation, cohesiveness, paternalism, and participation. What is more striking is that certain variables that had moderate zero-order r's with pay reduce to near zero when the main formal criteria of pay are held constant. This is the case for number of dependents (r = .48, but beta weight = .10); attendance (r = .36, beta = .07); performance (r = .29, beta = −.07); lifetime commitment (r = .31, beta = .05); and number of years in previous jobs (r = .24, beta = .03). These findings mean that employees' pay is not so much a function of how many dependents they have, their performance and attendance, their lifetime commitment, or the duration of their previous work experience, as of variables with which these are correlated. The formal criteria of pay—age, rank, and seniority—turn out to be the more basic determinants of pay.

The 17 independent variables together have a multiple R of .80 with pay. But age and seniority are themselves so highly correlated (r = .79) that we can drop seniority without losing any predictive power. Age alone has almost as high a correlation with pay (.70) as do all 17 variables together! The two best, non-redundant, independent variables—age and rank—together have as high a multiple R with pay (.80) as did the original 17 variables. It is also true that, as in Electric Factory, pay is the dependent variable that can be most fully predicted on the basis of our independent variables.

JOB CLASSIFICATION

Consider first the zero-order correlations in Table D.10. Job classification is most strongly related to seniority ($r = .81$), age (.80), sex (.64), and performance (.57). It is less correlated with the informational level of the section (.34), number of previous jobs (.31), size of community of origin (.20), participation (.15), lifetime commitment (.14), and father's occupation (.10); it varies independently of education (.01) and paternalism (.01). Thus, at the level of zero-order relationships, our hypotheses are generally confirmed: status variables are the best predictors (except for education), performance is moderately strongly related to job classification, and social origins and informal influences involving the social integration of the employee into the company are weakly related to job classification.

TABLE D.10
Multiple Regression of Job Classification in Electric Factory

Independent Variable	Standardized Partial Regression Coefficient: First Run	Standardized Partial Regression Coefficient: Second Run	Zero-Order Correlation[a]
84. Seniority	.44**	.77**	.81**
11. Age	.25**		.80**
24. Informational level of section	.16**	.22**	.34**
140. Performance	.14**		.57**
13. Education	−.07**		.01
148. Number of previous jobs	.06**		.31**
3. Sex	.04		.64**
123. Size of community of origin	.04*		.20**
143. Participation	.01		.15**
124. Father's occupation	−.01		.10**
144. Paternalism	−.01		.01
146. Lifetime commitment	−.01		.14**
Multiple R	.86	.83	
R^2	744	696	
N	762	1,032	

* Significant at the .05 level.
** Significant at the .01 level.
[a] For zero-order correlations between independent variables, see Appendix C.

When we examine the beta weights, seniority is seen to be by far the best predictor; the effect of age on job classification is greatly reduced when seniority is held constant; when the other variables are controlled, the informational level of the section emerges as the third best predictor of job classification. The low relationship between performance and job classification suggests that performance in fact has relatively little role vis à vis job classification, independent of seniority, age, and the section in which one works. All the other variables have virtually no relationship to job classification, either as zero-order *r*'s or as beta weights.

In the second run, all variables except the two strongest[7]—seniority and the informational level of the section—were excluded. These two variables have virtually as strong a multiple *R* with job classification (.83) as did all 12 of the original independent variables together (.86). Indeed, seniority alone predicts job classification almost as well as all the original variables together. Few other variables in our entire study have as much of their variance explained as do pay and job classification, no matter how many independent variables are used, let alone when only *one* independent variable is used.

RANK

Rank in Electric Factory is most strongly correlated with number of previous jobs (zero-order $r = .54$), age (.42), seniority (.38), sex (.27), and performance (.26) (see Table D.11). It is only weakly correlated with

TABLE D.11
Multiple Regression of Rank in Electric Factory

	Independent Variable	*Standardized Partial Regression Coefficient*		*Zero-Order Correlation*[a]
		First Run	*Second Run*	
84.	Seniority	.48**	.27**	.38**
148.	Number of previous jobs	.48**	.47**	.54**
13.	Education	.24	.22	.14**
11.	Age	−.19**		.42**
143.	Participation	.06*		.13**
144.	Paternalism	.05*		.09*
3.	Sex	−.05		.27**
24.	Informational level of section	−.02		.16**
123.	Size of community of origin	.02		.13**
146.	Lifetime commitment	.01		.10**
140.	Performance	.00		.26**
124.	Father's occupation	.00		.06
	Multiple *R*	.63	.64	
	R^2	.399	.411	
	N	762	1,033	

* Significant at the .05 level.
** Significant at the .01 level.
[a] For zero-order correlations between independent variables, see Appendix C.

[7] Following our general practice, age was dropped from the second run because it is so redundant with seniority.

education (.14), the informational level of the section (.16), size of community of origin (.13), and participation in company recreational activities (.13). As predicted, rank varies independently of father's occupation.

The beta weights reveal that the most important predictors of rank are seniority, number of previous jobs, education, and age. When other factors are controlled, employees are most likely to be in higher ranks if they have more seniority and education and have worked in other firms before coming to Electric Company, but are nevertheless still relatively young. Stepwise multiple regression analysis reveals that one can predict as much of the variance in rank on the basis of number of previous jobs, seniority, and education as when all 12 original variables are used (see the second run in Table D.11).

To summarize, after the three best predictors—number of previous jobs, seniority, and education—have been taken into account, the other nine independent variables add nothing further to the explanation of differences in rank. We have also seen that the major determinants of job classification are partly the same as and partly different from the main determinants of rank (see Table D.12). Whereas job classification is overwhelmingly determined by seniority, the influence of seniority on rank is less than that of number of previous jobs.

TABLE D.12
Comparison of the Multiple Regression of Job Classification and of Rank in Electric Factory

Independent Variable	*Standardized Partial Regression Coefficient with* Job Classification	Rank
84. Seniority	.77**	.27**
24. Informational level of section	.22**	
148. Number of previous jobs		.47**
13. Education		.22**
Multiple R	.83	.64

** Significant at the .01 level.

The results for Shipbuilding Factory appear in Table D.13. Rank is most strongly correlated with performance ($r = .37$), seniority (.30), and age (.23). It is weakly correlated with lifetime commitment (.19), education (.14), and size of community of origin (.11). It varies independently of section, participation, father's occupation, number of previous jobs, recruitment channel, and paternalism.

In the multiple regression, performance, education, and seniority emerge as the best predictors of rank. All other variables account for little

TABLE D.13
Multiple Regression of Rank in Shipbuilding Factory

Independent Variable	Standardized Partial Regression Coefficient		Zero-Order Correlation[a]
	First Run	*Second Run*	
125. Performance	.27**	.26**	.37**
95. Education	.23**	.31**	.14*
94. Seniority	.21**	.28**	.30**
105. Size of community of origin	.09*		.11*
93. Age	.09		.23**
106. Father's occupation	−.08		.04
128. Participation	.05		.10
131. Lifetime commitment	.05		.19**
98. Recruitment channel	.05		.07
129. Paternalism	.04		.07
108. Number of previous jobs	−.01		−.06
88. Section	.00		.08
Multiple *R*	.49	.48	
R^2	.244	.231	
N	324	512	

* Significant at the .05 level.
** Significant at the .01 level.
[a] For zero-order correlations between independent variables, see Appendix C.

or none of the variance in rank. When performance, education, and seniority are held constant, the influence of age on rank declines (because age is highly correlated with seniority); the influence of lifetime commitment on rank also declines somewhat. The variables that were unrelated to rank in the zero-order *r*'s continue to explain virtually none of the variance when other variables are held constant.

The multiple correlation between rank and all 12 independent variables is .49, but this is not a great improvement in predictability over that between rank and performance alone (.37). When rank is regressed on the three best predictors—performance, seniority, and education—each one is seen to exert a relatively equal influence on rank, when the other two are held constant. Together they account for as much of the variance in rank as all 12 of the original independent variables together (see Table D.13).

PERCEIVED PROMOTION CHANCES

Table D.14 presents the multiple regression of promotion chances on the 13 independent variables for Electric Factory. Perceived promotion chances are most strongly correlated with sex (.60), age (.43), informational level of section (.42), rank (.41), performance (.39), seniority (.38), participation in company recreational activities (.32), and education (.31).

TABLE D.14
Multiple Regression of Perceived Promotion Chances in Electric Factory

Independent Variable	Standardized Partial Regression Coefficient		Zero-Order Correlation[a]
	First Run	Second Run	
3. Sex	.44**	.44**	.60**
78. Rank	.25**	.25**	.41**
13. Education	.22**	.21**	.31**
11. Age	−.22**		.43**
84. Seniority	.15*		.38**
24. Informational level of section	.09**		.42**
140. Performance	.08**	.08**	.39**
143. Participation	.06*		.32**
148. Number of previous jobs	.05		.24**
124. Father's occupation	.03		.16**
146. Lifetime commitment	.02		.14**
144. Paternalism	.02		.06
123. Size of community of origin	−.00		.18**
Multiple *R*	.71	.67	
R^2	.506	.444	
N	743	972	

* Significant at the .05 level.
** Significant at the .01 level.
[a] For zero-order correlations between independent variables, see Appendix C.

Perceived promotion chances have weaker relationships with size of community of origin, number of previous jobs, father's occupation, and lifetime commitment. They vary independently of paternalism.

In the multiple regression of perceived promotion chances on all these independent variables, sex, status variables, and age continue to be the best predictors. When other variables are controlled, those males higher in rank, education, and seniority, and yet younger are more likely than their opposite numbers to think their promotion chances are good. When these variables are held constant, the influence of the informational level of one's section, performance, participation, number of previous jobs, and size of community of origin on perceived promotion chances is diminished.

Number of previous jobs, father's occupation, lifetime commitment, paternalism, and size of community of origin exert no significant independent influence on promotion chances. There is a multiple *R* of .71 between promotion chances and all 13 independent variables.

The variables with significant beta weights in the first run were rerun to obtain a more parsimonious explanation of promotion chances.[8] In the

[8] Because seniority is so strongly correlated with age, we dropped it, as the weaker of the two predictors, in the second run.

stepwise multiple regression analysis, we find that four of these independent variables—sex, rank, education, and performance—explain almost as much of the variance in perceived promotion chances (44.4 percent) as all 13 original variables (50.6 percent).

Turning now to Shipbuilding Factory, Table D.15 shows that, at the level of zero-order correlations, perceived promotion chances are most strongly correlated with rank ($r = .51$), performance (.28), participation (.26), recruitment channel (.25), and education (.20). Perceived promotion chances are weakly correlated with number of previous jobs (−.19), lifetime commitment (.17), and seniority (.12), and vary independently of age, section, paternalism, and social origins variables. The correlation with recruitment channel means that employees recruited via school recommendation perceive their promotion chances as better than those recruited by knowing someone in Shipbuilding Company, who in turn perceive their chances as better than those recruited on a more impersonal basis—want ads and government vocational centers.

In the multiple regression analysis, the best predictors of promotion chances, when other variables are controlled, are rank (beta = .43), age (−.30), and seniority (.25). The multivariate analysis suggests that the

TABLE D.15
Multiple Regression of Perceived Promotion Chances in Shipbuilding Factory

	Independent Variable	*Standardized Partial Regression Coefficient*		*Zero-Order Correlation*[a]
		First Run	*Second Run*	
91.	Rank	.43**	.49**	.51**
93.	Age	−.30**		−.05
94.	Seniority	.25**		.12*
131.	Lifetime commitment	.15**		.17**
98.	Recruitment channel	.14**	.21**	.25**
88.	Section	−.13**	−.13**	−.08
128.	Participation	.13**	.12**	.26**
95.	Education	.08		.20**
125.	Performance	.05		.28**
129.	Paternalism	.04		.08
108.	Number of previous jobs	.01		−.19**
105.	Size of community of origin	−.01		.04
106.	Father's occupation	.00		.06
	Multiple R	.63	.58	
	R^2	.396	.335	
	N	317	398	

** Significant at the .01 level.

[a] For zero-order correlations between independent variables, see Appendix C.

effect of age is hidden at the zero-order level: promotion chances appear to vary independently of age because both the youngest and the oldest employees see their chances as poor. But when rank and other variables are controlled, the effect of age is released. In a given rank, for example, the younger the employee, the better he perceives his promotion chances to be. Thus, the structural conditions that most conduce to optimism about promotion chances appear to be a combination of rank, age, and seniority: it is employees who have already achieved some promotions in rank and have accumulated seniority, and yet are still young relative to others in their present rank, who see their further opportunities for advancement through the rosiest glasses.

Some of the variance in perceived promotion chances is also explained in terms of lifetime commitment (beta = .15), recruitment channel (.14), participation (.13), and the section in which one works (−.13). The last finding may be interpreted as indicating that, when other variables are held constant, employees in Assembly, Design, and Staff sections see their chances of promotion as better than those in Machining. But this difference is slight.

Certain variables with at least some influence on promotion chances at the zero-order level are washed out when the main variables are held constant: this is true of performance, education, and number of previous jobs. It seems clear that whatever influence these variables may have on perceived promotion chances, they operate mainly through rank, age, and seniority.

The multiple R between perceived promotion chances and all 13 independent variables is .63. To obtain a more parsimonious explanation of promotion chances, a second regression was run, omitting all independent variables whose beta coefficients in the first run were non-significant; seniority was also dropped because it is highly correlated with age ($r = .79$). A stepwise regression analysis indicates that when rank, recruitment channel, section, and participation have been taken into account, in that order, the multiple R with promotion chances is .58 (see Table D.15); adding lifetime commitment and age only increases the R to .59.

EMPLOYEE COHESIVENESS

The strongest zero-order correlation for employee cohesiveness is with education; although this is in the predicted direction, it is weak ($r = -.24$; see Table D.16). There are also weak positive correlations between cohesiveness and job satisfaction (.18), lifetime commitment (.20), seniority (.12), number of previous jobs (.12), residence (.12), rank (.12), and job classification (.08), and a weak negative relationship with the informational level of the section (−.16). Cohesiveness varies independently

TABLE D.16
Multiple Regression of Employee Cohesiveness in Electric Factory

	Independent Variable	Standardized Partial Regression Coefficient		Zero-Order Correlation[a]
		First Run	Second Run	
13.	Education	−.20**	−.21**	−.24**
139.	Job satisfaction	.18**	.22**	.18**
84.	Seniority	.16		.12**
3.	Sex	−.16**		.00
146.	Lifetime commitment	.15**	.14**	.20**
109.	Perceived promotion chances	.13**		.07
24.	Informational level of section	−.12	−.15**	−.16**
11.	Age	−.11		.03
143.	Participation	.08		.03
148.	Number of previous jobs	.06		.12**
7.	Number of dependents	−.06		.07
2.	Residence[b]	.04		.12**
84.	Rank	.02		.12**
10.	Job classification	.01		.08*
	Multiple *R*	.41	.38	
	R^2	.164	.143	
	N	717	760	

* Significant at the .05 level.
** Significant at the .01 level.
[a] For zero-order correlations between independent variables, see Appendix C.
[b] Dichotomous dummy variable: whether employee lives in company housing or private residence.

of the other variables. It is not difficult to see why these relationships are so weak. With regard to lifetime commitment, it is likely that counterforces are at work. An employee with a high degree of lifetime commitment would assume that he would work with the same people for a long time to come, and that he might as well make the best of his relationships with them; thus, he would seek and encourage cohesiveness. But this cannot be more than a small part of the story, or there would be a higher positive relationship between lifetime commitment and cohesiveness. It may also be that an employee with *low* lifetime commitment stresses cohesiveness with his fellow employees as a compensatory, alternative source of satisfaction. And others high in lifetime commitment may avoid cohesiveness if they perceive their fellow employees as low in lifetime commitment.

The same kind of *ex post facto* "explanations" could be developed for the other variables that are only minimally correlated with cohesiveness, but we shall eschew this. Suffice it to say that when these variables are included in a multiple regression analysis, all 14 independent variables together yield a multiple *R* of only .41. A stepwise multiple regression analysis indicates that the four strongest variables—job satisfaction, education, lifetime commitment, and the informational level of the section—have a multiple R of .38 with cohesiveness. Adding promotion chances and sex, which also had significant beta weights in the first multiple regression run, only increases the multiple R to .39.

Thus, the closest we can come to stating the characteristics of employees who have greater cohesiveness is that they are middle school graduates working in low informational level (line production) sections, who are relatively satisfied with their job, and have somewhat more lifetime commitment than other employees. But these four variables together account for only 14 percent of the variance in employee cohesiveness. Because Electric Factory employees have a relatively high level of cohesiveness, it is the more difficult to find variables that discriminate between those who are more and those who are less cohesive.

Consider now the fate of these same hypotheses in Shipbuilding Factory. The zero-order correlations in Table D.17 indicate that cohesiveness varies positively with lifetime commitment (r = .31), job satisfaction (.28), participation (.16), and seniority (.12), but varies independently of age,

TABLE D.17
Multiple Regression of Employee Cohesiveness in Shipbuilding Factory

	Independent Variable	*Standardized Partial Regression Coefficient*		*Zero-Order Correlation*[a]
		First Run	*Second Run*	
131.	Lifetime commitment	.22**	.25**	.31**
124.	Job satisfaction	.20**	.17**	.28**
128.	Participation	.18**	.19**	.16**
93.	Age	−.18*		.09
94.	Seniority	.17*		.12*
102.	Residence[b]	−.14**		−.06
91.	Rank	−.13*	−.12**	−.02
95.	Education	−.09		−.08
89.	Job	.07		−.08
88.	Section	.06		.05
108.	Number of previous jobs	.03		−.03
101.	Number of dependents	.03		.05
73.	Perceived promotion chances	−.01		.04
	Multiple R	.43	.40	
	R^2	.186	.157	
	N	310	361	

* Significant at the .05 level.
** Significant at the .01 level.
[a] For zero-order correlations between independent variables, see Appendix C.
[b] Dichotomous dummy variable: whether employee lives in company housing or private residence.

job, education, rank, section, number of dependents, residence, number of previous jobs, and perceived promotion chances. When cohesiveness is regressed simultaneously on all 13 of these independent variables (Table D.17), lifetime commitment, job satisfaction, participation, age, and seniority continue to be the best predictors; residence and rank also have significant beta weights. The remaining variables continue to explain virtually none of the variance in cohesiveness.

To achieve a more parsimonious explanation of cohesiveness, we regressed it on the seven independent variables that had significant beta weights in the first run. Column 2 of Table D.17 shows that the four best predictors—lifetime commitment, participation, job satisfaction, and rank—account for almost as much of the variance in cohesiveness (15.7 percent) as all 13 of the original independent variables (18.6 percent). After these four variables have been introduced, adding seniority, age, and residence only marginally increases the amount of variance explained (to 17.1 percent).

PATERNALISM

When our hypotheses on paternalism are tested (see Tables D.18 and D.19), the zero-order correlations are low enough to indicate that paternalism varies almost independently of all of them. The beta weights for

TABLE D.18
Multiple Regression of Paternalism in Electric Factory

	Independent Variable	*Standardized Partial Regression Coefficient*		*Zero-Order Correlation*[a]
		First Run	*Second Run*	
84.	Seniority	−.17*		.02
13.	Education	−.13**	−.11**	−.07*
3.	Sex	.09		.04
8.	Pay	.09		.04
78.	Rank	.08*	.07*	.07*
143.	Participation	.08*	.07*	.06
146.	Lifetime commitment	.06*	.06*	.07*
2.	Residence[b]	.06		.06
123.	Size of community of origin	−.03		−.02
	Multiple R	.18	.14	
	R^2	.032	.020	
	N	801	815	

* Significant at the .05 level.
** Significant at the .01 level.
[a] For zero-order correlations between independent variables, see Appendix C.
[b] Dichotomous dummy variable: whether employee lives in company housing or private residence.

Electric Factory employees are also all low, ranging from −.17 for seniority and −.13 for education, when the other independent variables are held constant, to −.03 for the relationship between size of community of origin and paternalism when the other variables are controlled. The second multiple regression run shows that the four strongest predictors of paternalism—education, rank, participation, and lifetime commitment—have a multiple *R* with paternalism of only .14.

In Shipbuilding Factory, at the zero-order level, the only variables even weakly correlated with paternalism are education ($r = -.13$) and participation (.12). Employees' paternalism varies independently of pay, rank, seniority, lifetime commitment, size of community of origin, and residence. When paternalism is regressed on all these variables simultaneously, the best predictors are education (beta = −.18), participation (.15), and pay (.12). When other variables are controlled, employees are somewhat more paternalistic to the extent that they have lower education, participate more frequently in company recreational activities, and receive higher pay. The multiple *R* between paternalism and these eight independent variables is only .25, and one can predict paternalism virtually as well on the basis of participation, education, and pay as on the basis of all eight variables. When these three variables have been taken into account, lifetime commitment contributes virtually nothing further to the explanation of the variance in paternalism, and can therefore be eliminated as a best predictor.

TABLE D.19
Multiple Regression of Paternalism in Shipbuilding Factory

Independent Variable	Standardized Partial Regression Coefficient		Zero-Order Correlation[a]
	First Run	*Second Run*	
95. Education	−.18**	−.13**	−.13*
128. Participation	.15**	.15**	.12*
110. Pay	.12	.13**	.10
94. Seniority	−.08		.07
131. Lifetime commitment	.06		.11
91. Rank	.05		.08
105. Size of community of origin	−.04		−.04
102. Residence[b]	.00		.03
Multiple *R*	.25	.22	
R^2	.061	.048	
N	378	455	

* Significant at the .05 level.
** Significant at the .01 level.
[a] For zero-order correlations between independent variables, see Appendix C.
[b] Dichotomous dummy variable: whether employee lives in company housing or private residence.

The reasons for these weak findings appear to include a combination of the low reliability of our index of paternalism (only two items) and the high level of paternalistic preferences in the factories studied (as in Japan at large). The latter point reduces the variance in paternalism, thereby producing low correlation and regression coefficients in relation to the independent variables.

RESIDENCE

Tables D.20 and D.21 present the data that attempt to explain why some employees live in company housing, while the majority in both factories live in private residences.[9] Clearly, the best predictors of residence in Electric Factory are education and distance migrated; when these two variables are controlled, the remaining six are seen to have virtually no

TABLE D.20
Multiple Regression of Residence[a] in Electric Factory

Independent Variable	*Standardized Partial Regression Coefficient* First Run	Second Run	*Zero-Order Correlation*[b]
122. Distance of community of origin from Electric Factory	−.31**	−.35**	−.43**
13. Education	−.30**	−.31**	−.42**
84. Seniority	−.09		−.01
7. Number of dependents	.06		−.03
148. Number of previous jobs	.06*		−.00
78. Rank	−.03		−.12**
8. Pay	−.00		−.13**
144. Paternalism	.00		.03
Multiple *R*	.51	.54	
R^2	.264	.294	
N	957	1,020	

* Significant at the .05 level.
** Significant at the .01 level.
[a] Dichotomous dummy variable: whether employee lives in company housing or private residence. In this table negative relationships mean that the higher the score on the independent variable the more likely that one lives in company housing; positive relationships mean that the higher one's score on the independent variable the more likely one lives in a private residence.
[b] For zero-order correlations between independent variables, see Appendix C.

[9] The signs in Tables D.20 and D.21 are negative because the residence variable is coded (1) company housing, (2) private housing, whereas the independent variables are coded from low to high values on each variable.

TABLE D.21
Multiple Regression of Residence[a] in Shipbuilding Factory

Independent Variable	Standardized Partial Regression Coefficient		Zero-Order Correlation
	First Run	Second Run	
94. Seniority	.19**	.16**	.16**
95. Education	−.14**	−.17**	−.23**
104. Distance of community of origin from Shipbuilding Factory	−.13**	−.12**	−.14**
108. Number of previous jobs	.10*		.14**
91. Rank	−.10*	−.11**	−.10*
110. Pay	−.08		.03
101. Number of dependents	.00		.10*
129. Paternalism	−.00		.02
Multiple R	.32	.31	
R^2	.102	.094	
N	431	480	

* Significant at the .05 level.
** Significant at the .01 level.
[a] Dichotomous dummy variable: whether employee lives in company housing or private residence. For interpretation of signs, see footnote to table D.20.
[b] For zero-order correlations between independent variables, see Appendix C.

independent influence on housing. Two additional variables are important in Shipbuilding Factory: seniority and rank. Because most employees live in private housing, there is not much variance in housing to explain. Concerning the variance that exists, the variables of education and distance migrated provide a better explanation in Electric Factory than in Shipbuilding Factory, and this remains true even after two additional variables—seniority and rank—are introduced in Shipbuilding. In short, more of the variance in residence for Electric Factory employees can be explained on the basis of just two variables than can be explained in Shipbuilding on the basis of the four best variables, or even all eight independent variables.

Having considered the causes of residence, we next consider its alleged effects. The paternalism model implies that living in company housing has significant positive consequences for six areas of employee behavior and attitudes: participation in company recreational activities, performance in terms of company goals, job satisfaction, cohesiveness, paternalism, and lifetime commitment. This poses two questions which we can test, does company residence in fact increase each of these six phenomena, and if so, is this actually a result of company housing, independent of other variables that influence each of the aspects of behavior and attitudes? Consider the six dependent variables first in Electric Factory.

Participation in company recreational activities has a zero-order r of $-.29$ with residence: employees who live in company housing are more likely than those who live in private residences to participate frequently. This fits the paternalism model, but Table D.22 shows that when other variables are held constant, the best predictors of participation are education (beta = .29), sex (.26), seniority ($-.17$), and performance (.16). Participation was regressed on these four independent variables and residence. When education, sex, performance, and seniority are held constant, the influence of residence on participation is reduced to a beta of $-.10$. The multiple R between participation and education, sex, performance, and seniority is .46; adding residence as a fifth predictor does not increase the multiple R at all. Thus, the greater participation of employees who live in company housing is more a result of their high education, low seniority, high performance, and of their being men, than of their residence as such. In this sense, the paternalism model is disconfirmed.

Performance[10] varies virtually independently of residence ($r = -.11$). As shown in Table D.30 performance is much more a result of sex, organizational status, and knowledge of procedures (r = .54, .55, and .44, respectively) than of residence. In a multiple regression of performance on sex, organizational status, knowledge of procedures, and residence—not presented here—we found that the beta weights are .29 for sex, .26 for status, and .20 for knowledge of procedures; when these three variables are held constant, the beta for residence is .07. The multiple R between performance and the first three variables is .61; adding residence does not enable us to explain any more of the variance. Employees have high performance because they are male, have higher status in the company, and profess to know the procedures better, not because they live in company housing.

This second hypothesis of the model, then, is also disconfirmed, though in fairness to the model two points should be added. First, the model also recognizes that sex and status are important determinants of performance, and second, the model does not explicitly hypothesize that when sex and status are held constant, performance would still be higher among employees who live in company housing. The value of our analysis is to specify multivariate relationships that are vague and only implicit in the paternalism model.

Job satisfaction is somewhat higher among employees who live in company housing than among those who live in private residences ($r = -.17$). Although this fits the model, organizational theory suggests

[10] See Chapter 11 and Appendix B for details on the index of performance.

that job satisfaction is more directly a result of status in the company, which includes the kind of job one actually performs. The data we presented above (Table D.3) bear this out: the best predictors of job satisfaction are status (beta = .35), employee cohesiveness (.18), perceived promotion chances (.16), and sex (.14). When these four independent variables are held constant, the influence of residence on job satisfaction is reduced to nil (beta = .01). These results demonstrate that whatever influence living in company housing has on job satisfaction is in fact a result of status, cohesiveness, promotion chances, and sex. Having high job satisfaction is a result of having higher status, cohesiveness, and perceived promotion chances, and being male, and the relationship between residence and job satisfaction is essentially spurious. In this sense, our evidence again does not support the model.

Employee cohesiveness, as we saw in Table D.16, is virtually unrelated to residence, both in terms of the zero-order r and the beta weight. Insofar as we can explain cohesiveness at all, it appears to be due to such variables as job satisfaction, education, the informational level of the section, and lifetime commitment; there is no evidence that living in company housing makes for greater employee cohesiveness, which again disconfirms the model.

Paternalism presents a similar situation. We saw in Table D.18 that paternalisitic preferences are so widespread in Electric Factory that they vary relatively independently of all our hypothesized variables; clearly we find no indication that living in company housing makes one more favorable to paternalism, either in a direct way or as a result of interaction with other independent variables. Another expected relationship in the model fails to be confirmed.

Lifetime commitment will be shown (see Table D.28) to vary independently of residence (r = .00). When stronger predictors of lifetime commitment—job satisfaction, employee cohesiveness, and job autonomy—are held constant, lifetime commitment is seen to continue to vary independently of residence. The experience of living in company housing, then, does nothing to increase employees' lifetime commitment to the firm, either directly or in interaction with other independent variables. The sixth, and final, implication of the model is, like the other five, disconfirmed.

The same tests of the paternalism model in Shipbuilding Factory consistently disconfirm it. The zero-order correlations indicate that job satisfaction, cohesiveness, paternalism, participation, lifetime commitment, and performance all vary virtually independently of residence (r = .01, −.04, .02, −.05, .10, and .09, respectively). Nor can it be said that the effects of residence are suppressed, and become manifest only in the interaction between residence and other independent variables that influence each of these dependent variables. We regressed each of the latter—satisfaction, cohesiveness, paternalism, participation, lifetime commitment, and performance—on (1) residence and (2) the best predictors of each dependent variable, i.e., the second run variables in Tables D.4, D.17, D.19, D.23, D.29, and D.32. In each of the six multiple regressions, when the best predictor variables were held constant, residence consistently failed to exert any independent influence on the dependent variable. After the best predictors have explained as much of the variance as they can, adding residence as an independent variable fails to explain any of the remaining unexplained variance.

PARTICIPATION

In Electric Factory, the zero-order correlations shown in Table D.22 indicate that participation is moderately positively related to education ($r = .34$), perceived promotion chances (.32), sex (.30), and the informational level of the section (.29); personnel who live in company housing participate more than those who live at home (−.29). Participation is at least weakly positively related to distance of community of origin from Electric Factory, job satisfaction, performance, values, age, size of community of origin, and rank. That is, employees from larger and more distant communities of origin, older males, and those with higher rank, job satisfaction, performance, and work values are somewhat more likely than their opposite numbers to participate frequently. Participation varies virtually independently of the other variables.

When participation is regressed on all 17 independent variables simultaneously, the multiple R is .51. The explained variance in participation is due somwhat more to seniority (the influence of which is seen only when other variables are controlled: beta weight = −.25), sex (.20), education (.18), performance (.16), rank (.14), age (.13), number of previous jobs (−.13), and number of dependents (−.12) than to other variables, none of which explains more than a tiny portion of the variance. When variables such as seniority, sex, and education are held constant, the influence of perceived promotion chances, residence, values, informational level of the section, distance migrated, and job satisfaction is virtually zero.

Table D.23 shows for Shipbuilding Factory zero-order correlations that indicate participation varies positively, but weakly, with perceived promotion chances (.25), education (.19), cohesiveness (.14), performance (.13), and distance of community of origin from the factory (.11). It varies negatively, and again weakly, with age (−.18), number of previous jobs (−.17), number of dependents (−.15), and seniority (−.13). Participation varies independently of rank and section, job satisfaction and values, residence and paternalism, and size of community of origin.

TABLE D.22
Multiple Regression of Participation in Company Recreational Activities in Electric Factory

Independent Variable	Standardized Partial Regression Coefficient		Zero-Order Correlation[a]
	First Run	*Second Run*	
84. Seniority	−.25**	−.17**	.06
3. Sex	.20**	.26**	.30**
13. Education	.18**	.29**	.34**
140. Performance	.16**	.16**	.26**
78. Rank	.14**		.15**
11. Age	.13		.17**
148. Number of previous jobs	−.13**		.02
7. Number of dependents	−.12*		.07
141. Employee cohesiveness	.09**		.04
109. Perceived promotion chances	.08*		.32**
2. Residence[b]	−.07*		−.29**
123. Size of community of origin	.07*		.17**
144. Paternalism	.06*		.07
31. Values	.04		.21**
24. Informational level of section	.03		.29**
139. Job satisfaction	.03		.22**
122. Distance of community of origin from Electric Factory	−.01		.23**
Multiple R	.51	.46	
R^2	.261	.207	
N	740	976	

* Significant at the .05 level.
** Significant at the .01 level.
[a] For zero-order correlations between independent variables, see Appendix C.
[b] Dichotomous dummy variable: whether employee lives in company housing or private residence.

The first multiple regression run in Table D.23 enables us to assess the separate and joint effects of all 16 independent variables on participation. When other variables are controlled, perceived promotion chances, distance and size of community of origin, cohesiveness, performance, and education each exert a slight independent positive influence on participation. When these variables are held constant, the remaining 10 variables exert no significant independent influence on participation. The multiple R between participation and all 16 independent variables is .42.

When participation is regressed on the above six best predictors we find that once perceived promotion chances, cohesiveness, education, and distance of community of origin from Osaka have been taken into account—in that order—there is a multiple R of .32, and adding size of community of origin and performance only increases the multiple R to .34.

TABLE D.23
Multiple Regression of Participation in Company Recreational Activities in Shipbuilding Factory

	Independent Variable	Standardized Partial Regression Coefficient		Zero-Order Correlation[a]
		First Run	Second Run	
73.	Perceived promotion chances	.22**	.18**	.25**
104.	Distance of community of origin from Shipbuilding Factory	.19**	.11*	.11*
127.	Employee cohesiveness	.15**	.17**	.14*
125.	Performance	.13*		.13*
105.	Size of community of origin	.12*		−.03
95.	Education	.11*	.14**	.19**
101.	Number of dependents	−.09		−.15**
102.	Residence[b]	.08		−.02
129.	Paternalism	.08		.06
93.	Age	−.07		−.18**
88.	Section	.05		.03
124.	Job satisfaction	−.05		.02
94.	Seniority	−.05		−.13*
8.	Values	.04		.08
91.	Rank	−.04		.08
108.	Number of previous jobs	−.04		−.17**
	Multiple *R*	.42	.32	
	R^2	.179	.101	
	N	321	407	

* Significant at the .05 level.
** Significant at the .01 level.
[a] For zero-order correlations between independent variables, see Appendix C.
[b] Dichotomous dummy variable: whether employee lives in company housing or private residence.

COMPANY CONCERNS

The results of correlating company concerns with other variables for Electric Factory appear in Table D.24. What little variance there is in concerns varies independently of all the variables save rank, with which it is weakly positively correlated. When concerns are regressed on all 12 independent variables simultaneously, the strongest predictors are age (beta = −.23), pay (.15), rank (.13), lifetime commitment (−.12), sex (−.11), and job satisfaction (.10). However, none of these except lifetime commitment accounts for a significant portion of the variance in concerns. These results disconfirm all our hypotheses. Even the one variable with a significant beta, when the other variables are held constant—lifetime commitment—is related to concerns in the opposite direction from that

TABLE D.24
Multiple Regression of Company Concerns in Electric Factory

Independent Variable	Standardized Partial Regression Coefficient		Zero-Order Correlation[a]
	First Run	Second Run	
11. Age	−.23	−.12*	−.02
8. Pay	.15		.03
78. Rank	.13	.15**	.11*
146. Lifetime commitment	−.12*	−.11*	−.07
3. Sex	−.11		−.03
139. Job satisfaction	.10	.10	.06
140. Performance	.08		.05
148. Number of previous jobs	−.07		.02
109. Perceived promotion chances	.04		.06
2. Residence[b]	−.02		.00
84. Seniority	.02		−.01
13. Education	.01		.01
Multiple *R*	.22	.18	
R^2	.047	.032	
N	317	330	

* Significant at the .05 level.
** Significant at the .01 level.
[a] For zero-order correlations between independent variables, see Appendix C.
[b] Dichotomous dummy variable: whether employee lives in company housing or private residence.

predicted. It is employees who have less lifetime commitment who are more likely to express company concerns. All 12 independent variables together account for only five percent of the variance in concerns, and the multiple *R* is only .22.

In the second multiple regression run, pay was dropped because it is redundant with age ($r = .88$), and all variables with beta weights of .10 or less in the first run were also dropped.[11] One can explain concerns—to the extent they can be explained at all—almost as well on the basis of rank, age, lifetime commitment, and job satisfaction, as when all 12 independent variables are run.

Table D.25 shows that in Shipbuilding Factory, three hypotheses are confirmed: expressing company concerns varies significantly and positively with rank ($r = .28$), perceived promotion chances (.19), and lifetime commitment (.17). But concerns vary independently of other status variables, number of previous jobs, job satisfaction, performance, and

[11] Although age (as well as pay) was run in the second multiple regression, after job satisfaction, sex, and lifetime commitment are held constant, adding age increases the multiple *R* only to .20, and adding both age and pay increases it to .23.

TABLE D.25
Multiple Regression of Company Concerns in Shipbuilding Factory

Independent Variable	Standardized Partial Regression Coefficient First Run	Standardized Partial Regression Coefficient Second Run	Zero-Order Correlation[a]
91. Rank	.29**	.26**	.28**
131. Lifetime commitment	.21*	.11	.17*
108. Number of previous jobs	.14	.08	.11
124. Job satisfaction	−.12		.06
93. Age	−.10		.07
73. Perceived promotion chances	.08		.19**
125. Performance	−.07		.11
110. Pay	.04		.13
94. Seniority	−.04		.09
102. Residence[b]	.03		.00
95. Education	−.00		.02
Multiple R	.36	.32	
R^2	.126	.103	
N	178	190	

* Significant at the .05 level.
** Significant at the .01 level.
[a] For zero-order correlations between independent variables, see Appendix C.
[b] Dichotomous dummy variable: whether employee lives in company housing or private residence.

residence. When the concerns variable is regressed on all 11 independent variables simultaneously, the two best predictors are rank and lifetime commitment. When these and other variables are controlled, perceived promotion chances no longer accounts for a significant amount of the variance in concerns. The conditions under which one is more likely to express company concerns are having higher rank in the firm, higher lifetime commitment, and having had previous interfirm mobility. The concerns variable has a multiple R of .36 with all 11 independent variables, but can be predicted almost as well on the basis of three of these variables (.32). Of these three best predictors in the second multiple regression run, only rank has a significant beta weight. As in the case of other social integration variables, little of the variance in the concerns variable can be explained on the basis of our independent variables.

PERCEIVED RELATIVE ADVANTAGES OF FIRMS

Table D.26's zero-order correlations show that Electric Factory employees who perceive their company as offering unique advantages, relative to other firms, tend to have high cohesiveness with fellow employees, to be lower in status and perceived promotion chances, to be from nearby

TABLE D.26
Multiple Regression of Perceived Relative Advantages of Firms in Electric Factory

Independent Variable	Standardized Partial Regression Coefficient		Zero-Order Correlation[a]
	First Run	Second Run	
3. Sex	−.36**	−.32**	−.27**
139. Job satisfaction	.25**	.24**	.06
11. Age	.22**		−.09*
138. Organizational status	−.20*		−.15**
122. Distance of community of origin from Electric Factory	−.12**	−.15**	−.21**
141. Employee cohesiveness	.10**	.12**	.18**
2. Residence[b]	.09**		.23**
143. Participation	.07		−.06
148. Number of previous jobs	.04		.05
123. Size of community of origin	.03		−.09*
109. Perceived promotion chances	−.01		−.14**
Multiple R	.43	.38	
R^2	.183	.145	
N	785	849	

* Significant at the .05 level.
** Significant at the .01 level.
[a] For zero-order correlations between independent variables, see Appendix C.
[b] Dichotomous dummy variable: whether employee lives in company housing or private residence.

rather than from more remote parts of Japan, and to live in company housing. Men are more likely than women to say "many other firms have better conditions," but women are more likely to give "don't know" responses. The fact that those who live in private rather than company housing are more likely to see Electric Company as offering unique advantages suggests once more that company housing is not what makes the company's advantages appear unique. Younger employees and those from smaller communities of origin are somewhat more likely than their opposite numbers to think Electric Factory offers unique advantages.

When we shift from zero-order correlations to a regression of perceived relative advantages of firms on 11 independent variables, the best predictors are seen to be sex (beta weight = −.36), job satisfaction (.25), age (.22), and organizational status (−.20). On the basis of the first multiple regression run, then, we find that the best predictors of perceiving Electric Factory as offering unique advantages are: being older, more satisfied with one's job, lower in organizational status, from a nearby community of origin, having somewhat more cohesiveness with fellow employees, and living in a private residence. When these variables are held constant, the other independent variables—participation, number of previous jobs, size of community of origin, and perceived promotion chances—explain little of the remaining variance in perceived relative advantages of firms.

Perceived relative advantages of firms was regressed on the non-redundant independent variables that had a significant beta weight. A stepwise multiple regression program showed sex to be the most important variable. When job satisfaction is added as the second predictor, the multiple *R* is .33; with distance of community of origin added as a third predictor, the multiple *R* rises to .36; the fourth best predictor, employee cohesiveness, brings the multiple *R* to .38.[12] The negative relationship between sex and perceived advantages does not mean women are more likely than men to think Electric Company offers advantages few other firms can match; we have seen that what this actually means is that women are more likely to give the intermediate scale response, "I don't know if many other firms have as good conditions as Electric Company."[13]

Turning to Shipbuilding Factory, Table D.27's zero-order correlations show that employees with higher job satisfaction and cohesiveness are

TABLE D.27
Multiple Regression of Perceived Relative Advantages of Firms in Shipbuilding Factory

Independent Variable	Standardized Partial Regression Coefficient		Zero-Order Correlation[a]
	First Run	Second Run	
124. Job satisfaction	.22**	.23**	.22**
104. Distance of community of origin from Shipbuilding Factory	−.18**	−.07	−.09
105. Size of community of origin	−.10		.03
128. Participation	.06		.07
123. Organizational status	−.06		.02
127. Employee cohesiveness	.05		.13*
73. Perceived promotion chances	.05		.09
102. Residence[b]	.02		.02
93. Age	−.01		.03
108. Number of previous jobs	−.00		−.02
Multiple *R*	.28	.24	
R^2	.078	.055	
N	352	562	

* Significant at the .05 level.
** Significant at the .01 level.
[a] For zero-order correlations between independent variables, see Appendix C.
[b] Dichotomous dummy variable: whether employee lives in company housing or private residence.

somewhat more likely than those low in satisfaction and cohesiveness to think that only a few other firms offer as good conditions as Shipbuilding Company (r = .22 and .13, respectively). Perception of relative advantages of firms varies independently of all the other eight independent variables.

In the first multiple regression run, only job satisfaction and distance of community of origin account for any significant amount of the variance in perceived advantages of firms (beta = .22 and −.18, respectively). Perceiving only few other firms as offering the same advantages as Shipbuilding is somewhat more likely among personnel with higher job satisfaction and from nearby communities of origin. The multiple R between perceived advantages and all 10 independent variables is only .28, and the two best predictors can explain almost as much of the variance alone (R = .24). When perceived advantages is regressed on only job satisfaction and distance of community of origin (Table D.27), job satisfaction is clearly the better predictor (betas = .23 and −.07, respectively). But our variables explain virtually none of the variance in perceived relative advantages of firms.

LIFETIME COMMITMENT

Table D.28 presents the multiple regression of lifetime commitment on 12 hypothesized independent variables for Electric Factory males. In the first run, job satisfaction emerges as the most important predictor of commitment (beta weight = .19) when the other 11 variables are held constant. The only other variables significantly related to commitment, when other variables are controlled, are: employee cohesiveness (beta = .16) and perceived job autonomy (−.11). Even these are at best weak predictors of commitment. Commitment varies independently of age, organizational status, residence, perceived relative advantages of firms, perceived promotion chances, number of previous jobs, participation, paternalism, and size of community of origin. All 12 variables together explain only $(.33)^2$ or 11 percent of the variance in lifetime commitment.

Lifetime commitment was regressed on only those independent variables that were significantly related to it in the first run. The results of a stepwise multiple regression analysis appear as the second run in Table D.28. Job satisfaction was picked in the first step, with a beta weight

[12] Adding the fifth variable, residence, and the sixth, status, only increased the multiple R with perceived advantages of firms to .39.

[13] When the same independent variables are run against perceived relative advantages of firms for only male employees, rather than for all Electric Factory employees, the pattern of relationships is very similar. The best predictors continue to be cohesiveness, distance, and job satisfaction, and they have a multiple R of .41 with perceived advantages.

TABLE D.28
Multiple Regression of Lifetime Commitment Among Males in Electric Factory

	Independent Variable	*Standardized Partial Regression Coefficient*		*Zero-Order Correlation*[a]
		First Run	*Second Run*	
139.	Job satisfaction	.19**	.24**	.25**
141.	Employee cohesiveness	.16**	.16**	.23**
35.	Perceived job autonomy	−.11*	−.09*	−.07
11.	Age	.07		.10
72.	Perceived relative advantages of firms	.05		.16**
109.	Perceived promotion chances	.04		.12*
143.	Participation	.03		.07
148.	Number of previous jobs	−.03		.06
2.	Residence[b]	−.02		.00
123.	Size of community of origin	−.02		−.01
138.	Organizational status	.01		.11*
144.	Paternalism	.00		.05
	Multiple R	.33	.32	
	R^2	.111	.104	
	N	319	366	

* Significant at the .05 level.
** Significant at the .01 level.
[a] For zero-order correlations between *independent variables*, see Appendix C.
[b] Dichotomous dummy variable: whether employee lives in company housing or private residence.

of .24. The second variable added, employee cohesiveness, increased the the multiple R to .31. When the third variable, perceived job autonomy, is added, the R increases to .32. Thus, one can predict lifetime commitment virtually as well on the basis of job satisfaction, cohesiveness, and job autonomy as on the basis of all 12 of the original independent variables (multiple R = .32 vs. .33, respectively).

In testing the same hypotheses in Shipbuilding Factory, the following were confirmed (see Table D.29): lifetime commitment varies directly with job satisfaction (r = .50), age (.35), perceived relative advantages of firms (.31), cohesiveness (.28), organizational status (.20), perceived promotion chances (.16), perceived job autonomy (.13), recruitment channel (−.11), and paternalism (.10). The following hypotheses were disconfirmed: lifetime commitment varies inversely with size of community of origin (r = .04), inversely with number of previous jobs (−.02); lifetime commitment varies directly with participation in company recreational activities (.02), and directly with residence (.06). Finally, we had hypothesized that lifetime commitment varies with recruitment channel: it is highest when one was recruited on the basis of school recommendations, next highest when recruited on the basis of knowing someone in the company, and lowest when recruited in terms of impersonal media (vocational guidance centers or newspaper job ads). This hypothesis is also disconfirmed ($r = -.11$); the opposite is true, i.e., employees recruited via impersonal media have somewhat higher commitment than those recruited via personal contacts or school recommendations.

To examine the separate and joint effects of all 13 independent variables upon lifetime commitment, consider now the first multiple regression run in Table D.29. The best predictors of lifetime commitment, when other variables are held constant, are job satisfaction, age, perceived relative advantages of firms, and cohesiveness. When these variables are controlled, status, perceived promotion chances, job autonomy, paternalism, and recruitment channel cease to account for any significant portion of the variance in lifetime commitment. This indicates that the characteristics that most conduce to a higher level of lifetime commitment are being more

TABLE D.29
Multiple Regression of Lifetime Commitment Among Males in Shipbuilding Factory

	Independent Variable	*Standardized Partial Regression Coefficient*		*Zero-Order Correlation*
		First Run	*Second Run*	
124.	Job satisfaction	.32**	.34**	.50**
93.	Age	.26**	.24**	.35**
49.	Perceived relative advantages of firms	.18**	.18**	.31**
127.	Employee cohesiveness	.16**	.19**	.28**
108.	Number of previous jobs	−.09		−.02
73.	Perceived promotion chances	.07		.16**
129.	Paternalism	.05		.10*
102.	Residence[a]	.04		.06
123.	Organizational status	−.03		.20**
98.	Recruitment channel	−.02		.11*
14.	Perceived job autonomy	.02		.13*
105.	Size of community of origin	−.01		.04
128.	Participation	−.01		.02
	Multiple R	.61	.62	
	R^2	.367	.381	
	N	290	404	

* Significant at the .05 level.
** Significant at the .01 level.
[a] Dichotomous dummy variable: whether employee lives in company housing or private residence.

satisfied with one's job, older, more cohesive with one's fellow employees, and perceiving that few other firms offer as good conditions as Shipbuilding Company. Status has no independent influence on lifetime commitment once these variables have been held constant. Nor is this confounded by multicollinearity: all the relevant zero-order *r*'s among independent variables in this particular multiple regression analysis are only moderate or weak.

The multiple *R* of lifetime commitment and these 13 independent variables is .61, but the four best predictors—job satisfaction, age, cohesiveness, and perceived advantages of firms—alone predict as well ($R = .62$).[14]

PERFORMANCE

The zero-order correlations for Electric Factory in Table D.30 reveal that performance is most strongly related to status ($r = .55$), sex (.54),

TABLE D.30
Multiple Regression of Performance in Electric Factory

Independent Variable	Standardized Partial Regression Coefficient		Zero-Order Correlation[a]
	First Run	*Second Run*	
11. Age	.22**		.53**
3. Sex	.19**	.28**	.54**
39. Knowledge of procedures	.15**	.21**	.44**
139. Job satisfaction	.09*		.41**
31. Values	.07*		.35**
143. Participation	.06*		.23**
144. Paternalism	.04		.06
138. Organizational status	.04	.24**	.55**
123. Size of community of origin	−.04		.10**
109. Perceived promotion chances	.03		.41**
141. Employee cohesiveness	.02		.07
146. Lifetime commitment	.02		.17**
148. Number of previous jobs	−.01		.19**
Multiple *R*	.63	.61	
R^2	.394	.376	
N	667	1,005	

* Significant at the .05 level.
** Significant at the .01 level.
[a] For zero-order correlations between independent variables, see Appendix C.

[14] This analysis is based on the seven-item lifetime commitment index. The only difference when we use the four-item index (for comparability with Electric Factory) is that the four best predictors—job satisfaction, age, perceived relative advantages of firms, and cohesiveness—have a multiple *R* of .53 with lifetime commitment, and explain 28 percent of the variance, instead of 38 percent.

age (.53), knowledge of procedures (.44), perceived promotion chances (.41), and job satisfaction (.41). It is less closely related to work values (.35), participation (.23), number of previous jobs (.19), lifetime commitment (.17), and size of community of origin (.10). Performance varies independently of cohesiveness and paternalism.

Thus, performance is more strongly related to the sex, age, and status of the employee, to variables directly pertinent to his job (knowledge of procedures, job satisfaction), to work values, and to perceived promotion chances, than to either social origins and previous employment or to the degree of integration of the employee into the company, as measured by participation, cohesiveness, paternalism, and lifetime commitment.

In the first multiple regression run, shown in Table D.30, performance was regressed on all 13 independent variables simultaneously. No single independent variable emerges as a major predictor of performance. Instead, age and sex explain about the same amount of the variance in performance, followed by knowledge of procedures. When these variables are held constant, none of the other variables contributes more than marginally to the variance in performance.

It appears, then, that of the variables that had stronger zero-order correlations with performance, a few continue to exert at least some influence on performance when the others are controlled (age, sex, knowledge of procedures), while others have little influence independent of age, sex, and knowledge of procedures (job satisfaction, values, perceived promotion chances, and status). However, status and age are so highly correlated (.88) that they are measuring virtually the same thing; once either of these variables is entered in the multiple regression, the independent influence of the other is necessarily reduced to near zero. Accordingly, we regressed performance on two sets of the strongest predictor variables, one containing age but not status, the other containing status but not age (Table D.31). The other independent variables in each set were sex and knowledge of procedures.

As expected, the results are highly similar. The choice is therefore rather arbitrary. On theoretical grounds, we shall conclude that in Electric Factory performance is a result primarily of sex, status, and knowledge of procedures, though we note that the higher status employees—those with better performance—are disproportionately male and older.

Turning now to Shipbuilding Factory, the results of testing the same hypotheses appear in Table D.32.[15] The zero-order correlations indicate that performance is most strongly related to organizational status ($r = .37$), knowledge of procedures (.30), lifetime commitment (.29), perceived promotion chances (.27), and age (.26). It is more weakly correlated with job satisfaction, participation, and work values. It varies

[15] Except that sex is dropped and recruitment channel added.

TABLE D.31
Multiple Regression of Performance on Alternative Sets of Independent Variables in Electric Factory

Alternative I		Beta	Alternative II		Beta
3.	Sex	.28**	3.	Sex	.28**
138.	Organizational status	.24**	11.	Age	.25**
39.	Knowledge of procedures	.21**	39.	Knowledge of procedures	.19**
	Multiple *R*	.61			.62
	R^2	.376			.382
	N	1,005			1,006

** Significant at the .01 level.

TABLE D.32
Multiple Regression of Performance in Shipbuilding Factory

	Independent Variable	Standardized Partial Regression Coefficient		Zero-Order Correlations[a]
		First Run	Second Run	
123.	Organizational status	.23**	.25**	.37**
131.	Lifetime commitment	.18**	.23**	.29**
18.	Knowledge of procedures	.16**	.20**	.30**
73.	Perceived promotion chances	.09		.27**
98.	Recruitment channel	.08		.04
128.	Participation	.08		.12*
105.	Size of community of origin	−.07		−.01
108.	Number of previous jobs	.07		−.00
93.	Age	.07		.26**
127.	Employee cohesiveness	−.04		.03
8.	Values	.03		.15**
129.	Paternalism	.01		.08
124.	Job satisfaction	.00		.20**
	Multiple *R*	.50	.48	
	R^2	.249	.226	
	N	289	420	

* Significant at the .05 level.
** Significant at the .01 level.
[a] For zero-order correlations between independent variables, see Appendix C.

independently of the other variables. Employees with better performance, then, have higher status in the company, know procedures more fully, have greater lifetime commitment, think their promotion chances are relatively good, and are older. They are also somewhat more likely to have job satisfaction, to participate in company recreational activities, and to have "work is my whole life" values.

When performance is regressed on these 13 independent variables simultaneously, when other variables are controlled, the best predictors are status, lifetime commitment, and knowledge of procedures. Age, perceived promotion chances, and job satisfaction are seen to have little or no effect on performance independent of the effect of status, lifetime commitment, and knowledge of procedures. When Shipbuilding employees have high performance, it is due more to their higher status, better knowledge of procedures, and greater lifetime commitment to the company, than to their being older, more satisfied with their jobs, or more optimistic about their promotion chances. The remaining independent variables—involving the social integration of the employee into the firm, values, social origins, and recruitment—continue to explain virtually none of the variance in performance. There is a multiple R of .50 between all 13 independent variables and performance, but the second multiple regression run reveals that one can predict performance almost as well ($R = .48$) on the basis of only the three best predictors—status, lifetime commitment, and knowledge of procedures.

Figure D.1 states a human relations causal theory of performance: job satisfaction is a result of employee cohesiveness and participation in company recreational activities; all three of these variables in turn exert a positive influence on performance. The beta weights (path coefficients)

FIGURE D.1

A Human Relations Causal Theory of Performance[a]

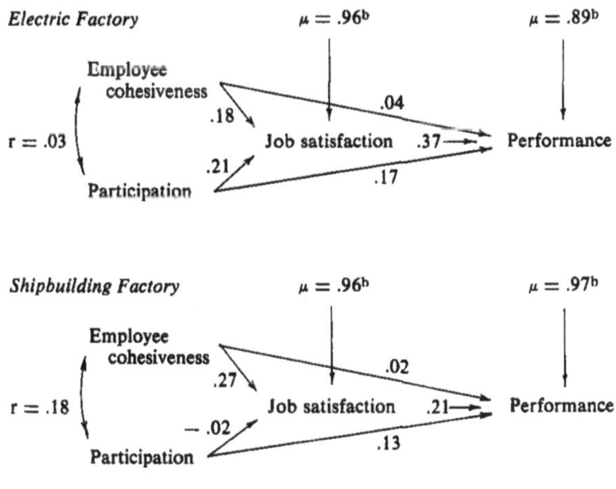

[a] The relationship between cohesiveness and participation is the Pearson product-moment r; all other relationships are path coefficients, which are the same as *beta* weights.

[b] $\mu =$ the estimate of the residual, i.e., the influence of all other variables than those stated, on the dependent variable. The assumption is that the total variation of each dependent variable (job satisfaction, performance) is completely determined by a linear combination of the specified independent variables and the residual, or error variables, μ. Since the independent variables and the residual variables are also assumed to vary independently of each other, $\mu = \sqrt{1-R^2}$ (Land 1969).

indicate that among Electric Factory employees, job satisfaction is about equally influenced by cohesiveness and participation, but performance is more strongly related to job satisfaction than to particpation, and more strongly related to participation than to employee cohesiveness. The multiple *R* between performance and these three human relations variables, shown in Table D.33, is .45, but job satisfaction alone predicts performance virtually as well ($r = .42$) as do all three variables together.

In Shipbuilding Factory, the beta weights show that job satisfaction is more a result of employee cohesiveness than of participation. Performance, as in Electric Factory, is more strongly related to job satisfaction than to participation, and more strongly related to participation than to cohesiveness. In Table D.34 we see that job satisfaction, participation, and

TABLE D.33
Multiple Regression of Performance on Human Relations Variables in Electric Factory

Independent Variable	Standardized Partial Regression Coefficient	Zero-Order Correlation[a]
139. Job satisfaction	.37**	.42**
143. Participation	.17**	.25**
141. Employee cohesiveness	.04	.12**
Multiple *R*	.45	
R^2	.204	
N	823	

** Significant at the .01 level.
[a] For zero-order correlations between independent variables, see Appendix C.

TABLE D.34
Multiple Regression of Performance on Human Relations Variables in Shipbuilding Factory

Independent Variable	Standardized Partial Regression Coefficient: First Run	Standardized Partial Regression Coefficient: Second Run	Zero-Order Correlation[a]
124. Job satisfaction	.21**	.20**	.22**
128. Participation	.13**	.14**	.14**
127. Employee cohesiveness	.02		.10*
Multiple *R*	.26	.26	
R^2	.066	.065	
N	427	482	

* Significant at the .05 level.
** Significant at the .01 level.
[a] For zero-order correlations between independent variables, see Appendix C.

cohesiveness all have significant, but weak zero-order correlations with performance. When all three of these central human relations variables act simultaneously on performance, they can explain only seven percent of the predictable variance.

RECRUITMENT

There are two variables involved in the recruitment of personnel into a firm that we seek to explain: previous interfirm mobility (number of firms for which one worked prior to entering one's present firm) and recruitment channel.

Tables D.35 and D.36 deal with the first of these. When the number of firms in which one has previously worked is regressed on the variables that precede them in time—age, sex, father's occupation, size of community of origin, and education—the best predictor among Electric Factory employees as a whole, and among Electric Factory males (table not shown), is age. Among Shipbuilding employees, age and education are the best predictors.

The second dependent variable in the recruitment process, recruitment channel, was coded from low to high, as follows: (1) recruited via government employment center or newspaper want ad; (2) knew someone in Shipbuilding Company; and (3) school recommendation. This ordering is based on the hiring preferences of Japanese employers. They believe employees recruited directly from schools will have the longest commitment to the firm and therefore prefer to hire on this basis. Those hired on the basis of knowing someone in the firm are somewhat less desirable

TABLE D.35
Multiple Regression of Previous Interfirm Mobility
in Electric Factory

	Independent Variable	*Standardized Partial Regression Coefficient*	*Zero-Order Correlation*[a]
11.	Age	.44**	.39**
3.	Sex	−.10**	.21**
13.	Education	−.03	.03
123.	Size of community of origin	.02	.07*
124.	Father's occupation	.01	.04
	Multiple R	.39	
	R^2	.150	
	N	947	

* Significant at the .05 level.
** Significant at the .01 level.
[a] For zero-order correlations between independent variables, see Appendix C.

TABLE D.36
Multiple Regression of Previous Interfirm Mobility in Shipbuilding Factory

Independent Variable	Standardized Partial Regression Coefficient		Zero-Order Correlation[a]
	First Run	*Second Run*	
93. Age	.26**	.25**	.32**
95. Education	−.23**	−.24**	−.31**
105. Size of community of origin	−.08*		−.07
106. Father's occupation	.00		−.04
Multiple *R*	.40	.39	
R^2	.159	.153	
N	474	493	

* Significant at the .05 level.
** Significant at the .01 level.
[a] For zero-order correlations between independent variables, see Appendix C.

because they have had previous work experience and may not stay in the firm as long. Those recruited on the basis of impersonal employment centers and newspaper want ads are the least desirable because they are believed to be the least likely to manifest lifetime commitment to the new firm. This variable was regressed on four variables that precede it in time as well as on the year recruited (see Table D.37). We find that when number of previous jobs and education are held constant, the social origins variables and the year in which one was recruited have no independent effect

TABLE D.37
Multiple Regression of Recruitment Channels in Shipbuilding Factory

Independent Variable	Standardized Partial Regression Coefficient		Zero-Order Correlation[a]
	First Run	*Second Run*	
108. Number of previous jobs	−.39**	−.39**	−.46**
95. Education	.22**	.22**	.34**
106. Father's occupation	−.04		.01
94. Cohort (year recruited into Shipbuilding Company)	−.02		−.11*
105. Size of community of origin	.01		.04
Multiple *R*	.51	.50	
R^2	.256	.253	
N	468	501	

* Significant at the .05 level.
** Significant at the .01 level.
[a] For zero-order correlations between independent variables, see Appendix C.

on recruitment channel. The multiple *R* between recruitment channel and all five independent variables is .51; however, the number of previous jobs variable alone has a correlation of −.46 with recruitment channel.

THE CONSEQUENCES OF PREVIOUS INTERFIRM MOBILITY

An inference one can draw from the lifetime commitment model is that an individual who voluntarily changes firms (and thereby violates lifetime commitment norms and values) will encounter negative sanctions in the new firm. As a result of these, we might expect at least five consequences. Compared to employees who have always worked for his new firm (and who thereby conform to lifetime commitment norms and values), and with other variables held constant, he will have (1) lower pay, (2) lower rank, (3) poorer perceived promotion chances, (4) lower cohesiveness with fellow employees, and (5) less frequent participation in company recreational activities.

Table D.38 shows the results of our tests of these inferences. The number of previous jobs variable has the lowest zero-order correlations seven out of 10 times. When other independent variables known to influence each of the five dependent variables are held constant, the number of previous

TABLE D.38
Relative Predictive Power of Selected Independent Variables[a] in Electric and Shipbuilding Factories

Electric Factory			*Shipbuilding Factory*		
1. Multiple Regression of *Pay*			1. Multiple Regression of *Pay*		
Independent Variable	*Beta*	*r*	*Independent Variable*	*Beta*	*r*
Seniority	.47**	.82**	Age	.69**	.72**
Number of dependents	.41**	.82**	Rank	.32**	.46**
Number previous jobs	.15**	.46**	Number previous jobs	−.09**	.11**
Multiple *R*	.89		Multiple *R*	.79	
R^2	.787		R^2	.628	
N	1,033		N	490	
2. Multiple Regression of *Rank*			2. Multiple Regression of *Rank*		
Independent Variable	*Beta*	*r*	*Independent Variable*	*Beta*	*r*
Number previous jobs	.47**	.56**	Performance	.29**	.38**
Seniority	.27**	.38**	Seniority	.25**	.28**
Education	.22**	.20**	Education	.24**	.17**
			Number previous jobs	−.01	−.04
Multiple *R*	.64		Multiple *R*	.47	
R^2	.411		R^2	.219	
N	1,033		N	460	

TABLE D.38 (*Cont.*)

Electric Factory

3. Multiple Regression of *Perceived Promotion Chances*

Independent Variable	*Beta*	*r*
Sex	.48**	.57**
Rank	.25**	.27**
Education	.21**	.27**
Number previous jobs	.03	.26**

Multiple *R*	.66
R^2	.441
N	981

4. Multiple Regression of *Employee Cohesiveness*

Independent Variable	*Beta*	*r*
Education	−.28**	−.26**
Lifetime commitment	.15**	.19**
Job satisfaction	.15**	.18**
Number previous jobs	.07*	.12**

Multiple *R*	.37
R^2	.137
N	761

5. Multiple Regression of *Participation in Company Recreational Activities*

Independent Variable	*Beta*	*r*
Education	.29**	.36**
Sex	.26**	.27**
Seniority	−.17**	.04
Performance	.16**	.24**
Number previous jobs	−.01	.04

Multiple *R*	.46
R^2	.207
N	976

Shipbuilding Factory

3. Multiple Regression of *Perceived Promotion Chances*

Independent Variable	*Beta*	*r*
Rank	.50**	.51**
Recruitment channel	.17**	.24**
Section	−.10**	−.06
Number previous jobs	−.08	−.17**

Multiple *R*	.56
R^2	.318
N	468

4. Multiple Regression of *Employee Cohesiveness*

Independent Variable	*Beta*	*r*
Lifetime commitment	.22**	.32**
Job satisfaction	.17**	.30**
Participation	.17**	.19**
Number previous jobs	−.03	−.06

Multiple *R*	.40
R^2	.156
N	339

5. Multiple Regression of *Participation in Company Recreational Activities*

Independent Variable	*Beta*	*r*
Employee cohesiveness	.18**	.19**
Perceived promotion chances	.17**	.23**
Education	.15**	.20**
Distance of factory from community of origin	.12**	.12**
Number previous jobs	−.06	−.15**

Multiple *R*	.36
R^2	.128
N	374

* Significant at the .05 level
** Significant at the .01 level
[a] For zero-order correlations between independent variables, see Appendix C.

jobs variable has the lowest beta weight nine out of 10 times, among the set of independent variables. We conclude that there is no evidence that changing firms brings about these negative sanctions and negative consequences for the individual in his new firm. Therefore the lifetime commitment model is again disconfirmed.

UNIFORMITY AND VARIATION AMONG FACTORIES

We next consider which aspects of factory social organization are most variable, and which most constant, across Electric, Shipbuilding, and Sake factories (see Table D.39). The findings are discussed on pp. 332–33.

TABLE D.39
Factory Comparisons of Male Employees on Aspects of Social Organization, Norms, Values, and Attitudes

	C/C_{max}
Sake > Shipbuilding: proportion of employees who live in company rather than private housing.	.83**
Sake > Electric: proportion of employees who live in company rather than private housing.	.74**
Electric > Shipbuilding: prefer job classification-based wage system because "it is rational that wages fit the job"; Shipbuilding > Electric: prefer seniority-based wage system because "it provides security to the worker as he gets older."	.53**
Shipbuilding > Electric: number of firms worked for prior to coming to present firm. Shipbuilding males' greater mobility also holds when age is controlled (see Table 12.1).	.51**
Sake > Electric: number of firms worked for prior to coming to present firm.	.46**
Electric > Sake: spontaneous expression of company (rather than personal, private) concerns.	.44**
Electric > Sake: seniority in the firm.	.44**
Shipbuilding > Sake: seniority in the firm.	.35**
Shipbuilding > Electric: job satisfaction index scores.	.34**
Electric > Shipbuilding: spontaneous expression of company (rather than personal, private) concerns.	.34**
Electric > Shipbuilding: prefer job classification wage system (Shipbuilding more likely to prefer seniority wage system).	.33**
Shipbuilding > Electric: perceived promotion chances poor because of age (too old); Electric > Shipbuilding: poor promotion chances because of lack of ability.	.32**
Sake > Shipbuilding: social relationships among employees are warm, family-like.	.31**
Sake > Electric: social relationships among employees are warm, family-like.	.31**
Electric > Shipbuilding: proportion of personnel who are in supervisory-managerial (rather than rank-and-file) positions.	.31**
Electric > Shipbuilding: participation in company recreational activities.	.30**
Electric > Shipbuilding: perceived promotion chances.	.28**
Electric > Shipbuilding: primacy of work values over family and pleasure values.	.27**
Electric > Shipbuilding: knowledge of procedures (bureaucratization).	.24**
Sake > Shipbuilding: perceived teamwork (harmony) among workers.	.21**
Electric > Shipbuilding: seniority in the firm.	.20**

TABLE D.39 (*Cont.*)

	C/C_{max}
Sake > Shipbuilding: agreement with ideology of lifetime commitment (if employee stays in one firm, the firm and he will both prosper).	.18**
Electric > Shipbuilding: proportion of employees who live in company rather than private housing.	.16**
Electric > Shipbuilding: perceived work load too heavy.	.15**
Electric > Shipbuilding: employee cohesiveness index scores.	.15**
Shipbuilding > Electric: lifetime commitment index scores.	.13**
Sake > Shipbuilding: favorable attitudes toward five aspects of one's job.	.06**
Electric > Shipbuilding: rank *education* higher than *number of dependents* as a factor that should determine pay.	.05**
For the following variables there are no significant differences between males in two (or all three) firms:	
I. Electric and Shipbuilding male employees are not significantly different with regard to:	
Rewards	
Legitimacy of the company's promotion system. (Ratio of reasons that attribute poor promotion chances to one's own limitations versus obstacles imposed by the company. The former include inadequate education, too old, not the right kind of personality, lack of ability, and poor record or poor reputation. The latter include no chance to get training within industry and "I'm not known to any of the factory managers.")	.03
Social Integration	
Preferring a paternalistic (rather than a functionally specific) relationship with superiors.	.02
Conflict	
Perception by supervisors and managers that their subordinates receive their communications with suspicion.	.27
Frequency with which managers and supervisors say they receive conflicting orders.	.17
Frequency of conflicts over authority and responsibility of the job, among managers and supervisors.	.09
II. Electric and Sake male employees are not significantly different with regard to:	
Perceived teamwork (harmony) among employees.	.12
Preferring a paternalistic (rather than a functionally specific) relationship with superiors.	.02
Favorable attitudes toward five aspects of one's job.	.01
III. Shipbuilding and Sake male employees are not significantly different with regard to:	
Spontaneous expression of company (rather than personal, private) concerns.	.18
Number of firms worked for prior to coming to present firm.	.10
Preferring a paternalistic (rather than a functionally specific) relationship with superiors.	.02

* Significant at the .05 level.
** Significant at the .01 level.

The next way of considering uniformity and variation across factories is to ask whether the causes of these aspects of factory social organization are similar or different in Electric and Shipbuilding. For each of 14 social organization variables and the performance variable we want to identify those independent variables that produce significant amounts of change in the dependent variable, when other hypothesized independent variables are held constant. Unlike the earlier multiple regression analyses in Appendix D, however, we shall now shift from the standardized partial regression coefficient (beta weight) to the unstandardized partial regression coefficient (*b*).[16]

The independent variables included in the causal diagrams to follow are not necessarily the same as those in the second run of the corresponding multiple regression tables (Tables D.1 to D.37). Where the variables differ, the source of the difference is twofold. First, the multiple regression tables already presented referred to all employees in Electric and Shipbuilding factories, whereas the causal diagrams in Figure D.2 refer only to males in the two factories, and somewhat different causal processes operate with regard to each sex. Second, the *first* run in many of the preceding multiple regression tables contained some multicollinearity, the most important example of which was the .86 correlation between status and age in Electric Factory. Multicollinearity was essentially removed in the *second* multiple regression run in Tables D.1 to D.37, but the possibility remains that had the multicollinearity been eliminated in the *first* run of each table—for example, by dropping either age or status—a different set of "best predictors" might have emerged and therefore a different set of independent variables might have emerged in the second run. To examine this possibility systematically, in preparing Figure D.2 we re-ran the first and second multiple regression runs for each dependent variable, this time removing the multicollinearity even from the first run. Thus, the

[16] Standardized coefficients are appropriate when one wants to compare different beta weights within a specific population, or to generalize to a specific population–either Electric Factory or Shipbuilding Factory, which is what we have done thus far in this study. But we now want to compare these two populations directly to see if the underlying causal processes are basically similar for given dependent variables. Unstandardized regression coefficients are more appropriate for this task of describing causal laws, stated in hypothetical "if-then" form. These *b*'s measure the concrete contribution of an independent variable in an absolute sense. Each *b* represents the average amount of change in a dependent variable that can be associated with a one-unit change in one of the independent variables, with the remaining independent variables held constant. Thus, in Shipbuilding Factory, when we say the promotion chances variable has a *b* of 0.77 in relation to the dependent variable, job satisfaction, we mean: a one-unit increase in perceived promotion chances causes a 0.77 unit increase in job satisfaction; i.e., if employees see their promotion chances as changing from "not too good" to "fairly good"—a one-unit change—their job satisfaction can be expected to increase by 0.77 points on the 20-point scale with which we measure satisfaction. By comparison, among males in Electric Factory, a similar one-unit increase in perceived promotion chances, on the average, produces a 0.56 unit increase in the level of job satisfaction. See Blalock 1972:450, 452; Schoenberg 1972.

causal structures diagrammed in Figure D.2 have been specified through a process which is less "contaminated" by the initial multicollinearity among independent variables, than in the case of Tables D.1 through D.37.

Note that to be considered a common cause, a variable does not have to have the same relative importance with regard to a given dependent variable in both factories; it only has to be one of the statistically significant independent variables in both factories. Had we used relative importance as a second criterion of common causal processes, the effect would have been to reduce further the ratio of common to unique causes.

Figure D.2 has at least three uses. First, it summarizes our main quantitative findings for Electric and Shipbuilding factories (at least for male employees). Second, it provides evidence with which to question the lifetime commitment model's uniformity of social organization hypothesis, since it indicates that even in these two factories, the causes of social organization are more often different than the same. Third, it provides the most concise answer to the question: which causal variables are the most robust under varying conditions? These variables, from among the larger set with which we began this study, are the ones most likely to prove important in future research that seeks to explain similarities and differences among Japanese (and other nations') firms or other organizations. In the end we have a more parsimonious explanation of aspects of social organization than the one with which we began, deriving as it did from the paternalism–lifetime commitment model of the Japanese factory.

FIGURE D.2

Causal Models

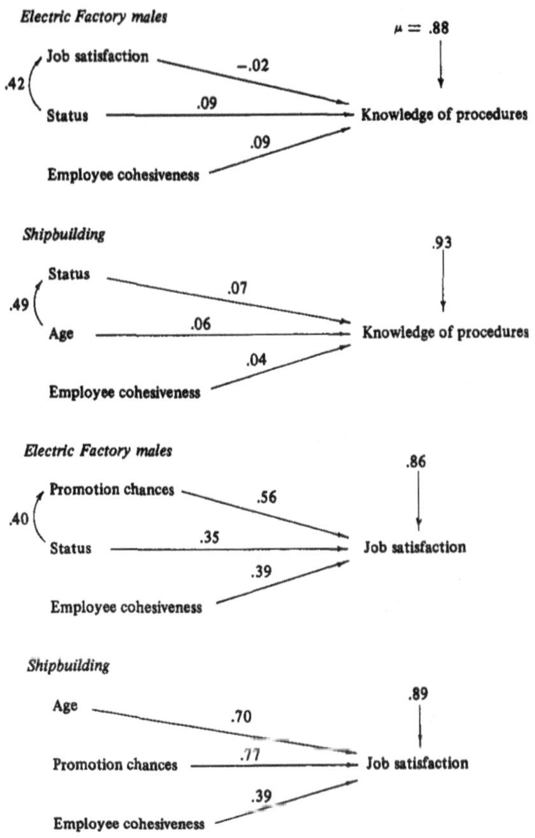

Note: Straight lines indicate unstandardized partial regression coefficients, *b*; curved lines zero-order correlations between independent variables. Only independent variables whose *b* is statistically significant (one-tailed t-value) at the .01 level are included in these diagrams. The .01 level is used because, given our large *N*'s, even weak relationships tend to be significant at the .05 level. Only *r*'s for relationships between independent variables that are .30 or higher are shown in these diagrams. A broken line indicates no theoretical causal link; the correlation is probably spurious, a result of some third variable. Thus, it is not that the number of one's dependents causes one's rank, but that both are a function of a third variable, probably age. See footnote b, Fig. D.1, p. 407, for the meaning of μ.

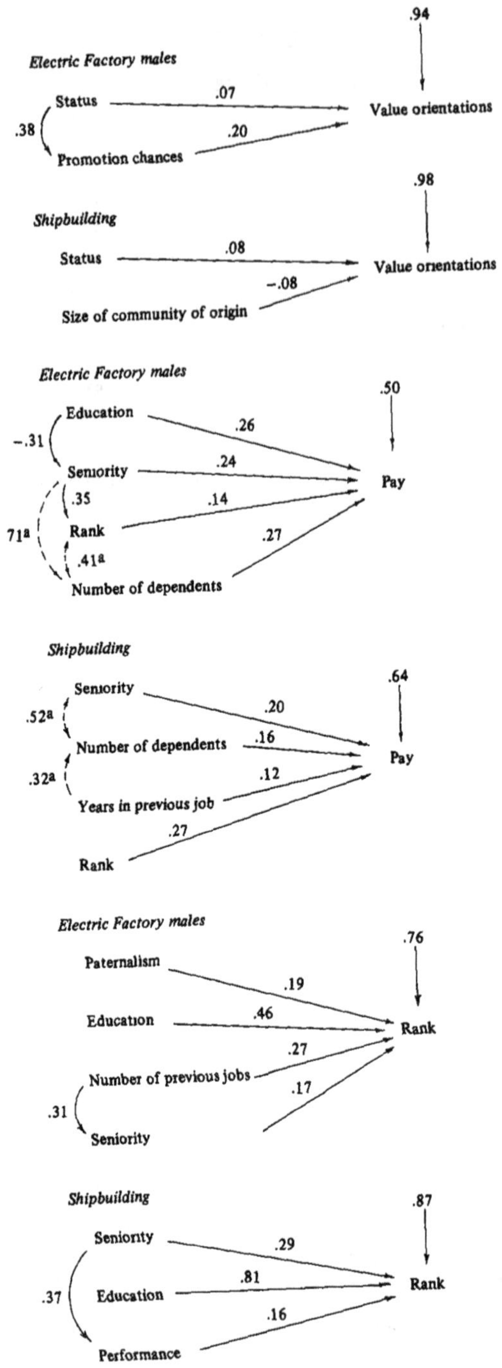

Electric Factory males
.94
Status
.07
Value orientations
.38
.20
Promotion chances
Shipbuilding
.98
Status
.08
Value orientations
−.08
Size of community of origin
Electric Factory males
.50
Education
.26
−.31
Seniority
.24
Pay
.35
.14
71[a]
Rank
.27
.41[a]
Number of dependents
Shipbuilding
.64
Seniority
.20
.52[a]
Number of dependents
.16
Pay
.32[a]
.12
Years in previous job
.27
Rank
Electric Factory males
.76
Paternalism
.19
Education
.46
Rank
.27
Number of previous jobs
.17
.31
Seniority
Shipbuilding
.87
Seniority
.29
.81
.37
Education
Rank
.16
Performance

Electric Factory males

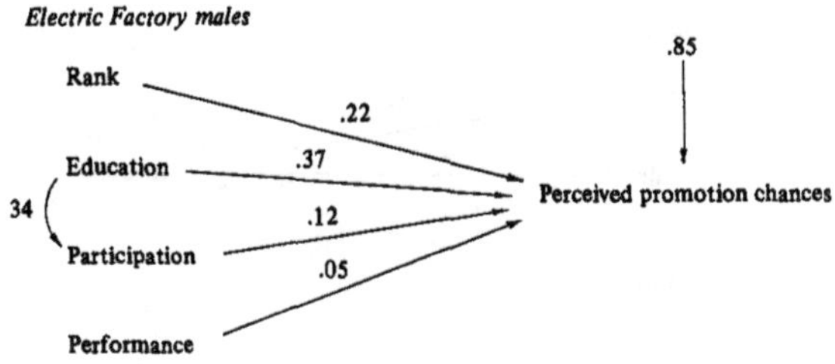

Shipbuilding

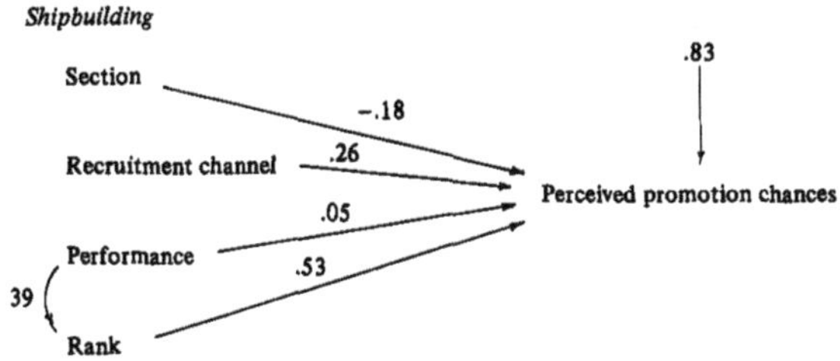

Electric Factory males

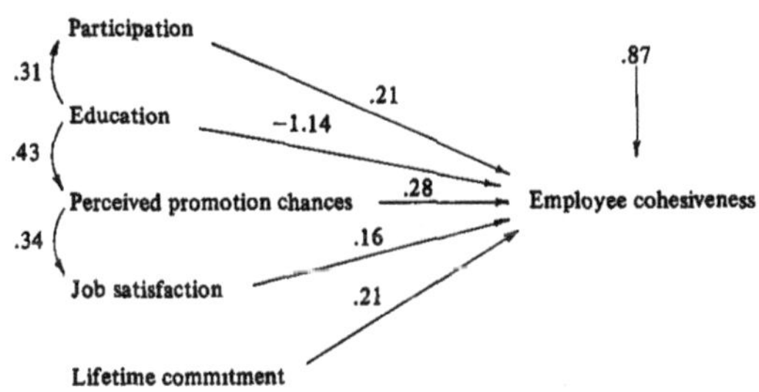

Shipbuilding

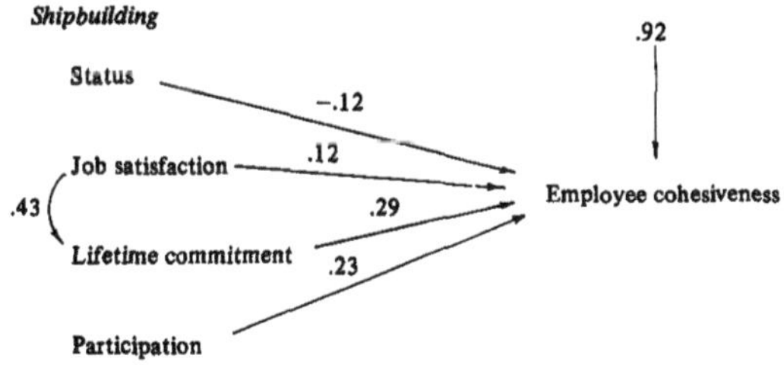

Electric Factory males

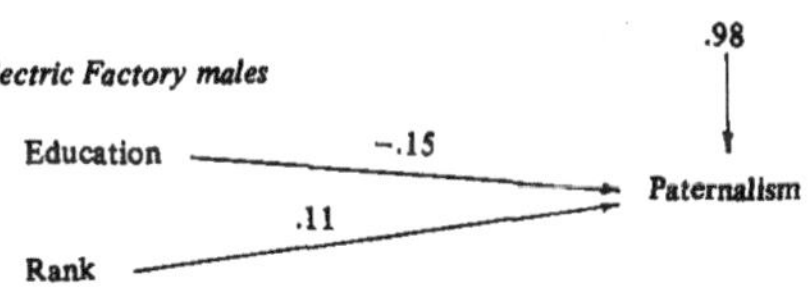

Shipbuilding

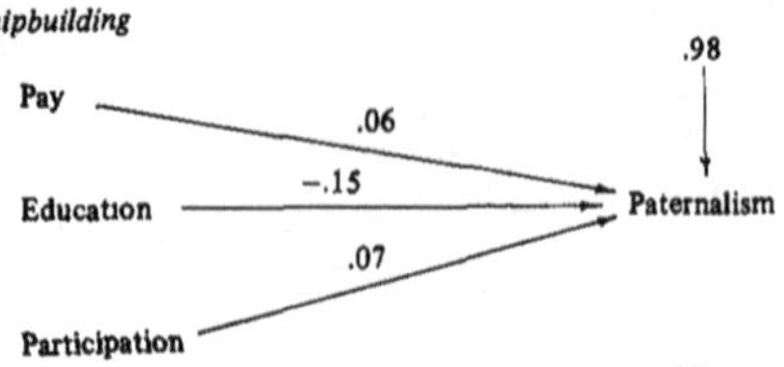

Electric Factory males

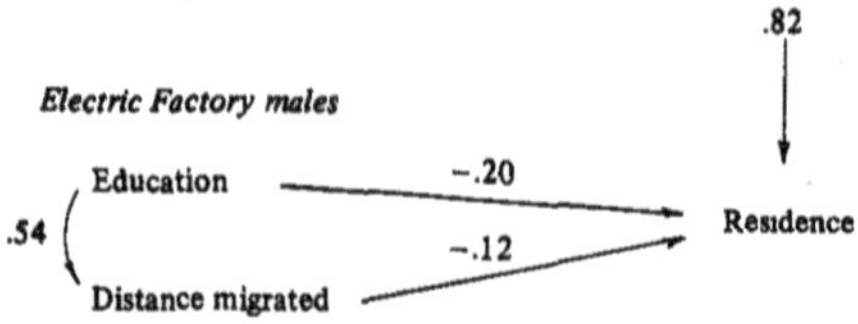

Shipbuilding

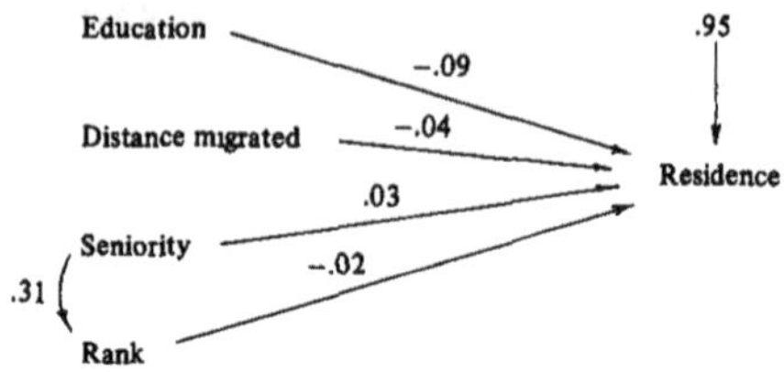

Electric Factory males

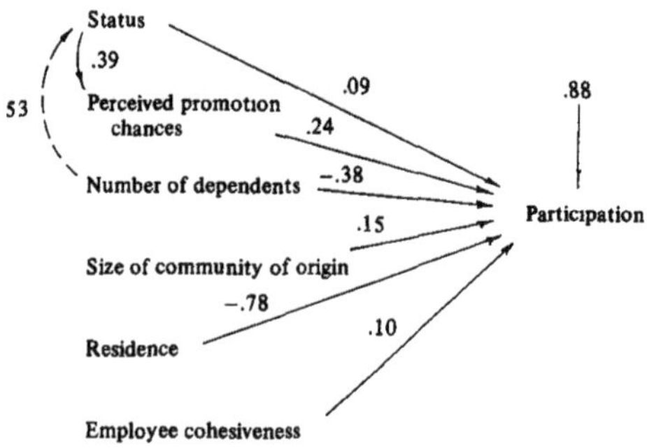

Shipbuilding

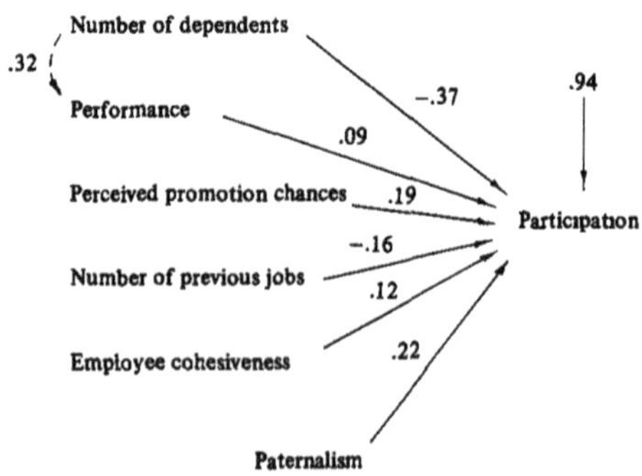

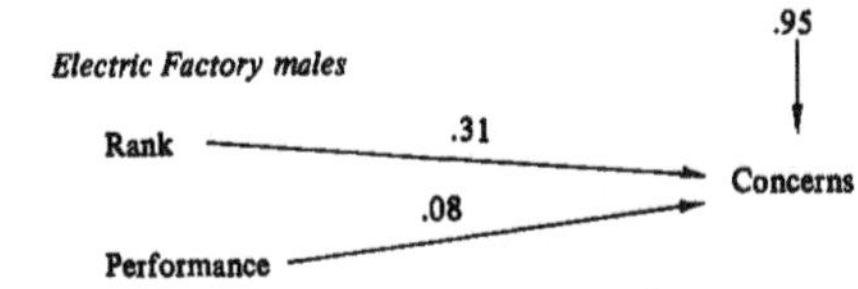
Electric Factory males
.95
Rank
.31
Concerns
.08
Performance

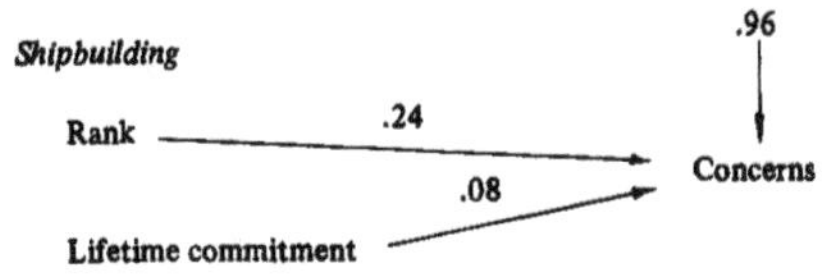
Shipbuilding
.96
Rank
.24
Concerns
.08
Lifetime commitment

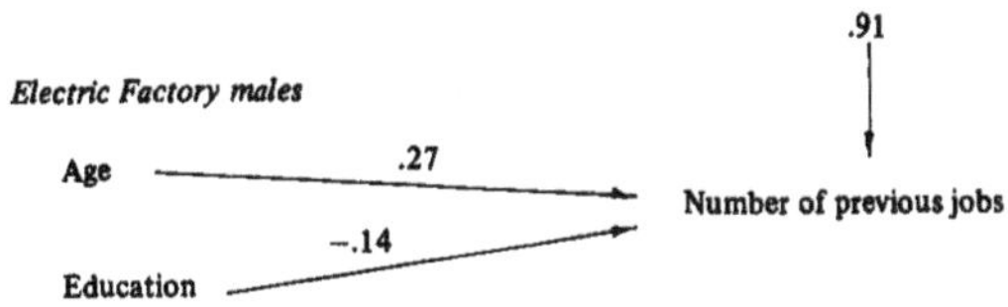
.91
Electric Factory males
Age
.27
Number of previous jobs
−.14
Education

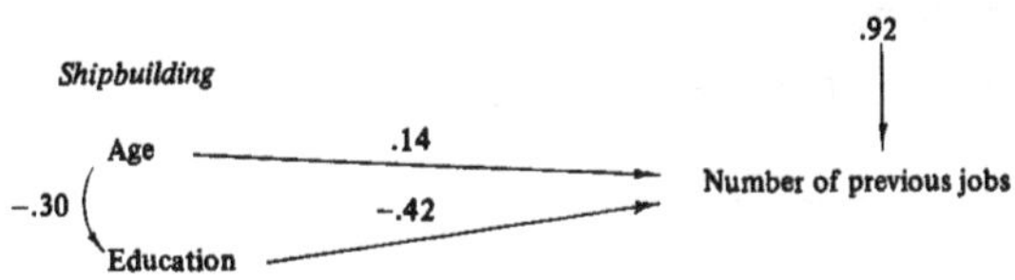
.92
Shipbuilding
Age
.14
−.30
Number of previous jobs
−.42
Education

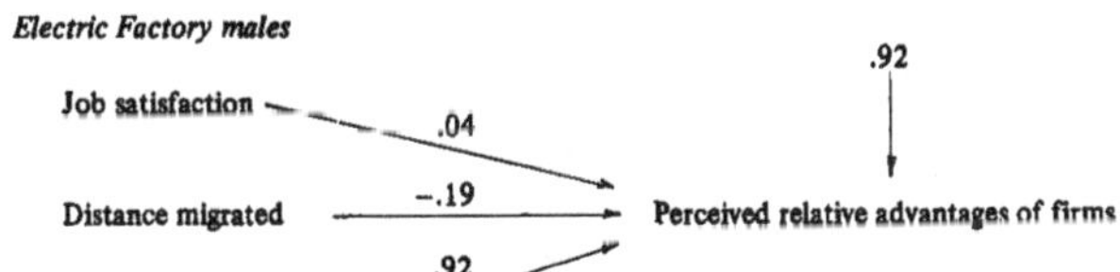
Electric Factory males
.92
Job satisfaction
.04
Distance migrated
−.19
Perceived relative advantages of firms
.92
Employee cohesiveness

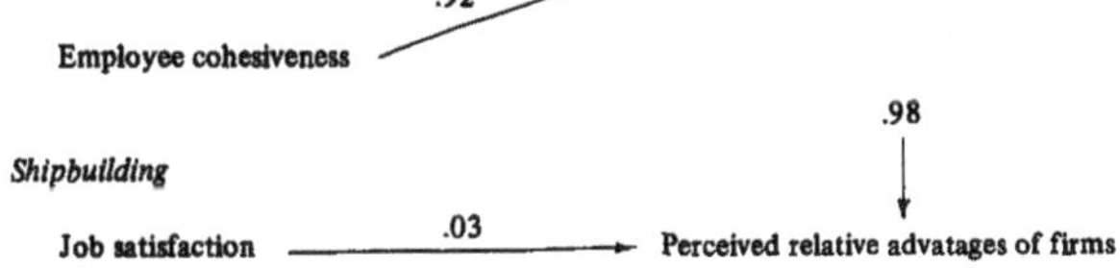
.98
Shipbuilding
Job satisfaction
.03
Perceived relative advatages of firms

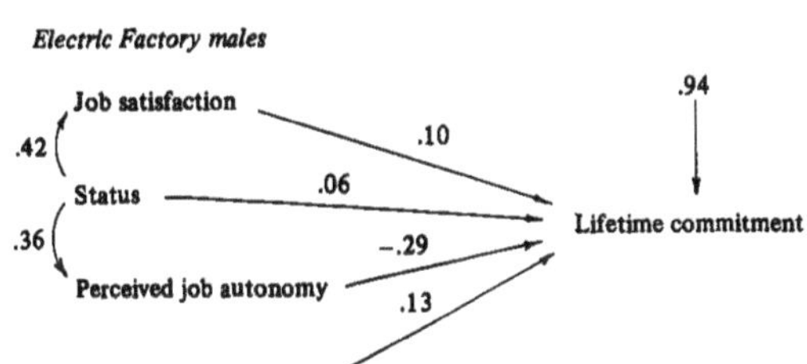
Electric Factory males
.94
Job satisfaction
.42
.10
Status
.06
Lifetime commitment
.36
−.29
Perceived job autonomy
.13
Employee cohesiveness

Shipbuilding

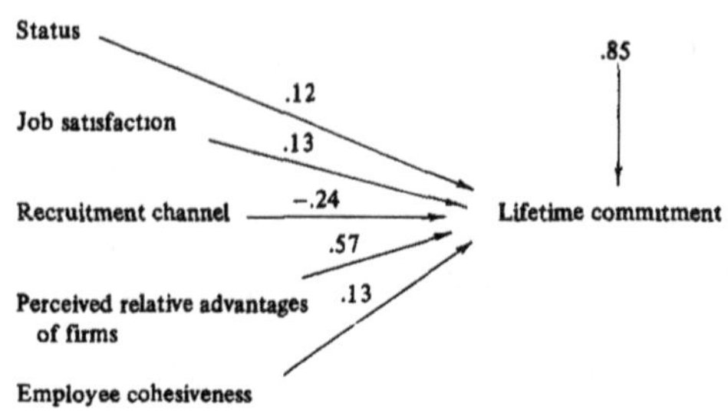

Electric Factory males

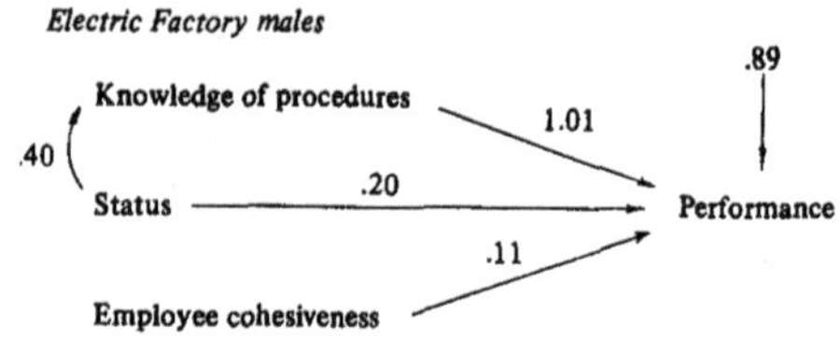

Shipbuilding

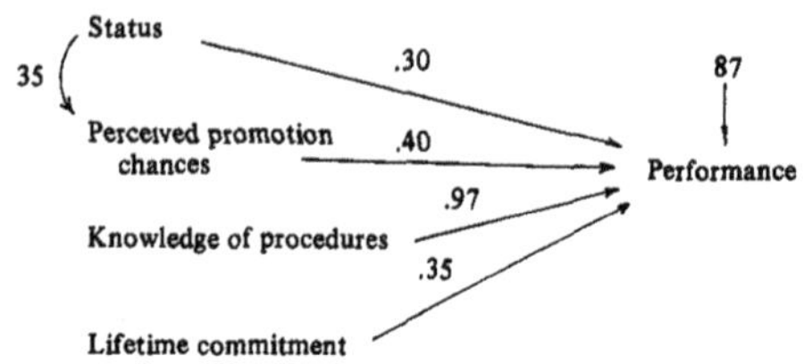

BIBLIOGRAPHY

Abegglen, James C., 1958. *The Japanese Factory*. Glencoe: The Free Press.

———, 1969. "Organizational Change." In *The Japanese Employee*, edited by Robert J. Ballon, pp. 99–119. Rutland, Vermont: Charles E. Tuttle.

Aiken, Michael and Hage, Jerald, 1968. "Organizational Interdependence and Intra-Organizational Structure." *American Sociological Review* 33 (December): 912–30.

Allen, G. C., 1965. *Japan's Industrial Expansion*. London: Oxford University Press.

Asahi Evening News, 1970a, "Young Salary Earners Serious in their Work but Also in Play." March 27.

———, 1970b. "Japan's Youth Today." March 28.

Asahi Shimbun, 1970. "Atarashii sarariman," March 26, section 2:21–23.

Azumi, Koya and Hage, Jerald, 1972. *Organizational Systems*. Lexington: D. C. Heath.

Azumi, Koya and McMillan, Charles J., 1973. "Worker Sentiments in the Japanese Factory: Organizational Determinants." Unpublished paper prepared for seminar on "Japan by 1980," Yale University, February 15.

Bales, Robert F., 1953. "The Equilibrium Problem in Small Groups." In T. Parsons, R. F. Bales, and E. A. Shils, *Working Papers in The Theory of Action*, Ch. 4. Glencoe: The Free Press.

Bellah, Robert N., 1957. *Tokugawa Religion*. Glencoe: The Free Press.

Bennett, John W., 1967. "Japanese Economic Growth: Background For Social Change." In *Aspects of Social Change in Modern Japan*, edited by R. P. Dore, pp. 411–53. Princeton: Princeton University Press.

———, 1968. "Paternalism." In *International Encyclopedia of the Social Sciences*, edited by David Sills, vol. 2, pp. 472–77. New York: Macmillan.

Bennett, John W. and Ishino, I., 1963. *Paternalism in the Japanese Economy*. University of Minnesota Press.

Bennis, Warren G. and Slater, Philip E., 1968. *The Temporary Society*. New York; Harper & Row.

Berrien, F. Kenneth, 1966. "Japanese Values and the Democratic Process." *The Journal of Social Psychology* 168:129–38.

Blalock, Hubert M., Jr., 1972. *Social Statistics*. New York: McGraw-Hill.

Blau, Peter M. and Schoenherr, Richard A., 1971. *The Structure of Organizations*. New York: Basic Books.

Blauner, Robert, 1964. *Alienation and Freedom*. Chicago: The University of Chicago Press.

———, 1966. "Work Satisfaction and Industrial Trends in Modern Society." In *Class, Status and Power: Social Stratification in Comparative Perspective*, edited by Reinhard Bendix and S. M. Lipset, pp. 473–87. New York: The Free Press.

Burns, T. and Stalker, G. M., 1961. *The Management of Innovation*. London: Tavistock Publications.

Carey, Alex, 1967. "The Hawthorne Studies: A Radical Criticism." *American Sociological Review* 32 June:403–16.

Casal, U. A., 1940. "Some Notes on the *Sakazuki* and on the Role of *Sake* Drinking in Japan." *The Transactions of the Asiatic Society of Japan*, Second Series, XIX:1–186.

Caudill, Willian and Scarr, Harry A., 1962. "Japanese Value Orientations and Cultural Change." *Ethnology* 1.1 (January):53–91.

Chao, Kang, 1968. "Labor Institutions in Japan and Her Economic Growth." *Journal of Asian Studies* 28 (November):5–17.

Cole, Robert E., 1967. *Japanese Blue Collar Workers: A Participant-Observation Study*. Ph.D. dissertation, Department of Sociology, University of Illinois.

———, 1971a. *Japanese Blue Collar: The Changing Tradition*. Berkeley: University of California Press.

———, 1971b. "The Theory of Institutionalization: Permanent Employment and Tradition in Japan." *Economic Development and Cultural Change* 20 (October): 47–70.

———, 1972. "Permanent Employment in Japan: Facts and Fantasies." *Industrial and Labor Relations Review* 26 (October):615–30.

Coser, Lewis, 1956. *The Functions of Social Conflict*. Glencoe: The Free Press.

Crozier, Michel, 1964. *The Bureaucratic Phenomenon*. Chicago: University of Chicago Press.

Dahrendorf, Ralf, 1959. *Class and Class Conflict in Industrial Society*. Stanford: Stanford University Press.

Dator, James Allen, 1966. "The 'Protestant Ethic' in Japan." *The Journal of Developing Areas* 1 (October):23–40.

Davis, Louis E., 1971. "Job Satisfaction Research: The Post-Industrial View." *Industrial Relations* 10 (May):176–93.

Dean, Lois R., 1954. "Union Activity and Dual Loyalty." *Industrial and Labor Relations Review* 7 (July):526–36.

Dore, Ronald P., 1973. *British Factory: Japanese Factory*. Berkeley: University of California Press.

Downing, George D., 1967. "The Changing Structure of a Great Corporation." In *The Emerging American Society* I, edited by William L. Warner, pp. 158–240. New Haven: Yale University Press.

Dubin, R.; Homans, G. C.; Mann, F. C.; and Miller, D. C.; 1965. *Leadership and Productivity*. San Francisco: Chandler.

Ducommun, Rosalie, 1969. *Measuring Labor Productivity, Studies and Reports*, New Series, No. 75. Geneva: International Labour Office.

Dunlop, John, 1958. *Industrial Relations Systems*. New York: Holt.

Eisenstadt, S. N., ed., 1968. *The Protestant Ethic and Modernization: A Comparative View*. New York: Basic Books.

Etzioni, Amitai, 1964. *Modern Organizations*. Englewood Cliffs: Prentice-Hall.

Ford, Robert N., 1969. *Motivation Through the Work Itself.* New York: American Management Association.

Form, William H., 1969. "Occupational and Social Integration of Automobile Workers in Four Countries: A Comparative Study." *International Journal of Comparative Sociology* 10 (March and June):95–116.

Fortune, 1963–1973. "The Fortune Directory: The 200 (or 300) Largest Foreign Industrial Companies." Annual August issue.

Fullan, Michael, 1970. "Industrial Technology and Worker Integration in the Organization." *American Sociological Review* 35 (December):1028–39.

Georgopoulous, Basil S., and Tannenbaum, Arnold S., 1957. "A Study of Organizational Effectiveness." *American Sociological Review* 22 (October):534–40.

Goldthorpe, John H., et al., 1968. *The Affluent Worker: Industrial Attitudes and Behaviour.* Cambridge: Cambridge University Press.

Hall, Richard H.; Haas, J. Eugene; and Johnson, Norman J., 1967. "Organizational Size, Complexity and Formalization." *American Sociological Review* 32 (December):903–12.

Hazama, Hiroshi, 1960. "The Logic and the Process of Growth on the 'Familistic Management in Japan' (*keiei kazokushugi*)." *Shakaigaku hyoron* 11 (July):2–18, 137–38.

———, 1964. *Nihon rōmu kanrishi kenkyū.* Tokyo: Diamond.

Herbst, P. G., 1967. "Generalised Behavior Theory: Behaviour Under Conditions of Outcome Uncertainty." *Acta Sociologica* 10.3–4:201–57.

Heydebrand, Wolf, V., ed., 1973. *Comparative Organizations: The Results of Empirical Research.* Englewood Cliffs: Prentice-Hall.

Imai, Masaaki, 1968. *Tenshoku no susume.* Tokyo: Saimaru.

Institute of Statistical Mathematics, Research Committee of Japanese National Character, 1961. *Nihonjin no kokuminsei.* Tokyo: Shiseido.

Japan Institute of Labor, 1964a, 1964b, 1966, 1968, 1971a, 1971b, 1971c, 1972. *Japan Labor Bulletin.* Tokyo: Japan Institute of Labor: New Series. 3.8 (August): 3–4; 3.11 (November):2; 5.9 (September):7–8; 7.8 (August):3; 10.3 (March):2; 10.5 (May):2; 10.9 (September):2–3; 11.9 (September):7.

Japan, Office of the Prime Minister, 1971. "Shakai ishiki." *Getsukan yoron-chōsa.* (November):2–49.

Japan Report, 1972. "Attitudes of Japanese Workers Place Emphasis on 'Life Congenial to their Taste,'" 18.14 (July 16):6–7.

Kahn, Robert L. et al., 1964. *Organizational Stress.* New York: Wiley.

Kaneko, Yoshio, 1970. "Employment and Wages." *The Developing Economies.* VIII (December):445–74.

Karsh, Bernard and Cole, Robert E., 1968. "Industrialization and the Convergence Hypothesis: Some Aspects of Contemporary Japan." *Journal of Social Issues* 24:45–64.

Kawasaki, Ichiro, 1969. *Japan Unmasked.* Tokyo: Tuttle.

Kluckhohn, Florence and Strodtbeck, Fred L., 1961. *Variations in Value Orientations.* New York: Harper and Row.

Kobayashi, Yoshihiro, 1973. "Zosen." In *Nihon no sangyō soshiki 1*, edited by Hisao Kumagai, pp. 183–234. Tokyo: Chuokoronsha.

Komiya, Ryutaro et al., 1973. "Kateidenki." In *Nihon no sangyō soshiki 1*, edited by Hisao Kumagai, pp. 15–82. Tokyo: Chuokoronsha.

Koshiro, Kazutoshi, 1969. "Industrial Relations in the Japanese Electrical Machinery Industry I, II." *Japan Labor Bulletin* 8.1 (January):4–8; 8.2 (February): 4–8.

———, 1974. "Wage Hikes and the Tax Burden." *Japan Labor Bulletin* 13.2 (February):4–8.

Kurihara, Kenneth K., 1971. *The Growth-Potential of the Japanese Economy.* Baltimore: Johns Hopkins Press.

Land, Kenneth C., 1969. "Principles of Path Analysis." In *Sociological Methodology 1969*, edited by Edgar Borgatta, pp. 3–37. San Francisco: Jossey-Bass.

Lawler, Edward E., 1971. *Pay and Organizational Effectiveness.* New York: McGraw-Hill.

Levine, Solomon B., 1958. *Industrial Relations in Postwar Japan.* Urbana: University of Illinois Press.

Levy, Marion J., Jr., 1953. "Contrasting Factors in the Modernization of China and Japan." *Economic Development and Cultural Change* 2.3 (October):161–97.

———, 1966. *Modernization and the Structure of Society.* Princeton: Princeton University Press.

Likert, Rensis, 1967. *The Human Organization.* New York: McGraw-Hill.

Lockwood, William, 1954. *The Economic Development of Japan: Growth and Structural Change 1868–1938.* Princeton: Princeton University Press.

The Mainichi Daily News, 1969. "Economic White Paper for Fiscal 1968." July 17:5.

Manganji, Issaku, 1934. "Kinsei shuzōgyō no seisan-kikō." *Rekishigaku kenkyū* 1.6 (April):411–55.

Mannari, Hiroshi; Maki, Masahide; and Seino, Masayoshi, 1967. "Sake tsukuri no rōdō no soshiki: sangyō shakaigakuteki kenkyū." *Kwansei gakuin daigaku shakaigakubu kiyo* 15 (December): 1–32.

Mannari, Hiroshi and Marsh, Robert M., 1970. "Shūshin kōyō no kōzo kino bunseki." *Kwansei gakuin daigaku shakaigakubu kiyo* 21 (November): 67–77.

———, 1971a. "Nihon no sangyō soshiki ni okeru shūshin kōyō sei no saikentō." *Shakaigaku hyoron* 21.4 (March):37–50, 106–07.

———, 1971b. "Nihon no sangyō no teichaku to ido." *Nihon rōdō kyokai zasshi* 13 (November):64–74.

March, James G., ed., 1965. *Handbook of Organizations.* Chicago: Rand McNally.

Marsh, Robert M. and Mannari, Hiroshi, 1971. "Lifetime Commitment in Japan: Roles, Norms and Values." *American Journal of Sociology* 76 (March):795–812.

———, 1972. "A New Look at 'Lifetime Commitment' in Japanese Industry." *Economic Development and Cultural Change* 20 (July):611–30.

———, 1973. "Japanese Workers' Responses to Mechanization and Automation." *Human Organization* 32 (Spring):85–93.

Matsushima, Shizuo, 1951. *Rōdō shakaigaku josetsu.* Tokyo: Fukumura-shoten.

———, 1962. *Rōmukanri no nihonteki tokuishitsu to hensen.* Tokyo: Diamond.

———, 1968, "Labour-Management Relations." In *Comparative Perspectives on Stratification: Mexico, Great Britain, Japan,* edited by Joseph A. Kahl, pp. 223–35. Boston: Little Brown.

Matsushima, Shizuo and Nakano, Taku, 1958. *Nihon shakai yoron*. Tokyo: Tokyo University Press.

Merton, Robert K., 1968. *Social Theory and Social Structure*. New York: The Free Press.

Moore, Wilbert E. and Feldman, Arnold S., 1960. *Labor Commitment and Social Change in Developing Areas*. New York: Social Science Research Council.

Mott, Paul, 1972. *Characteristics of Effective Organizations*. New York: Harper and Row.

Nakamura, Hajime, 1964. *Ways of Thinking of Eastern Peoples: India-China-Tibet-Japan*. Honolulu: East-West Center Press.

Nakano, Takashi, 1964. *Shokadōzokudan no kenkyū*. Tokyo: Miraisha.

Nakayama, Ichiro, 1964. *Industrialization of Japan*. Honolulu: East-West Center Press.

Neuhauser, Duncan, 1971. *The Relationship Between Administrative Activities and Hospital Performance*. Chicago: Center for Health Administration Studies Research Series #28, University of Chicago Press.

Nihon Ginkō Chōsa-Kyoku, 1931. *Nada no seishu*. Tokyo: Nihon Ginkō.

Nihon Jimbun Kagaku-Kai, ed., 1963. *Gijutsukakushin no shakaiteki eikyō*. Tokyo: Todai-shuppankai.

Nihon Keieisha Dantai Renmei, ed., 1971. *Waga kuni rōmukanri no gensei: dai san kai rōmukanri seido chōsa*. Tokyo: Nihon keieisha dantai renmei.

Nihon Keizai Shimbunsha, ed., *Kaisha nenkan*. Tokyo: Nihon keizai shimbunsha, various years.

———, ed., 1964, 1965, 1968, 1969, 1970. *Kaisha sōkan: mijojo kaisha han*. Tokyo: Nihon keizai shimbunsha.

Nippon Kōgyō Ginkō, 1972. "The Shipbuilding Industry." *Quarterly Survey of Japanese Finance and Industry* 24.4 (October–December): 17–35.

Noda, Kazuo, 1963. "Traditionalism in Japanese Management." *Oyo shakaigaku hyoron* 6:127–70.

Nunnally, Jum, 1967. *Psychometric Theory*. New York: McGraw-Hill.

Odaka, Kunio, 1950. "An Iron Workers' Community in Japan: A Study in the Sociology of Industrial Groups." *American Sociological Review* 15 (April):186–95.

———, 1953. "Jūgyōin no keiei to kumiai ni taisuru kizokuishiki no sokutei." In *Sangyō ni okeru ningenkankei no kakaku*, pp. 311–47. Tokyo: Yuhikaku.

———, 1963. "Traditionalism, Democracy in Japanese Industry." *Industrial Relations* 3.1 (October):95–104.

———, ed., 1952. *Imono no machi*. Tokyo: Yuhikaku.

Ōhara Shakai Mondai Kenkyūjo, ed., 1966. *Nihon rōdō nenkan*, v. 36. Tokyo: Rōdō jumpōsha.

Okamoto, Hideaki, 1970. "Manpower Policy at the Enterprise Level, II." *Japan Labor Bulletin* 9.11 (November):7–10.

———, 1971a. "The Growing Middle Class Identification of Industrial Workers." *Japan Labor Bulletin* 10.5 (May):7–10.

———, 1971b. "Work and Leisure in Japan, I, II." *Japan Labor Bulletin* 10.10 (October):5–8; 10.11 (November):4–10.

———, 1972. "Industrialization: Apprenticeship and Manpower Development, I, II, III." *Japan Labor Bulletin* 11.6 (June):4–8; 11.7 (July):5–8; 11.8 (August):5–8.

Okochi, Kazuo, 1972. "Waga kuni ni okeru roshikankei no tokushitsu." In *Roshi kankeiron no shiteki hatsuten*, pp. 177–202. Tokyo: Yuhikaku.

Okochi, Kazuo et al., 1959. *Rōdōkumiai no kōzō to kinō*. Tokyo: Todai Shuppankai.

Okochi, Kazuo; Karsh, Bernard and Levine, Solomon B., eds., 1973. *Workers and Employers in Japan: The Japanese Employment Relations System*. Princeton: Princeton University Press.

Okochi, Kazuo and Sumiya, Mikio, eds., 1955. *Nihon no rōdōshakaikyū*. Tokyo: Tōyōkeizai shimpōsha.

Ono, Tsuneo, 1971. *Wage Problems and Industrial Relations in Japan*. Tokyo: The Japan Institute of Labor.

Oriental Economist, ed., 1969, 1971. *Japan Economic Yearbook 1969, Japan Economic Yearbook 1971*. Tokyo: The Oriental Economist.

Oshikawa, Ichiro; Nakayama, Ichiro; Arisawa, Hiromi; and Isobe, Kiichi, eds., 1962. *Chūshō-kigyō no hatten, dai ni-ji chūshō kigyō kenkyū jū-ichi, kinsei shuzogyō no seiritsu*. Tokyo: Tōyō keizai shimposha.

Ōuchi, Hyoē et al., 1971. *Nihon keizai zusetsu*. Tokyo: Iwanami,

Parnes, Herbert S., 1968. "Labor Force: Markets and Mobility." In *International Encyclopedia of the Social Sciences*, edited by David Sills, vol. 8, pp. 481–87. New York: Macmillan.

Parsons, Talcott, 1956. "Suggestions for a Sociological Approach to the Theory of Organizations I and II." *Administrative Science Quarterly* 1 (June and September):63–85, 225–39.

———, 1964. *Essays in Sociological Theory*. Rev. ed. New York: The Free Press.

———, 1966. *Societies: Evolutionary and Comparative Perspectives*. Englewood Cliffs: Prentice-Hall.

Parsons, Talcott et al., eds., 1961. *Theories of Society*. New York: The Free Press.

Perrow, Charles B., 1970. *Organizational Analysis: A Sociological View*. Belmont, California: Brooks/Cole Publishing Co.

President, 1968, 1970. "The Largest 200 Manufacturing and Mining Companies in the World." Tokyo: Diamond. August issues.

Price, James L., 1968. *Organizational Effectiveness, An Inventory of Propositions*. Homewood, Illinois: Richard D. Irwin.

Richardson, Stephen A., 1956 "Organizational Contrasts on British and American Ships." *Administrative Science Quarterly* 1 (September):168–206.

Rōdō Shō, ed., 1968, 1969. *Rōdō hakusho*. Tokyo: Okura sho.

———, 1965, 1966, 1968. *Rōdō tōkei nempō*. Tokyo: Nihon rōsei kyōkai, v. 17, 1964: v. 18, 1965: v. 20, 1967.

Rōdō Shō rōdō tōkei chōsa bu, 1969. *Rōdō tōkei yoran*. Tokyo: Okura sho.

Rueschemeyer, Dietrich, 1969. "Partielle Modernisierung." In *Theorien des sozialen Wandels*, edited by Wolfgang Zapf, pp. 382–96. Köln: Kiepenheuer and Witsch.

Sakaguchi, Kinichiro, 1964. *Nihon no sake*. Tokyo: Iwanami.

Sato, Morihiro, 1964. "Gijutsukakushin-ka no daikigyō rōdōsha." In *Gijutsukakushin to ningen no mondai* edited by Kunio Odaka. Tokyo: Diamond.

Schoenberg, R., 1972. "Strategies for Meaningful Comparison." In *Sociological Methodology 1972*, edited by Herbert L. Costner, pp. 1–35. San Francisco: Jossey-Bass.

Seashore, Stanley E. and Bowers, David G., 1963. *Changing the Structure and Functioning of an Organization*. Monograph No. 33. Ann Arbor: Survey Research Center, University of Michigan.

Shepard, Jon M., 1970. "Functional Specialization, Alienation and Job Satisfaction." *Industrial and Labor Relations Review* 23.2 (Jan.):207–19.

———, 1971. *Automation and Alienation. A Study of Office and Factory Workers*. Cambridge: MIT Press.

Shiba, Kimpei and Nozue, Kenzo, 1969. "The Brains Behind Japan's Big business." *Asahi Evening News*, November 5, p. 6.

Shirai, T., 1967. "Labor Relations in the Japanese Shipbuilding Industry." *Japan Labor Bulletin* 6.1 (January):5–10.

Smelser, Neil J., 1968. *Essays in Sociological Explanation*. Englewood Cliffs: Prentice-Hall.

Smith, Thomas C., 1958. "Old Values and New Techniques in the Modernization of Japan." *The Far Eastern Quarterly* 14 (May). 355–63.

Stone, P. B., 1969. *Japan Surges Ahead*. New York: Praeger.

Sumiya, Mikio, ed., 1959. *Sangyō to rōdōkumiai*. Tokyo: Diamond.

Sumiya, Mikio et al., 1967. *Nihon shidon shugi to rōdō mondai*. Tokyo: Todai Shuppankai.

Suzuki, Tatsuzo, 1966. "A Study of the Japanese National Character Part III." *Annals of the Institute of Statistical Mathematics* Supplement IV:15–64.

———, 1970. *A Study of the Japanese National Character*, Part IV. Tokyo: Research Committee on the Study of Japanese National Character.

Taira, Koji, 1962. "Characteristics of Japanese Labor Markets." *Economic Development and Cultural Change* 10:150–68.

———, 1970. *Economic Development and Labor Markets in Japan*. New York: Columbia University Press.

Takeuchi, H., 1966. *Electrical Machinery Industry*. Tokyo: Tōyōkeizai.

Takezawa, Shin-ichi, 1961. "Nihon no keiei to ningen kankei." *Shakaigaku hyoron* 12 (September):20–29, 110–11.

———, 1969. "The Blue-Collar Worker in Japanese Industry." *International Journal of Comparative Sociology* 10 (March and June):178–93.

Tannenbaum, Arnold S., 1968. *Control in Organizations*. New York: McGraw-Hill.

Tominaga, Ken'ichi, 1961. "Industrial Organization and Social Structure in Japan." *Shakaigaku hyoron* 12 (September): 30–45.

———, 1962. "Occupational Mobility in Japanese Society: Analysis of Labor Market in Japan." *The Journal of Economic Behavior* (Japan) 2 (April):1–37.

Tōyō Keizai, ed., 1969, 1970, 1971, 1973. *Chingin soran*. Tokyo: Tōyō keizai shimpōsha, 1968 ed.: 146–200; 1969 ed.: 156–66; 1970 ed.: 156; 1972 ed.: 168.

———, 1967, 1972, *Tōkei geppō*. Tokyo: Tōyō keizai shimpōsha, v. 27, (December) and v. 32 (August):8, 48.

Tsuda, M., 1965. *The Basic Structure of Japanese Labor Relations*. Tokyo: The Society for the Social Sciences, Musahi University.

———, 1973. "Personnel Administration at the Industrial Plant Level." In *Workers and Employers in Japan*, edited by Kazuo Okochi, Bernard Karsh, and Solomon B. Levine, pp. 399–440. Princeton: Princeton University Press.

United Nations, 1970. *Monthly Statistical Bulletin.* New York: United Nations.

———, 1971, *Statistical Yearbook, 1970.* New York: Statistical Office of the United Nations.

Vogel, Ezra F., 1963. *Japan's New Middle Class.* Berkeley: University of California Press.

Weber, Max, 1946. *From Max Weber: Essays in Sociology.* New York: Oxford University Press.

———, 1950. *The Protestant Ethic and the Spirit of Capitalism.* New York: Scribners.

Whitehill, Arthur M., Jr., and Takezawa, Shin-ichi, 1968. *The Other Worker: A Comparative Study of Industrial Relations in the U.S. and Japan.* Honolulu: East-West Center Press.

Wilensky, Harold L., 1960. "Work, Careers, and Social Integration." *International Social Science Journal* 12 (Fall): 543–60.

Woodward, Joan, 1965. *Industrial Organization: Theory and Practice.* London: Oxford University Press.

Yoshino, M. Y., 1968. *Japan's Managerial System: Tradition and Innovation.* Cambridge: The M.I.T. Press.

INDEX

Library of Congress Cataloging in Publication Data

Marsh, Robert Mortimer.
Modernization and the Japanese factory.

Bibliography: p.
Includes index.
1. Industrial sociology—Japan. 2 Personnel management—Japan. 3. Machinery in industry.
I. Mannari, Hiroshi, 1925- joint author.
II. Title.
HD6957.J3M37 658.4'00952 75-3466
ISBN 0-691-09365-2
ISBN 0-691-10037-3 pbk.

GPSR Authorized Representative: Easy Access System Europe - Mustamäe tee 50, 10621 Tallinn, Estonia, gpsr.requests@easproject.com

www.ingramcontent.com/pod-product-compliance
Lightning Source LLC
LaVergne TN
LVHW020646300926
849161LV00004B/834